Simulating the Evolution

Springer

London
Berlin
Heidelberg
New York
Barcelona
Hong Kong
Milan
Paris
Singapore
Tokyo

Angelo Cangelosi and Domenico Parisi (Eds)

Simulating the Evolution of Language

 Springer

Angelo Cangelosi, Laurea, PhD
Centre for Neural and Adaptive Systems, School of Computing,
University of Plymouth, 7 Kirkby Place, Plymouth, PL4 8AA, UK

Domenico Parisi, Laurea, MA
Institute of Psychology, National Research Council, Viale Marx 15, 00137 Rome, Italy

British Library Cataloguing in Publication Data
Cangelosi, Angelo
    Simulating the evolution of language
    1.Historical linguistics - Computer simulation
    2.Evolutionary computation
    ITitle II.Parisi, Domenico
    417.7'0113
    ISBN 1852334282 ✓

Library of Congress Cataloging-in-Publication Data
Cangelosi, Angelo, 1967-
    Simulating the evolution of language / Angelo Cangelosi and Domenico Parisi.
        p. cm.
    Includes bibliographical references and index.
    ISBN 1-85233-428-2 (alk. paper)
        1. Language and languages--Origin--Data processing. 2. Language and
    languages--Origin--Simulation methods. I. Parisi, Domenico. II. Title.
    P116 .C36 2001
    417'.7'0285--dc21   2001049624

ISBN 1-85233-428-2 Springer-Verlag London Berlin Heidelberg
a member of BertelsmannSpringer Science+Business Media GmbH
http://www.springer.co.uk

Typesetting: Camera ready by editors
Printed and bound at the Athenæum Press Ltd., Gateshead, Tyne and Wear
34/3830-543210 Printed on acid-free paper SPIN 10791784

to Allegra and Virginia – AC
to Cristina and Elene – DP

Acknowledgements

The editors wish to thank all those that have contributed to the realization of this volume. First, we would like to thank all the authors for providing interesting and high quality chapters. The success and impact of this book is due to their contribution. We are also grateful to those that have helped review the chapters. These include some of the authors that have cross-reviewed the chapters, and other external reviewers such as Guido Bugmann, Dan Joyce, and Liz Stuart. We also thank the two anonymous referees of the whole book for their constructive comments. A special thank you goes to Allegra Cattani for her help in preparing the Subject and Author Indices. Finally, we are grateful to the editorial assistant team of Springer Verlag London, in particular to Karen Borthwick, Melanie Jackson, Rosie Kemp and Catherine Drury. They have been very prompt and responsive to any request that we have made.

Angelo Cangelosi also wishes to thank the School of Computing of the University of Plymouth for the time and resources provided during the preparation of this volume. Angelo's work on this book and his research on the simulaton of language evolution and acquisition have been partially supported by The Nuffield Foundation (grant: SCI/180/97/116/G), and the UK Engineering and Physical Science Research Council (grants: GR/N01118 and GR/N38145/01).

Table of Contents

List of Contributors

Michael A. Arbib, Director USC Brain Project, University of Southern California, Los Angeles CA 90089-2520, USA. arbib@pollux.usc.edu

Bart de Boer, Center for Mind, Brain and Learning, University of Washington, Box 357988, Seattle WA 98195, USA. bdb@u.washington.edu

Seth Bullock, Information Research Institute, School of Computing, University of Leeds, Leeds LS2 9JT, UK. seth@scs.leeds.ac.uk

Angelo Cangelosi, Centre for Neural and Adaptive Systems and Plymouth Institute of Neuroscience, School of Computing, University of Plymouth, Drake Circus, PL4 8AA Plymouth, UK. acangelosi@plymouth.ac.uk

Morten H. Christiansen, Department of Psychology, Uris Hall, Cornell University, Ithaca NY 14853, USA. mhc27@cornell.edu

Christopher M. Conway, Department of Psychology, Uris Hall, Cornell University, Ithaca NY 14853, USA. cmc82@cornell.edu

Rick A.C. Dale, Department of Psychology, Uris Hall, Cornell University, Ithaca NY 14853, USA. rad28@cornell.edu

Ezequiel Di Paolo, School of Cognitive and Computing Sciences, University of Sussex, Falmer, Brighton BN1 9QH, UK. ezequiel@cogs.susx.ac.uk

Michelle R. Ellefson, Cognitive Science Lab, Department of Psychology, Southern Illinois University, Carbondale IL 62901-6502, USA. ellefson@siu.edu

Alberto Greco, Department of Anthropological and Psychological Sciences, University of Genoa, vico S. Antonio 5, 16126 Genoa, Italy. greco@nous.unige.it

Stevan Harnad, Centre for Cognitive Science, Department of Electronics and Computer Science, University of Southampton, Southampton SO17 1BJ, UK. harnad@cogsci.soton.ac.uk

Takashi Hashimoto, School of Knowledge Science, Japan Advanced Institute of Science and Technology (JAIST), 1-1 Tatsunokuchi, Ishikawa, 923-1292, Japan. hash@jaist.ac.jp

Brian Hazlehurst, Kaiser Permanente Center for Health Research, 3800 N. Interstate Avenue, Portland Oregon 97227-1098, USA. Brian.Hazlehurst@kpchr.org.

James R. Hurford, Language Evolution and Computation Research Unit, Department of Linguistics, University of Edinburgh, 40 George Square, EH8 9LL Edinburgh, UK. jim@ling.ed.ac.uk

Edwin Hutchins, Department of Cognitive Science, University of California at San Diego, 9500 Gilman drive, La Jolla CA 92093-0515, USA. hutchins@cogsci.ucsd.edu

Simon Kirby, Language Evolution and Computation Research Unit, Department of Linguistics, University of Edinburgh, 40 George Square, EH8 9LL Edinburgh, UK. simon@ling.ed.ac.uk

Natalia L. Komarova, Institute for Advanced Study, Einstein Drive, Princeton, NJ 08540 USA, and Department of Applied Mathematics, University of Leeds, LS2 9JT Leeds, UK. natalia@ias.edu

Daniel Livingstone, Department of Computing and Information Science, University of Paisley, High Street, PA1 2BE Paisley, Scotland, UK. daniel.livingstone@paisley.ac.uk

Jason Noble, Information Research Institute, School of Computing, University of Leeds, Leeds LS2 9JT, UK. jasonn@comp.leeds.ac.uk

Martin A. Nowak, Head of Program of Theoretical Biology, Institute for Advanced Study, Einstein Drive, Princeton NJ 08540, USA. nowak@ias.edu

Domenico Parisi, Institute of Psychology, National Research Council, viale Marx 15, 00137 Rome, Italy. parisi@ip.rm.cnr.it

Michael Tomasello, Max Planck Institute for Evolutionary Anthropology, Inselstrasse 22, D-04103 Leipzig, Germany. tomas@eva.mpg.de

Huck Turner, Centre for Neural and Adaptive Systems, School of Computing, University of Plymouth, Drake Circus, PL4 8AA Plymouth, UK. h.turner-2@plymouth.ac.uk

Luc Steels, Director SONY Computer Science Lab (Paris), and VUB Artificial Intelligence Laboratory, Vrije Universiteit Brussel, Pleinlaan 2, B-1050 Brussels, Belgium. steels@arti.vub.ac.be

Part I

INTRODUCTION

Chapter 1

Computer Simulation: A New Scientific Approach to the Study of Language Evolution

Angelo Cangelosi and Domenico Parisi

Introduction

Language is such an important human characteristics that we would like to know how it first came into existence and how it managed to reach its present form. If we could go back in time for a sufficient number of generations we would encounter human ancestors that did not have language. In a succession of generations the descendants of those non-linguistic ancestors came to possess the ability to speak and to understand the speech of others. What made the transition possible? How did the transition occur? Were there intermediate forms of language in the sense of communication systems different from known animal communication systems but also different from language as we know it and as is spoken today by all humans?

These questions constitute the scientific study of language origins and language evolution. The study of language origins does not directly include historical linguistics since this focuses on how human languages change historically and how new languages emerge from pre-existing languages, nor does it directly include the study of how language is acquired by children since they learn in an environment in which language already exists. The study of language origins and evolution is the study of how language emerges from a situation in which there is no language. However, both historical linguistics and the developmental psychology of language are concerned with causes, mechanisms, and processes of language change and with primitive forms of language such as pidgin, creole, and children's languages. These disciplines can indirectly contribute to the study of language origins and evolution.

The study of language origins and evolution clearly is an interdisciplinary field (Knight, Studdert-Kennedy and Hurford, 2000; Hurford, Studdert-Kennedy and

Knight, 1998; Harnad, Steklis and Lancaster, 1976). Evolutionary biology and the neurosciences are involved in this study because the human capacity of language has a genetic and a neural basis. Furthermore, as we have seen, historical linguistics and developmental psycholinguistics and, more generally, linguistics and psychology, may have something to say about language evolution. Finally, paleoanthropology and cultural anthropology are clearly relevant since language has evolved in human ancestors and, although it can have an evolved genetic basis, is culturally transmitted, i.e., learned from others, and it changes culturally (historically) from one generation to the next.

The rich interdisciplinary basis of the study of language evolution can be an asset but also a liability since the various disciplines that contribute to this study tend to have different theoretical and methodological orientations and practices which are difficult to reconcile. But the real problem that makes the study of language origins and evolution a difficult and even problematic research field is that it is a field with a multitude of theories but a very limited, or even non-existent, empirical basis. Language has emerged from non-language in the distant past and there is no direct trace of the event. Furthermore, the origin of language is an event that does not repeat itself and therefore cannot be observed. We can observe a child's first acquisition of language and the emergence of pidgin/creole languages but these are different events, although somewhat related.

How can we reconstruct the origin and evolutionary stages of human language given these difficulties? Since language is so important for defining what a human being is, human beings have a natural tendency to propose theories and evolutionary scenarios for its origin and evolution. But these theories inevitably tend to remain speculative because we don't have the empirical evidence to confirm or disconfirm them or to choose among them. In the nineteenth century this situation persuaded the *Société Linguistique de Paris* to emit its famous ban on research and publication on the subject. Today we know much more about language, about mind and the brain, and about human ancestors than was known in the nineteenth century. But still the variety of different theories and interpretations of language origin and evolution that have at least initial plausibility is so great and the empirical basis for choosing among these different theories and interpretations so restricted that the study of language origins and evolution remains a somewhat dubious research field. Science is the production of theories about causes, mechanisms, and processes underlying the phenomena of reality and explaining them but a mature science cannot stop at theories. To be called scientific a theory must generate specific empirical predictions and these predictions must be compared with directly or indirectly observed facts.

Theories of language origins and evolution not only are difficult to test empirically but they tend to be stated in such vague and general terms that these theories are unable to generate specific and detailed empirical predictions. Empirical predictions are the "interface" between theories and empirical reality. Science makes progress when this "interface" is extended and detailed. This is not the case for theories of language evolution. Hence, the problem of the objective scarcity of empirical evidence for such theories becomes even more serious because the theories do not generate specific and detailed empirical predictions and they do not help the researcher to look for empirical evidence.

It is this very problematic aspect of the study of language evolution which computer simulations can help us to overcome. Computer simulations are theories of the empirical phenomena that are simulated. Traditionally, scientific theories are expressed verbally or with the help of formal and mathematical symbols. Simulations are a novel way to express theories in science. Simulations are scientific theories expressed as computer programs. The program incorporates a set of hypotheses on the causes, mechanisms, and processes underlying the simulated phenomena and, when the program runs in the computer, the results of the simulations are the empirical predictions derived from the theory incorporated in the simulation. All this contributes to the development of a new approach to the study of the origins and evolution of language.

A New Approach to the Study of Language Evolution: Computer Simulation

The use of computer modeling and simulation in the natural and human sciences has greatly progressed in recent decades. This progress has been possible because of the availability of very efficient and low cost computers. The new methodological approach has also facilitated interdisciplinary research because the great memory and processing speed of computers make it possible to incorporate in one and the same simulation data and ideas from a variety of disciplines. For example, in cognitive science and artificial life the computer simulation methodology plays a major role in bringing together scientists from different disciplines. In the specific field of language evolution, simulation provides a complementary methodology that can help researchers to develop detailed and precise hypotheses on language origins and evolution and to test these hypotheses in the virtual experimental laboratory of the simulation. Various bits of empirical evidence such as observations and experiments on animal communication, fossil record analysis, genetics studies, and experimental psycholinguistic data can lead to the formulation of theories and predictions that can be verified through computer simulations.

As we have said, a simulation is the implementation of a theory in a computer. A theory is a set of formal definitions and general assertions that describe and explain a class of phenomena. Examples of language evolution theories are those that identify a specific ability as the major factor explaining the origins of language, such as gestural communication (Armstrong, Stokoe and Wilcox, 1998), tool making (Parker and Gibson, 1979), vocalization and speech (Lieberman, 1994), or symbol acquisition abilities (Deacon, 1997).

Theories expressed as simulations possess three characteristics that may be crucial for progress in the study of language origins and evolution[1]. First, if one expresses one's theory as a computer program the theory cannot but be explicit, detailed, consistent, and complete because, if it lacks these properties, the

---

[1] Some of these characteristics will be discussed in more detail in the section on the advantages of simulation.

theory/program would not run in the computer and would not generate results. This is crucial for theories of language evolution because, as we have said, these theories tend to be stated in vague and general terms and, therefore, it is virtually impossible to determine if they are internally consistent and complete, that is, sufficient to explain the phenomena. Second, a theory expressed as a computer program necessarily generates a large number of detailed predictions because, as we have said, when the program runs in the computer, the simulation results are the predictions (even predictions not thought of by the researcher) derived from the theory. This is important for theories of language evolution because a large and diverse set of empirical predictions broadens and strengthens the usually restricted and weak interface between theory and data which characterizes these theories. Third, simulations are not only theories but also virtual experimental laboratories. As in a real experimental laboratory, a simulation, once constructed, will allow the researcher to observe phenomena under controlled conditions, to manipulate the conditions and variables that control the phenomena, and to determine the consequences of these manipulations. This is crucial for the study of language evolution not only because experimental manipulations yield a richer and more detailed set of observed empirical phenomena but because these manipulations allow the researcher to test hypotheses and ideas experimentally, which is otherwise entirely impossible if one is interested in long past events and phenomena such as those concerning the origins and evolution of language.

Simulations not only implement general theories but they can implement a specific model derived from a theory, or some specific hypothesis. A model is a simplification of a general theory, or focuses on a specific aspect of the phenomena of interest. For example, there are formal models that seek to explain the necessary conditions for the evolution of syntactic communication (Nowak, Plotkin and Jansen, 2000), or that focus on the role of imitation of motor skills and the development and specialization of Broca cortical areas for language (Rizzolatti and Arbib, 1998). A computer simulation is a clearer and more *practical* way of expressing what a theory or a model says. This makes a theory or model more verifiable, because the execution of a computer program can easily identify the problems, inconsistencies, and incompleteness of the theory or model. Simulations also permit the discovery of new predictions that can (and must) be derived from a theory and the search of new (real) empirical data that can verify these predictions.

Computer simulations are also important because they tend to be inherently quantitative research tools. Quantitative predictions derived from theories, quantitative descriptions of the empirical data, quantitative manipulations of variable and parameters controlling the observed phenomena, are critical features of a mature science. The natural sciences have made great progress in the last 3-4 centuries not only because they have created an intense and detailed dialogue between theories and experiments but because empirical observations, predictions derived from theories, and experimental manipulations are all expressed in precise, quantitative terms. Quantification makes ideas and observations more objective and makes it possible for other people to replicate observations and experiments. Computer simulations tend to be quantitative. The simulated phenomena tend to be described and analyzed in quantitative terms and the parameters and conditions controlling the phenomena that are manipulated by the researcher also tend to be quantitative. Furthermore, simulations can be repeated ad libitum, varying the initial conditions, and they make possible all sorts of statistical and reliability analyses.

But simulations are not only a novel way of expressing scientific theories and not only virtual experimental laboratories. Simulations re-create reality, they *are* artificial reality. This is a property of simulations which can be very important for the study of language origins and evolution. The study of language origins and evolution is the study of long past and non-repeating phenomena that it is impossible for us to observe today. There is some empirical evidence available from which we can reconstruct these past events but it is only very indirect. From archeological evidence about ancient stone tools we can learn about the tool-making skills of our ancestors (Davidson, 2000) and from the study of cranial bones we can speculate on their vocalization abilities (Lieberman, 1994). From observations of the communicative behavior of species that are closely related to humans, we can hypothesize about the communicative abilities of our ancestors (Savage-Rumbaugh, 1998) and reflect on the differences between human language and animal communication systems (Hauser, 1996). But as we have said, this is all indirect empirical evidence. The phenomena and events of language origin and evolution elude us. Computer simulations allow us to *re-create* these phenomena and events. To some extent, we can simulate the process of natural selection (Holland, 1975) and re-create possible evolutionary scenarios (natural environments, social groups, survival strategies) to directly test the validity of possible evolutionary explanations of the origins of language. We can observe the primitive form that this simulated language tends initially to assume and the changes which occur in successive forms due to both biological and cultural evolution. We can systematically study the adaptive advantages of different communication systems and the pre-conditions that make the evolutionary emergence of these communicative systems possible.

With computer simulations researchers tend to adopt a synthetic strategy (Steels, 1997), which is quite different from classical scientific methodologies based on the analytic approach. In the natural sciences such as biology a top-down approach is often used. The organism is analyzed, i.e., divided up, into organs: brain, heart, lungs, etc. An organ is then divided up into different tissues, tissues are divided into cell types, and so on until the molecular and purely physical level is reached. In linguistics, language is analyzed, i.e., decomposed, into lexicon and syntactic principles, then lexical entries are divided up into morphemes and phonemes, and so on. A synthetic strategy uses a bottom-up and constructive approach. The researcher decides the basic components of a system, the rules by which these components interact and, possibly, the environment in which the components interact with each other or with the environment itself. The computer program will then simulate the interactions among the components and as a result of these interactions the researcher should be able of observing the emergence of the various higher-level entities with their properties and the different phenomena that involve these entities. With a simulation, the feasibility and validity of assumptions regarding components, their interaction rules, and the environment can be tested. Wrong, incomplete, or inadequate assumptions will make it impossible to observe the emergence of higher-level entities or of entities that do not have realistic properties and do not exhibit realistic phenomena. The bottom-up approach of synthetic simulations also permits the study of problems and phenomena that are analytically intractable, such as those of complex and non-linear systems – and language and evolution are typical complex systems (De Jong, 1999).

The synthetic methodologies that can be used to study the origins and evolution of language include evolutionary computation (Goldberg, 1989), neural networks (Rumelhart and McClelland, 1986; Levine, 2000), artificial life and synthetic ethology (Langton, 1995; Parisi, 1997) and rule-based agents (Kirby, 1999). In this book we are not restricting the use of the term simulation only to software implementation of models. We also include *embodied* simulations, which are hardware-based models such as robots. In fact, various robotic models for studying the evolution of communication and language have been developed (Steels and Vogt, 1997; Steels and Kaplan, 1998; Billard and Dautenhahn, 1999). All these methods will be described in detail in Chapter 2. In the next paragraph we look at the advantages and limitations of simulation.

Advantages of simulation

1. Simulations as virtual experimental laboratories

Simulations are tools for the development of a variety of "experiments". They can be considered as virtual experimental laboratories that allow researchers to run *realistic, impossible*, and *counterfactual* experiments. Simulations can be used to replicate empirically *realistic* laboratory experiments with the added advantage that manipulation of conditions and variables is easier and can be extended to conditions and variables that cannot be manipulated in real experiments. With simulations it is possible to replicate old experiments and to execute new experiments for the first time. In the first case, the simulated experiments can be useful for systematically monitoring and manipulating previously ignored or overlooked variables. In the second case, the virtual experiment can be sometimes used as a pilot study or as a substitute for a real experiment.

Simulations can also be virtual laboratories for conducting empirically *impossible* experiments, i.e., for experimentally studying phenomena that it would be difficult to bring into the real laboratory. These are phenomena that are long past and do not recur, or that are too big or last too long to do real experiments on them. This is often the case for research on evolutionary phenomena. It is generally impossible to run real controlled evolutionary experiments except for simple and short-lived species such as microscopic creatures and fruit flies, or for isolated molecules and proteins (Weiss and Sidhu, 2000). Instead, there are modeling techniques such as genetic algorithms that can be used to simulate some of the processes of natural selection and make it possible to run virtual evolutionary experiments on more complex and longer-lived species. These techniques are commonly used in language evolution models. In some simulations genetic algorithms are used as simple optimization techniques, while in other models they simulate some evolutionary mechanisms and the interaction between ontogenetic learning and evolution (Nolfi and Floreano, 1999). For example, researchers can first define the genotype and phenotype of virtual organisms able to develop language and other behavioral skills. Then organisms are selected and reproduce according to their behavioral performance. The manipulation and close monitoring of evolutionary and behavioral variables can allow researchers to understand the adaptive role of communication and language.

Other types of impossible experiments are those that are impossible for ethical reasons. For example, in simulations of linguistic behavior one can examine the internal functioning of neural networks that control the linguistic behavior of organisms in ways that would be impossible to replicate with the real nervous system of human beings. Or one can lesion these neural networks and observe the consequences of the lesions for the linguistic behavior of the artificial organisms, which of course would be impossible to do for ethical reasons with real human organisms.

Finally, simulations can be used to run *counterfactual* experiments, i.e., experiments that are characterized by conditions and parameter values that we know are not those of reality. For example, one could simulate how a communicative system with the same complexity as human language might emerge in a population of organisms with very different sensory and motor capacities than those of humans. Even if the results of these simulated experiments by definition cannot be compared with real empirical data, still counterfactual simulations can be useful to enlarge that range of phenomena to which the hypotheses incorporated in the simulation can apply. By running both real and counterfactual simulations our theories and hypotheses can be tested with both real phenomena and possible phenomena, thereby enriching and making more explicit their empirical content.

A further advantage of simulations as virtual experimental laboratories is that many more factors can be considered at the same time in the same experiment. In real laboratories only one or a few variables can be manipulated in any given experiment. In a simulation of language origins evolutionary variables (e.g., mutation and crossover rate, selection strategies), environmental factors (e.g., abundance of resources, geographical separations), individual variables (e.g., learning abilities, neural control of behavior), and social variables (patterns of social interaction, parent-child and peer-to-peer communication) can be controlled and studied in the same experiment. This type of simulation corresponds more to a piece of "field" research than to a laboratory-based experiment.

There are also practical advantages in the use of simulations as virtual experimental laboratories. First, computer simulation software is generally cheap to develop, and only requires the possession of normal programming skills. The time required to implement a simulation model can be relatively short. Moreover, simulation software tools can be obtained at affordable prices, or for free, from non-profit research and academic organizations. For example, Swarm (Luna and Stefansson, 2000) is a freely available program that can be used for the execution of evolutionary and multi-agent simulations. The availability of standard simulation software also makes it easier to replicate simulation experiments.

A second practical advantage of simulation experiments is that they are cheap to run. A standard personal computer is generally sufficient to run simulations of language evolution. The design and implementation of the model could require some time, but then the execution of experiments is a relatively automatic process.

The simulation can sometime require long execution times, especially when genetic algorithms are combined with other computationally expensive techniques such as neural networks, but the execution of experiments can be fully automated. Of course, the analysis of large quantities of experimental results will require sufficient time.

2. Simulations as tools for testing the internal validity of theories

The main goal of simulation models is to generate empirical predictions that can be verified through existing empirical data (Tomasello, this volume), real experiments, or further simulations. This makes the internal validation of theories more objective and dependent on clear and unambiguous definition of variables and assumptions. A very important function of simulations is to establish if a prediction actually and consistently derives from a theory. If a prediction embodied in a simulation is not confirmed, that is, if the simulation results do not match the empirical phenomena that the theory was intended to explain, then the theory must be rejected as false, inadequate or incomplete.

The generation of practical predictions through simulation models is made possible, and is actually facilitated, by the fact that computer programs require the detailed specification of any aspect that must be implemented. In experimental research the specification of the meaning of variables is called *operationalization*. The operational definition of a variable is its expression in terms of observable and quantifiable factors and, as will be remembered, simulations are typically quantitatively expressed theories.

Simulations are exceptionally good tools for testing the internal validity of theories because they make empirical predictions explicit and this is critical for theories of language origins and evolution. Since language is such a complex phenomenon, and its origins are long past events that cannot be directly observed, theories of language origins and evolution tend to be speculative, making it difficult to decide whether they actually explain the phenomena they are intended to clarify. Is a specific theory of how language has originated sufficiently complete, detailed, and non-contradictory that, assuming that the conditions specified by the theory are realized, language would actually emerge? Such a question may be impossible to answer to the satisfaction of both proponents and opponents of the theory, with the result that theories of language origins can be discussed ad infinitum but never confirmed or abandoned.

Simulations can make a critical difference from this point of view. If a theory of language origins and evolution is sufficiently well defined, explicit, and detailed that the theory can be incorporated in a simulation, the results of the simulation can actually show to the satisfaction of everyone whether the theory actually generates the intended phenomena: given a population of organisms initially lacking language, the population goes through a process of biological and/or cultural changes at the end of which the members of the population actually exhibit a type of communicative behavior possessing the properties of human language.

Once the simulation has been realized it can generate known facts about human language and also predictions about facts and properties of language which have not been previously observed. These novel predictions can then be used to search for new empirical evidence and to look for real data, for example using fossil records, animal behavior observations and experiments, population genetics analyses, and psycholinguistic experiments. An example of a study based on the match between data on population genetics and predicted patterns of linguistic change is Cavalli-Sforza (2000; see also Piazza, 1996).

Finally, simulations can help testing the internal validity of theories when used as *thought experiments* (Di Paolo, Noble and Bullock, 2000; Parisi, 2001). A thought experiment is a simplified situation which does not exist in reality. It focuses on some problematic (and often paradoxical) aspect of reality which is critical for deciding between alternative interpretations. For example, Searle's Chinese Room thought experiment (Searle, 1982) is often used in the literature on the origin and grounding of the meaning of words (Harnad, 1990). Thought experiments could only be conducted "in the head" prior to computers and therefore one could only appeal to one's intuitions in assessing the validity of such experiments. Computer simulations can be used to implement some of these hypothetical situations and to test their consequences in a more objective and systematic way.

3. Simulations as tools for studying language as a complex system

One further advantage of using simulation methods to study language evolution is that simulations allow us to clearly see the nature of language as a complex system. Complex systems are made up of a large number of entities that by interacting locally with each other give rise to global properties that cannot be predicted or deduced from an even complete knowledge of the entities and of the rules governing their interactions. Complex systems are very sensitive to initial conditions, tend to change in unpredictable ways, to react to external disturbances in ways that are not commensurate with the importance of the particular external disturbance. In many cases they are adaptive systems, that is, they tend to change in ways that depend on the particular environment in which they exist. All of these properties are exhibited by language and by language evolution. Linguistic behavior is a property of individuals, but language-using individuals, by interacting with each other and with the external environment, give rise to global properties of their language that make language a complex, self-organizing, and adaptive system. In time, the properties of individuals (their linguistic behavior during individual development) and the global properties of language (language evolution, language change) change in adaptive ways. Complex systems also tend to be organized in a hierarchy of levels, with a number of entities at one level determining the properties of a single entity at the next higher level. Hierarchically organized complex systems exhibit further unpredictable properties, like the fact that changes at one level of the hierarchy can remain hidden, with no resulting visible change until a small further change at the lower level suddenly produces a very visible change. This hierarchical organization is also typical of language in which, for example, neurons interact in the brain to produce the linguistic behavior of an individual and individuals interact linguistically with each other to produce the global properties of language.

Traditional research methods, i.e., laboratory experiments, data collection and analysis, systematic observation, are not fully adequate to study language as a complex system. Laboratory experiments are more appropriate for studying simple systems that can be isolated from their environment and that have properties which are the effect of a limited number of causes that can be manipulated by the experimenter one at a time. Data collection and analysis and systematic observation constitute the empirical basis for testing theories but, by themselves, they tell us little directly about the mechanisms and processes underlying the observed phenomena. Computer simulations allow the researcher to create scenarios in which a large number of entities are assigned a set of properties and the computer causes these entities to interact in such a way that the researcher can observe (literally, see on the computer screen) the global properties that emerge from their interactions. These global properties may change in time due to the adaptation of the system to its environment and the properties assigned to the individual entities themselves can change. Furthermore, the enormous memory and processing speed of a computer allows the researcher to introduce and manipulate all sorts of different variables and conditions and to observe the effects of these manipulations on the (simulated) phenomena.

Computational methodologies using this agent-based, bottom-up approach, can contribute significantly to our understanding of complex adaptive systems and in particular to our understanding of language and language evolution as complex systems. In other disciplines that are concerned with complex systems, such as chemistry, biology, and the social and economic sciences, it has already been shown that the bottom-up approach of synthetic techniques can generate important insights (Langton, 1995).

Limits of simulation

Simulation is not immune from limitations and difficulties. Some of these limitations are due to the current limits of computer technology. Models that simulate populations of complex communicating agents for a large number of generations may be computationally expensive. Often researchers have to simplify their models to avoid long simulation runs. Simulation is also limited by the creativity of researchers and their inability to exploit the full potential of current simulation methodologies and platforms. Sometimes, innovative but simple models can capture important phenomena but innovations may be difficult to arrive at. A final source of difficulty for simulation models depends on the general problems associated with language evolution research and the limited direct evidence available.

In the following sections we discuss some of the major limits of simulation of language origins and evolution.

1. Simplification

Most current computational models, both in language origin research and in other disciplines, are *toy models*. A toy model is a highly simplified and idealized model of a real phenomenon. It tries to capture only certain, very restricted, features of a system. For example, current artificial neural networks (Rumelhart and McClelland, 1986; Levine, 2000) are toy models of the brain. Neural network units are only loosely inspired by real neurons and artificial neural networks leave out many potentially important aspects of the brain. The functioning of an artificial neural network mimics a highly simplified network of neurons, mainly by simulating the distributed and parallel processing of information. In computational neuroscience (Churchland and Sejnowski, 1994; Arbib, 1995; Koch and Segev, 1998), however, more detailed and realistic models of real neurons are used.

Language evolution models are mostly toy models because only a limited subset of the factors affecting the origins and evolution of human language are simulated and explicitly encoded in any particular model. For example, the process of natural selection is simulated through genetic algorithms (Holland, 1975). These evolutionary computation techniques offer a simplified representation of the main mechanisms of evolution such as selective reproduction using a fitness formula, genetic crossover, and random mutation of genotypes. In a specific simulation, only one selection strategy tends to be implemented, among a possible choice of different, and plausible, strategies (e.g., elitism, roulette wheel).

The level of simplification of natural selection using genetic algorithms is relatively satisfactory, however, and a standard approach is commonly chosen. Unfortunately, this is not true for many other aspects of language evolution models. For example, the behavior of communicating organisms can be simulated in a variety of different and not directly comparable ways: artificial neural networks (Cangelosi, and Parisi, 1998; Hutchins and Hazlehurst, 1995), abstract and simplified meaning-signal probability tables (Steels, 1996; Oliphant and Batali, 1997), and rule systems (Kirby, 1999).

Most of these toy models do not deal with the problem of scaling up. For example, in artificial neural network research the problem of scaling up concerns the difficulty of constructing networks that scale from a few tens of neurons up to the billions of neurons of real neural systems. Work has been done to directly deal with this problem, but no convincing solutions have been proposed yet. The problem of scaling up in language evolution models is exemplified by the limited complexity of syntax that most of the models have been able to simulate or the limited number of items in lexicons. Most of the models simulate simple syntactic rules, such as subject-verb-object, not the rich morphological and grammatical features of human languages, and use lexicons with a few words as opposed to thousands of words comprising the lexicons of natural languages.

Although oversimplification and scaling up are problems that language evolution models must be able to solve, it must be noted that the aim of these models, at least currently, is not to replicate reality as it is. The main aim of simulation models is to generate predictions that can be empirically tested through further simulations, or through experiments and empirical data collection. As we have already noted, simulation models are implementations of theories, and as theories they aim at describing reality at some essential and necessarily simplified level because in science it is simplification that produces understanding. Of course, what is crucial is that researchers make the right simplifications.

2. Arbitrariness of assumptions and details

Researchers have to make subjective and arbitrary choices when they decide the assumptions of a simulation model and define the details of the implementation. Assumptions are choices that will affect the whole simulation. In a model that focuses on the evolution of linguistic recursion, the researcher can assume that a minimum of syntactic knowledge is already available to individuals (e.g., the grammatical rules that generate subject-verb-object sentences). If recursion can be shown to emerge in the simulation, the generalization value of the simulation results will depend on the correctness of such an assumption. In fact, it could well be that recursion and the basic subject-verb-object sentence structure are more strictly interrelated and they may have evolved together. Therefore, it is important that general assumptions, that will affect the whole of the simulation, are clearly stated and their plausibility is explicitly discussed and justified.

The implementation of a simulation model into a computer program requires a lot of arbitrary choices for the definition of the model's details. For example, the programming of a genetic algorithm requires detailed decisions on the number of individuals in the population, the mutation rate of genotypes, the total number of generations, and many other parameters that are known to affect the evolutionary process (Goldberg, 1989). Due to the large number of parameters that must be set in a computer simulation, researchers should try to take as many explicit, justified, and plausible choices as possible. The running of systematic batteries of simulations permits the monitoring of the model's details which is not easy to control. This systematic testing can highlight the specific biases of certain factors in the outcome of the simulation, and allow a more objective interpretation and generalization of the simulation results.

Among the various synthetic methodologies that can be used to study language origins, some techniques offer a better approach to problems of simplification and arbitrary details. This is the case of robotic models of the emergence of communication (Steels and Kaplan, 1998). The implementation of a model in an embodied system, such as a robot, releases the researcher from the task of making too many subjective decisions on the model features. Initially, the architecture and functional procedures controlling the robot must be decided, hopefully according to plausible principles. Then, the overall functioning of the robotic system will depend on these initial decisions, and no more technical details have to be defined.

3. Difficult external validation

Simulations generally produce large amounts of data and results. The analysis of these data can be quite difficult and time consuming. Direct comparison between simulation and real data are not often possible. First, because there is not much direct evidence available for the evolution of language and in general for evolutionary studies. Second, even when some comparisons between simulation data and available real data are in principle possible, direct and quantitative evaluations are difficult to perform.

Various heuristics and techniques have been proposed for the analysis of data and simulation results within models of the evolution of language. Except for few studies, not much work has been done implying comparisons between simulation results and real data. For example, in a model of the emergence of vowel systems, de Boer (2000; this volume) directly compares the simulated vowel systems with real language data. In simulations that use neural networks, various aspects of these artificial neural systems can be systematically analyzed (Hanson and Burr, 1990) and compared with experimental brain processing data. In a recent model of the evolution of simple syntactic languages (Cangelosi and Parisi, 2001; Parisi and Cangelosi, this volume) the relationships between the network's motor representation and the internal representations of different word classes have been studied in detail. The results appear to show that internal representations of verbs have a stronger motor component while nouns reflect sensory information. These results may be directly related to existing work on brain imaging techniques for word processing (Martin *et al.*, 1995). Other neural network models were direct comparisons are made with experimental data are discussed in detail in the chapter on sequential learning (Christiansen *et al.*, this volume)

The contribution of simulation to an understanding of language origins and evolution should rely more often, and more directly, on the comparison and validation of model results with real-world data. Without a more direct integration between simulation data and evidence collected from other disciplines and methodologies, the predictive and explanatory value of language evolution simulations remains limited (Di Paolo *et al.*, 2000).

Simulation Approaches to the Evolution of Language

The study of the origins and evolution of language and communication has a broad range of issues to address. These research questions spread along different continua. First, the differentiation between language and communication can be used to stress different levels of complexity in the communication medium. The study of the evolution of *communication* mostly focuses on simple animal communication systems, such as monkeys' calls or honeybees' dance. It investigates, for example, the adaptive role of communication in groups of animals. The evolution of *language* mainly refers to human languages, and focuses on their inherent complexity, such as syntax. However, this distinction is not always a very significant one, because of the large overlap between research aims in the evolution of language and of communication. Indeed, the study of language is inherently connected to the evolution of animal communication. Language/communication issues lie along the same continuum.

Secondly, some differences exist between the study of the origins of language and that of the evolution of language. Research on the *origins* of language concentrates on the particular spatial (e.g., geographical), temporal (e.g., historical), individual (e.g., neural, cognitive) and social conditions that might have favored the beginning of language. The issue of the *evolution* of language is a broader one, and deals with the continuity in animal communication systems and human language, the historical changes of languages and differentiation of language families, and the adaptive, behavioral, neural, and social aspects of language. In fact, evolution includes the study of the origins of language, along with all other evolutionary and historical aspects.

Current simulation models of language evolution have dealt with various research issues along these continua. This book presents a variety of approaches that focus on these issues. The volume is organized into seven parts. The first part, which includes this chapter, gives a general introduction to computer simulation in the study of language evolution and to the various methods and techniques. The various research questions will be addressed in the following five parts: (Part II) the evolution of simple signaling systems; (Part III) the emergence of syntax and syntactic universals; (Part IV) the sensorimotor grounding of symbols in evolving language systems; (Part V) the neural and behavioral factors in language use, evolution, and origins; (Part VI) the role of auto-organization and dynamical factors in the emergence of linguistic systems. In the final part, the concluding chapter suggests ways in which current studies of primate communication, including humans, may broaden the empirical base for future simulation approaches to language evolution.

Introduction to simulation methods
In Chapter 2 Huck Turner introduces the reader not familiar with computational methods to the different simulation techniques used in language evolution. Methods are grouped into two sections, one for the representation of the individual agents, and one for the learning and interaction algorithms. The agent representation methods serve to control the behavior of the agents. These include symbolic rules and neural networks. The interaction amongst agents is implemented through learning techniques (rule generalization, obverter, imitation, self-understanding), evolutionary algorithms (genetic algorithms, game theory, synthetic ethology) and robotics. In the second part of the chapter the author gives some comments on the implications of these methods for language evolution issues such as innateness and adaptive benefits.

The evolution of signaling systems: adaptive, phonetic, and diversity factors
The chapters in Part I focus on the evolution of simple signaling systems. The emergence of shared communication in animal-like groups can be seen as a preliminary stage towards the emergence of more complex human languages. The simulation of simple signaling systems can help the understanding of various factors in language evolution. In particular, the three chapters look (a) at the adaptive factors in signaling systems, (b) at the emergence of a shared speech systems, and (c) at the formation of dialects.

Chapter 3 investigates the selective pressures which affect the origin, organization, and maintenance of shared signaling systems. Jason Noble, Ezequiel Di Paolo and Seth Bullock present a series of simulations which model the evolution of populations of individuals interacting in shared environments. This evolutionary simulation modeling approach is presented as a way of augmenting conventional game theoretic and mathematical models, allowing modelers to explore aspects of signaling system evolution that would otherwise remain cryptic. First, evolutionary simulation models address *trajectories* of evolutionary change, revealing how model populations change over evolutionary time, perhaps reaching one equilibrium or another, perhaps cycling endlessly, etc. Second, these models help us to understand the relationship between the high-level phenomena exhibited by signaling systems and the low-level behaviors of the individual signalers and receivers that constitute these systems.

The three main questions addressed by Noble *et al.* are the role of ecological feedback, the handicap principle in honest signaling, and the use of communication in animal contests. Ecological factors such as the spatial distribution of individuals and the presence of noise are shown to have non-trivial implications for the evolution of altruistic communication. In a first model, action-response games in which signalers and receivers suffer a conflict of interest are simulated. Results show that evolutionary pressures bring about a spatial organization of the population which itself alters the selective pressures of the model such that altruism is supported. A second simulation investigates Zahavi's (1975) handicap principle, where extravagant displays such as colorful tail feathers are thought to be adaptive because their costliness guarantees trustworthiness. In particular, the authors investigate the evolutionary attainability of handicap signaling equilibria from appropriate initial conditions. Simulations suggest that although there exist scenarios in which signaling systems with any degree of exaggeration are evolutionarily *stable*, only those signaling systems which exhibit high degrees of exaggeration are likely to be evolutionarily *attainable* by populations evolving from reasonable initial conditions. In a final model, Noble *et al.* describe a continuous-time simulation of animal contests. This tests Enquist's hypothesis (1985) that weak animals cannot afford to risk commencing a fight and communicate this honestly. Simulation results reject Enquist's explanation because it is based on an implausible discrete model of animal contests.

A quite different aspect of the evolution of signaling systems is investigated in Chapter 4. Bart de Boer focuses on the *form* of signaling systems, i.e., speech sounds, rather than on the content and reliability of communication. He first reviews the different approaches to modeling speech systems and discusses the advantages of a modeling approach that combines genetic algorithms and language games. Subsequently, de Boer presents a population-based language game model of the emergence of a shared vowel system. Simulation results show that there is a significant bias towards the emergence of optimal vowel systems that closely reflects the frequency and structure of human language vowel systems.

The author favors an explanation of the evolution of language speech universals based on the functional optimizations of communication over noisy channels. For example, it is shown that the frequency of vowel occurrence in human language reflects the optimization of acoustic distinctiveness. This functional explanation is proposed as an alternative to the view that speech universals are a reflection of innate human linguistic abilities, and to the historical explanation of relatedness between human languages.

The evolution of dialect and linguistic diversity is presented in Chapter 5. Daniel Livingstone uses simulation models to inform the debate on the two existing opposite explanations of the role of dialect diversity. The first hypothesis is that dialects cannot be an accident of natural selection and, therefore, must have a purpose. The second explanation argues that the processes of language learning and transmission provide sufficient means to explain language diversity.

Livingstone first reviews the different modeling approaches to the development of linguistic diversity, including both mathematical and computer simulation models. In particular, he criticizes Nettle's models of dialect formation (Nettle, 1999; Nettle and Dunbar, 1997) for the use of language acquisition rules based on explicit averaging and thresholding processes. Livingstone argues that these explicit mechanisms bias Nettle's results on the evolution of diversity and its explanation based on social status influences.

The author describes simulations in which communicating agents are spatially organized, but no social relationships exist between them. In a first simulation, agents learn an abstract signaling system. This model shows that dialect continua form depending on the patterns of spatial organization of individuals. For instance, larger neighboring sizes reduce the number of distinct dialects. The second model is an extension of de Boer's simulation of evolving vowel system (see Chapter 4). This new simulation again shows the emergence of vowel dialect continua, due to the geographically limited interactions allowed between agents. It supports the general view that language diversity arises as part of the auto-organization process of language evolution and transmission.

The emergence of syntax and syntactic universals

The three chapters in Part II investigate the emergence and use of syntactic structures. Most of the chapters in this book deal with the emergence of syntax and syntactic structures, along with other aspects of the evolution of language such as lexicon formation. The three included in this section mainly focus on the emergence of syntactic universals such as compositionality, the population dynamics of grammar acquisition, and the role of sequential learning. Moreover, the authors employ a variety of simulation approaches also in conjunction with other formal and experimental methods.

The emergent properties of syntactic universals such as compositionality are explored in Chapter 6. Simon Kirby and Jim Hurford give an overview of the Iterated Learning Model (ILM) that they have used with other collaborators to simulate the emergence of linguistic structure. The ILM is presented as a general approach to simulate the process of glossogenesis, i.e., the cultural transmission of language in populations of interacting agents. Each agent is able to learn a meaning-signal mapping (1) from being exposed to the training data (E-language) produced by other agents and (2) through its ability to identify regularities in the language via neural networks or rule inference systems. In ILM simulations the agents are initialized without any linguistic system whatsoever and after many (thousands) learning iterations a stable (but potentially dynamic) language system is reached

The authors present various simulations of the ILM. In a first model neural network learning agents are used for the emergence of expressive and stable meaning-signal mappings. Symbolic rule-based agents are used to simulate the emergence of compositionality and recursion "out of learning". A final simulation studies the interaction between the process of glossogenesis and that of phylogeny and shows that functional pressures bias the initial setting of grammatical Principles. Overall all these simulations show that much of the structure of language is emergent. In particular, most fundamental features of human language can be explained as by-products of the pressures on language transmission, without necessary reference to communication.

A mathematical model of the population dynamics of the acquisition of grammar is presented in Chapter 7. Natalia Komarova and Martin Nowak use a rigorous definition of grammar and formalize the process of grammar learning in evolving populations. Three factors are manipulated to study the evolution of shared coherent grammars: search space, learning mechanism, and learning examples. The search space refers to the "pre-formed linguistic theory" of a child, or the "universal grammar" that, according to Chomsky, specifies the form of possible human grammars. The learning mechanism is the algorithm that children use to evaluate the available linguistic data in order to deduce the appropriate grammatical rules; since the actual learning mechanism in humans is unknown, two model grammar learning mechanisms are used, namely memoryless and batch learning; they provide the lower and the upper bound on the efficiency of the mechanism used by children. The effects of varying the number of learning interactions are also studied.

The key result of this model is the finding of a "coherence threshold". It turns out that in order to facilitate successful communication, i.e., to maintain grammatical coherence, the learning accuracy of children must be sufficiently high. Below this threshold no common grammar can evolve in a population. Above the threshold, coherent grammatical systems manifest themselves as stable equilibria. The analyses of the threshold condition relate the maximum complexity of the grammars' search space to the number of learning interactions and the performance of the learning mechanism. Only a universal grammar that operates above this threshold can induce and maintain successful communication in a population. Komarova and Nowak also extend their model to include variants of universal grammars and natural selection amongst them.

The interaction between general cognitive limitations in sequential learning and their resulting linguistic by-products, such as word order constraints, is discussed in Chapter 8. Morten Christiansen and collaborators present a multi-methodological approach to the investigation of the role of sequential learning in language acquisition and evolution. Sequential learning is defined as the acquisition of hierarchically organized structure in which combinations of primitive elements can themselves become primitives for further higher-level combinations. The authors propose artificial language learning as a complementary paradigm for testing hypothesis about language evolution. This method permits the investigation of language learning abilities in infants, children, adults, and also computational models such as artificial neural networks. Subjects are trained on artificial language with particular structural constraints on word order, and then their knowledge of language is tested.

The results of these experimental and modeling studies demonstrate how constraints on basic word order and complex question formation are a by-product of underlying general cognitive limitations on sequential learning. In general this suggests how many constraints on language development are a consequence of limitations on cognitive abilities. This view is further supported by converging evidence from studies with aphasic patients, literature on primate cognition, and other computational models.

The grounding of symbols in evolving languages

The issue of the sensorimotor grounding of symbols is of central interest in language evolution research. It involves questions such as: How do symbols acquire their meaning? How do different individuals autonomously ground sufficiently similar representations that allow them to share a common lexicon? Is there a single mechanism for the symbol grounding, whereby all symbols are directly grounded to object in the world during language acquisition? Or are there other others complementary mechanisms for the transfer of grounding?

Language evolution models differ in the way they approach (or avoid) the problem of grounding symbols into sensorimotor representations and objects in the world. Some simulations do not include any direct grounding of evolving symbols, and assume symbols are virtually grounded. The modeler has the interpretative power of linking the symbols of an evolving language with external objects. Other simulations have directly confronted symbol grounding. In these models the strategy for grounding symbols in reality is an integral part of the system and becomes one of the various factors in how language evolves. Two models that directly deal with the problem of symbol grounding in evolving language systems are included in the book. They show the importance of sensorimotor grounding and its potential contribution to the explanation of the origins of language.

In Chapter 9, Angelo Cangelosi, Alberto Greco and Stevan Harnad discuss the different definitions of a symbol and explain their implications for the symbol grounding problem in cognitive and linguistic models. The authors propose categorical perception (the compression of perceived similarities within categories and the expansion of perceived differences between categories as a side-effect of learning) as a unifying theoretical and computational framework for understanding the cognitive mechanisms of category learning and the acquisition of grounded symbols. Category learning and symbol grounding are based on a series of processes by which individuals (i) first transform categorical representations (i.e., filtered features of internal sensorimotor projections of objects) into grounded low-level symbols (i.e., the names of basic categories), and (ii) subsequently learn new higher-level symbols through Boolean combination of already-grounded symbols. All these processes are shown to be possible in a neural network model for the categorization and naming of geometrical shapes.

A simulation model that tests the "symbolic theft" hypothesis of the origins of language is also presented. The "symbolic theft" strategy refers to the linguistic acquisition of new categories by hearsay. New categories are defined as propositions consisting of Boolean combination of already-grounded categories. In contrast, the "sensorimotor toil" strategy for acquiring new categories is based on direct, real-time trial-and-error experience with the objects, guided by corrective feedback from the consequences of categorization and miscategorization. In competition, symbolic theft is always found to outperform sensorimotor toil. The picture of language origins and evolution that emerges from this hypothesis is that of a powerful hybrid symbolic/sensorimotor capacity. Initially, organisms evolved an ability to detect some categories of the world through direct sensorimotor toil. Category "names" were, originally, whatever differential instrumental action it was adaptive to perform on them (e.g., eat, mate, attack, carry, nurse, etc.). These analog actions may then have done some double duty in communication and been simplified by social convention into arbitrary names. Subsequently, some organisms may have experimented with stringing combinations of the names of those categories and discovered the advantage of this new way of acquiring and conveying categories: "stealing" them via hearsay instead of doing it the hard way. This hypothesis is tested in a computational model where two populations of organisms can learn new categories through sensorimotor toil or symbolic theft. The results clearly show the adaptive superiority of category learning by theft over toil, as well as the effect of categorical perception phenomena in language learning. This helps to explain the adaptive advantage of symbolic theft over direct sensorimotor toil and its role in the origins of language.

A slightly different approach to the evolution of grounded symbols is proposed by Luc Steels in Chapter 10. He presents a complex and innovative approach to the problem of sensorimotor symbol grounding in groups of robots and Internet agents. The robotic setup consists of a set of "Talking Heads" connected through the Internet. Each Talking Head features a rotating video camera, a computer for the cognitive processing of perception, categorization, and language, and devices for audio input and output. Agents can load themselves in a physical Talking Head and teleport themselves to another Head by traveling through the Internet.

Steels stresses the importance of representations used by a cognitive agent that are directly grounded in external reality through a sensorimotor apparatus. These representations need to be sufficiently similar to those used by other agents in the group. They will enable coordinated actions and shared communication. In this system, each agent autonomously acquires internally grounded representations. Steels argues that language plays a crucial role in the learning of similar grounded representations because it is a source of feedback and constrains the degrees of freedom of the representations used in the group. Evolutionary language games are introduced as a framework for concretizing the structural coupling between concept formation and symbol acquisition. An example experiment with physical robots is discussed in detail to show how grounded representations emerge.

Behavioral and neural factors in language use, evolution, and origins

Some theories and models of the evolution of language focus on the interaction between linguistic abilities and other behavioral and cognitive skills. For example, it has been hypothesized that there is a strict interdependence between the origins of language and those of other motor skills, such as hand signing, or the ability to imitate motor actions from others. Moreover, in these motor theories of language origins great importance is given to the neural control of motor behavior. In other theories, the evolution of language and cognitive skills are hypothesized to be strictly interdependent. The chapters in this section will present two of these theories that show the strict interaction between the evolution of language and other behavioral abilities, and the neural control of both systems.

In Chapter 11 Michael Arbib proposes the view that the human brain and body have been evolutionarily shaped to be language-ready. As a consequence, the variety of human languages evolved culturally, as a more or less cumulative set of inventions. In particular, the author proposes the Mirror System hypothesis of the evolution of language. That is, speech derived from an ancient gestural system based on the mirror mechanism: the link between observer and actor became, in speech, a link between the sender and the receiver of messages.

Arbib introduces computational models of monkey mechanisms for the control of grasping and for the mirror system of grasping. The proposed FARS model of visually-directed grasping in monkeys is used to ground the study of the evolution of the human brain upon a detailed understanding of the brain of monkeys and of our common ancestors. The chapter offers a very detailed exposition of the mirror system hypothesis. Arbib has postulated that the progression from grasp to language through primate and hominid evolution proceeded via seven stages: grasping, a mirror system for grasping, a "simple" imitation system, a "complex" imitation system, a manual-based communication system, protospeech, and language. During the discussion of such transitions between stages the author goes "beyond the mirror" to offer more detailed hypotheses about the evolution of the language-ready brain.

In Chapter 12, Domenico Parisi and Angelo Cangelosi suggest that a unified computer simulation scenario can be used to try to answer three research questions: (i) What is the origin and past history of the ability to speak in the individual? (ii) What is the origin and past history of the ability to speak in the species? (iii) What is the origin and past history of the particular language spoken by the social group of which the individual is a member? That is, simulations should simultaneously include phenomena of language development, evolution, and historical changes. The proposed scenario is based on the interaction between the process of biological evolution and adaptation (simulated through evolutionary computation algorithms), the process of ontogenetic acquisition of language (through neural network learning), and the process of cultural transmission of linguistic knowledge (through cultural learning between organisms). This approach also permits investigation of the effects of behavioral, cognitive, and neural factors in the evolution, learning, and transmission of language.

To show how such a scenario can be used to study language evolution, the authors discuss three simulations that include lexicons of various complexity (from single-signal communication to compositional languages) and learning and interaction patterns of increasing complexity (from genetically-inherited signals to culturally-learned words). In particular, attention is placed on the interaction between behavioral and neural factors in the evolution of language. For example, results show that the evolution of good categorization abilities (to discriminate essential foraging stimuli) favors the emergence of optimal lexicons that discriminate between these categories. At the same time, language has a positive effect on categorization because it improves the neural network's categorization by optimizing its internal representation of categories. Moreover, in compositional languages using verb-noun structure, verbs have a larger positive effect on performance than nouns and this is reflected in their neural representation.

Auto-organization and dynamic factors

This part contains two chapters that discuss in depth the role of auto-organization and dynamical processes in the emergence of language structure. Language is seen as a complex dynamical system whose functioning depends on the interaction of autonomous individuals. The communicative behaviors of these individuals determine the dynamic auto-organization of complex linguistic phenomena. The communication between language users affects the concurrent and interlaced development of categories (meanings) and words (forms).

The processes of the auto-organization and emergence of shared language structure such as lexicons and grammars are thoroughly discussed in Chapter 13 by Ed Hutchins and Brian Hazlehurst. Emergent linguistic structures result from the organized interactions of patterns that are present in the initial conditions or are produced in the history of interactions within the models. Positive feedback loops control this process, which is capable of creating novel and complex structures. However, researchers can take different stances regarding the elements of language (Meanings, Referents and Forms) and the relations among those elements (e.g., arbitrariness in Forms-Referents associations). Hutchins and Hazlehurst consider three major frameworks for modeling the emergence of shared lexicons and grammars: (a) Expression/Induction, (b) Form-Tuning and (c) Embodied guessing games. For example, in models of the emergence of lexicons, researchers that use the Expression/Induction approach start with a closed set of forms and meanings to evolve shared form-meaning mappings. Following the Form-Tuning framework, employed by the same authors, the set of forms is not pre-determined and is made dependent on the organization and sharing of sensory-motor experiences. With the Embodied language game approach the most complex representation of meanings is used together with a set of closed and arbitrary forms.

Hutchins and Hazlehurst argue that Expression/Induction models are limited because these assume that complex structured meanings simply exist prior to the language phenomena that emerge later. This assumption is inherited from the physical symbol system hypothesis which posits that intelligent processes (including language) are realized through internal manipulation of language structure. The private language of thought (meanings) arises prior to, and independently of, the public language. In contraposition, the Form-Tuning and the Embodied frameworks view the evolutionary development of language and cognitive abilities to be strictly interdependent on each other. This supports a cultural symbol system hypothesis in which public symbols arise concurrently with the internal meaning structures with which they are coordinated.

Language is characterized as a complex system in the work of Takashi Hashimoto (Chapter 14). As such, language is regarded as an essentially dynamic system. There are four dynamic processes that are of interest in the study of language evolution: (a) the origin of first linguistic systems, (b) the evolution of various languages and language structures, (c) the development and acquisition of language in children and adults, and (d) the sense-making process of giving meanings to words during communication. In addition, the language dynamics depend on the subjectivity of individual language users. Complex linguistic structures will emerge from the interaction of such dynamical and subjective processes.

To understand such a complex dynamical system the constructive approach of computer simulation is proposed. This permits the modeling of the interaction of the individual activities of speaking, listening and understanding that contribute to the emergence of dynamical language structure. Hashimoto presents various simulations using this constructive approach. For example, he models the sense-making activity as the formation of a web of relationships between words through conversation between individuals. The analysis of the dynamics of the formation of word clusters (categories) shows the coexistence of global stability (closed categories with rigid links) and local adaptability (prototype categories with changeable, gradual links) depending on the use of words during communication. Subsequently, the integration of word-web categorizations into a system with evolving grammatical structures causes the emergence of a bigger variety of word-web categories. Hashimoto's work supports the view that the essence of language relies on a series of internal mental processes and on a series of conversations.

Simulation and empirical research

In the concluding chapter Michael Tomasello suggests a direction for computer simulations of language evolution. He acknowledges that there is not much directly comparable evidence on the ancient past *facts* of the origins and evolution of language. However, there are a number of *processes* that are currently ongoing in the real world that were very likely to have been involved in the evolution of language. These processes regard non-human primate communication, child language acquisition, imitation and cultural learning, and language change mechanisms such as pidginization and creolization. Most of these processes are studied empirically, with observational and experimental techniques. Some of the researchers using computer simulations have already attempted to benefit from recent research, sometimes by directly comparing simulation research with empirical data. Following this approach Tomasello suggests that computer simulations "can only make real progress with at least a little bit of back-propagation from facts about the way these processes operate in the real-world". This can create a "virtuous circle": research-informed simulations will produce new predictions and insight into the evolution of language; experimental studies will verify these predictions leading to new insights that will then produce better simulations, and so on.

Tomasello also claims that experimental and simulation studies on the origins and evolution of language require attention to three distinct time frames: evolutionary, historical, and ontogenetic. For each of these dimensions, he proposes a series of ideas. In evolution, to investigate human ability to use language we must look at human ability to understand others as intentional agents. In historical terms, processes such as grammaticalization suggests that constructions of individual languages have been built up by groups of people communicating with one another under the general constraints of human cognition, communication, and vocal-auditory processing. In ontogeny, young children themselves create linguistic abstractions, such as word classes, by using general cognitive and social-cognitive skills.

Other approaches

This book contains a collection of some of the current simulative work in the evolution of language. However, it is not meant to cover *all simulation approaches* to the evolution of language and *all issues* addressed through computational modeling. The editors have tried to cover broadly the full range of simulation methodologies, and the range of different research issues. However, we want to acknowledge explicitly other work that has not been possible to include in the book. Among the many computational models, we would like to mention the early simulations on the evolution of communication by Hurford (1991), MacLennan (1992), Werner and Dyer (1994), Ackley and Littman (1994), more recent models by Saunders and Pollack (1996), Arita and Taylor (1996), Oliphant (1996; 1999), Wagner (2000), Grim, St Denis and Kokalis (2001), the recent models of the evolution of syntax (Briscoe, 2000; Teal and Taylor, 2000), the robotic approach of Billard and Dautenhahn (1999), the computational modeling of primate social intelligence (Worden, 1996), the connectionist simulation of historical changes in English morphology (Hare and Elman, 1995), and other models of language diversity (Nettle, 1999; Nettle and Dunbar, 1997). The reader will find references to such models in the various book chapters.

Conclusions

Among the advantages of computer simulation in language evolution research we have stressed the feasibility of synthetic methodologies as new scientific tools for studying language as a complex dynamical system. The theme that language is a complex system, whose functioning is due to bottom-up and non-linear interactions of local components, is common to the various chapters in this book.

Concepts typical of the study of complex systems, such as that of emergence, auto-organization, and interaction will be used over and over again. A first common thread that links the variety of approaches, models and theories presented in the chapters is that of the *emergence of language* (see also Knight *et al.*, 2000). Most of the chapters claim and show that linguistic behaviors, structures and processes emerge due to the interaction between the components of complex systems. These components are constituted by interacting autonomous agents, their neural, sensorimotor, cognitive and communication abilities, and their social and physical environment. Computer simulation permits fine analyses for the investigation of the role of such components in the process of the evolutionary emergence of language. This volume, and in general the simulation research on the evolution of language, provide a long series of examples of the analysis of emergent linguistic behavior and structure. Shared vowel systems are shown to emerge because of their optimality and discriminatory features (Chapter 4), dialect continua emerge due to spatial organization of agents (Chapter 5), shared representations for sensorimotor grounding emerge through interacting language games between simulated and robotic agents (Chapters 9, 10 and 13), compositional linguistic structures emerge due to learning bottleneck (Chapter 6), and shared word categories emerge from internal mental processes and conversational games (Chapter 14).

A second common thread that connects most chapters in this volume is the strict *interaction and interdependence between language and the other non-linguistic abilities* and characteristics of the organisms and their environment. The study of the origins of language needs to look at the way other non-linguistic and non-communicative factors can have affected language evolution. Various chapters show that general cognitive constraints such as sequential learning can explain the evolution of word order constraints (Chapter 8), that our ancestors' behavioral and neural structures, such as the mirror system for grasping, can have shaped the evolution of the language-ready brain (Chapter 11), that sensorimotor and categorization abilities favor the evolution of language and contribute to the formation of word classes (Chapter 12). The interaction between language and general adaptive and population dynamics factors are analyzed in simulations on the evolution of simple signaling systems (Chapter 3) and in a mathematical model of the evolution of universal grammar (Chapter 7).

The final common theme of this volume is the *relation between simulation studies and empirical research*. At present, not many computer models make direct comparisons between simulation results and available empirical evidence, although simulations tend to be constrained on empirical research. In this volume, only few chapters systematically compare simulation and experimental data. For example, the evolved vowel system of simulations on the emergence of speech is compared with that of humans (Chapter 4), and the performance of neural networks in artificial language learning is directly compared with human subjects' data (Chapter 8). All other chapters look at available empirical evidence to define the parameters of models and to make some indirect comparisons between simulation results and published literature data. As Tomasello highlights in the final chapter, the simulation approach to language evolution can make real and significant progress if it directly looks at known facts related to language evolution, such as non-human primate communication, child language acquisition, imitation and cultural learning, and language change processes.

References

Ackley DH, Littman ML (1994) Altruism in the evolution of communication. In: Brooks R, Maes P (eds) *Proceedings of the Fourth Artificial Life Workshop*. MIT Press, Cambridge

Arbib MA (ed) (1995) *The handbook of brain theory and neural networks*, A Bradford Book/The MIT Press, Cambridge

Arita T, Taylor CE (1996) A simple model for the evolution of communication. In: Fogel LJ, Angeline PJ, Bäck T (eds) *Evolutionary Programming V*, MIT Press, Cambridge

Armstrong DF, Stokoe WC, Wilcox SE (eds) (1998). *Gesture and the nature of language*. Cambridge University Press

Billard A, Dautenhahn K (1999) Experiments in learning by imitation: Grounding and use of communication in robotic agents. *Adaptive Behavior*, 7: 415-438

Briscoe EJ (2000) Grammatical acquisition: Inductive bias and coevolution of language and the language acquisition device. *Language*, 76: 245-296

Cangelosi A, Parisi D (1998) The emergence of a 'language' in an evolving population of neural networks. *Connection Science*, 10: 83-97

Cangelosi A, Parisi D (2001). How nouns and verbs differentially affect the behavior of artificial organisms. In Moore JD, Stenning K (eds), *Proceedings of the 23rd Annual Conference of the Cognitive Science Society*, Lawrence Erlbaum Associates, pp 170-175

Cavalli Sforza LL (2000) *Genes, peoples and languages*. Penguin Press.

Churchland PS, Sejnowski TJ (1994) *The computational brain*. MIT Press

Davidson I (2000) Tools, language and the origins of culture. In: Dessalles JL, Ghadakpour L (eds) *Proceedings of the 3rd International Conference on the Evolution of Language* de Boer B (2000), Self organization in vowel systems. *Journal of Phonetics*, 28: 441-465

de Jong ED (1999) Analyzing the evolution of communication from a dynamical systems perspective. In: *Proceedings of the European Conference on Artificial Life ECAL'99*. Springer-Verlag, Berlin, pp 689-693

Deacon TW (1997) *The symbolic species: The coevolution of language and human brain*. Penguin, London.

Di Paolo EA, Noble J, Bullock S (2000) Simulation models as opaque thought experiments In: Bedau MA, McCaskill JS, Packard NH, Rasmussen S (eds) *Artificial Life VII: Proceedings of the Seventh International Conference on Artificial Life*. MIT Press, Cambridge MA, pp 497-506

Goldberg DE (1989) *Genetic algorithms in search, optimization, and machine learning*. Addison-Wesley

Grim P, St. Denis P, Kokalis T (2001) Learning to communicate: The emergence of signaling in spatialized arrays of neural nets. Technical Report, Department of Philosophy, SUNY at Stony Brooks

Hanson SJ, Burr DJ (1990) What connectionist models learn: Learning and representation in connectionist networks. *Behavioral and Brain Sciences*, 13(3): 471-518

Hare M, Elman JL (1995) Learning and morphological change. *Cognition*, 56: 61-98

Harnad S (1990) The symbol grounding problem. *Physica D*, 42: 335-346

Harnad S, Steklis HD, Lancaster J (eds) (1976) *Origins and evolution of language and speech*. National Academy of Sciences.

Hauser MD (1996) *The evolution of communication*. MIT Press, Cambridge MA

Holland JH (1975) *Adaptation in natural and artificial systems*. Univ. of Michigan Press.

Hurford J, Studdert-Kennedy M, Knight C (eds) (1998) *Approaches to the evolution of language*. Cambridge University Press, Cambridge UK

Hurford JR (1991) The evolution of the critical period for language acquisition. *Cognition*, 40: 159-201.

Hutchins E, Hazlehurst B (1995) How to invent a lexicon: the development of shared symbols in interaction. In: Gilbert N, Conte R (eds) *Artificial Societies: The computer simulation of social life*. UCL Press

Kirby S (1999) *Function, selection and innateness: the emergence of language universals*. Oxford University Press, Oxford

Knight C, Studdert-Kennedy M, Hurford J (eds) (2000) *The evolutionary emergence of language: Social function and the origins of linguistic form*, Cambridge University Press

Koch C, Segev I (eds) (1998) *Methods in neuronal modeling: From ions to networks (2nd edition)*, A Bradford Book/The MIT Press, Cambridge MA

Langton CG (1995) *Artificial life: An overview*. MIT Press, Cambridge MA

Levine DS (2000) *Introduction to neural and cognitive Modeling*. LEA, Mahwah, NJ

Lieberman P (1993) *Uniquely human: The evolution of speech, thought, and selfless behavior*. Harvard University Press.

Luna F, Stefansson B (eds) (2000) *Economic simulations in Swarm: Agent-based modelling and object oriented programming*. Kluwer Academic Publishers

MacLennan B (1992) Synthetic Ethology: An approach to the study of communication. In: Langton CG, Taylor CDF, Rasmussen S (eds) *Artificial Life II: SFI studies in the science of complexity*. Addison-Wesley, Redwood City, pp 631-658

Martin A, Haxby JV, Lalonde FM, Wiggs CL, Ungerleider LG (1995) Discrete cortical regions associated with knowledge of color and knowledge of action. *Science*, 270: 102-105

Nettle D (1999) Using social impact theory to simulate language change. *Lingua*, 108: 95-117

Nettle D, Dunbar RIM (1997) Social markers and the evolution of reciprocal exchange. *Current Anthropology*, 38: 93-99

Nolfi S, Floreano D (1999) Learning and evolution. *Autonomous Robots*, 7(1): 89-113

Nowak MA, Plotkin JB, Jansen VAA (2000) The evolution of syntactic communication. *Nature*, 404: 495-498

Oliphant M (1996) The dilemma of saussurean communication. *Biosystems*, 37: 31-38

Oliphant M (1999) The learning barrier: Moving front innate to learned systems of communication. *Adaptive Behavior*, 7: 371-383

Oliphant M, Batali J (1997) Learning and the emergence of coordinated communication. *Center for Research on Language Newsletter*, 11(1)

Parisi D (1997) An Artificial Life approach to language. *Mind and Language*, 59: 121-146.

Parisi D (2001) *La simulazione*. Il Mulino, Bologna.

Parker ST, Gibson KR (1979). A developmental model for the evolution of language and intelligence in early hominids. *Behavioral and Brain Sciences*, 2(3): 367-407

Piazza A (1996) Genetic histories and patterns of linguistic change. In: Velichkovsky BM, Rumbaugh DM (eds) *Communicating meaning: The evolution and development of language*, LEA Publishers, Mahwah NJ

Rizzolatti G, Arbib MA (1998) Language within our grasp. *Trends in Neurosciences*, 21(5): 188-194

Rumelhart DE, McClelland JL (eds) (1986) *Parallel Distributed Processing: Explorations in the microstructure of cognition*. MIT Press, Cambridge MA

Saunders GM, Pollack JB (1996) The evolution of communication schemes over continuous channels. In: *Proceedings of the SAB'96 conference on the simulation of adaptive behavior*. MIT Press, Cambridge MA

Savage-Rumbaugh ES (1998) *Apes, language, and the human mind*. Oxford U. Press

Searle JR (1982) The Chinese room revisited. *Behavioral and Brain Sciences*, 5: 345-348

Steels L (1996) Self-organizing vocabularies In: Langton CG (ed) *Proceedings of Artificial Life V*. Nara, 1996

Steels L (1997) The synthetic modelling of language origins. *Evolution of Communication* 1: 1-34

Steels L, Kaplan F (1998) Situated grounded word semantics. In: *Proceedings of IJCAI-99*. Morgan Kauffman Publishing, Los Angeles, pp 862-867

Steels L, Vogt P (1997) Grounding adaptive language games in robotic agents. In: Husband P, Harvey I (eds) *Proceedings of the Fourth European Conference on Artificial Life*. MIT Press, Cambridge MA, pp 474-482

Teal TK, Taylor CE (2000) Effects of compression on language evolution. *Artificial Life*, 6: 129-143

Wagner K (2000) Cooperative strategies and the evolution of communication. *Artificial Life*, 6(2): 149-179

Weiss GA, Sidhu SS (2000) Design and evolution of artificial M13 coat proteins. *Journal of Molecular Biology*, 300(1): 213-9

Werner GM, Dyer MG (1994) BioLand: a massively parallel environment for evolving distributed forms of intelligent behavior. In: Kitano H (ed) *Massively parallel artificial intelligence*. MIT Press, Cambridge MA

Worden RP (1996) Primate social intelligence. *Cognitive Science*, 20: 579-616

Zahavi A (1975) Mate selection - A selection for a handicap. *Journal of Theoretical Biology*, 53: 205-214

Chapter 2

An Introduction to Methods for Simulating the Evolution of Language

Huck Turner

This chapter introduces some of the basic technical and theoretical background necessary to understand current computational approaches to studying the evolution of language. It is divided into two halves. The first introduces some of the specific technical details required to understand the major methods referred to in this book. The second half goes beyond specific implementation details to place these approaches in the context of relevant theoretical issues to get a sense of how they can and cannot be used to inform debate.

Simulations have been used to investigate many different aspects of the evolution of language and in many different ways. The models covered here differ with respect to the linguistic subject they attempt to illuminate and the simulation methods they utilize for the task. In terms of the linguistic subject, the major models fall into a number of categories. Cangelosi and Harnad (in press), Cangelosi and Parisi (1998), Hutchins and Hazlehurst (1995), MacLennan and Burghardt (1994) and Steels and Vogt (1997) have modeled the emergence of symbols and simple lexicons, while others have concentrated on the emergence of various syntactic properties. These include regular compositionality (Batali, 1998; Kirby and Hurford, 1997; Steels, 1998), recursion (Batali, 2000; Christiansen and Devlin, 1997; Kirby, 1999), syntactic selection (Cangelosi, 1999) and syntactic universals (Briscoe, 2000; Christiansen, Dale, Ellefson and Conway, this volume). Steels (1998) has also produced a composite model in which both symbols and simple syntax emerge. Others have modeled the emergence of coordinated communication (Di Paolo, 2000; Noble, 2000), self-organization of sound-systems for communication (de Boer, 1997), the dynamics of language evolution (Hashimoto, this volume) and aspects of historical change such as the formation of dialects (Livingstone and Fyfe, 1999).

These simulation approaches to modeling language evolution can be contrasted with mathematical models like that of Komarova, Niyogi and Nowak (2001).[1] It is beyond the scope of the present review to compare the relative merits of mathematical and computational approaches, but the reader should be aware of such alternatives. In their mathematical model, Komarova *et al.* (2001) calculated the learning accuracy required for a grammar to conventionalize in a population and related the number of example sentences that would be required to accurately induce the grammar of a target language to the size of the search space of grammars specified by a universal grammar.

Simulation Methods

In nearly all of the simulations discussed here, communicating organisms are modeled using populations of communicating agents. These agents are represented in different ways in different studies and come to possess linguistic abilities via different processes. The following outlines a categorization of these differences within which the major simulation methods are located (including all of those referred to in the present volume). Some introductory concepts are covered to make it accessible to readers from a broad range of disciplines. Those familiar with these concepts can probably skip many of the details, but may nevertheless gain some perspective by following the discussion.

Agent representations

Rules
The control of an agent's behavior can be modeled using finite state machines, which are essentially tables or sets of rules that, given a current state of a system and a valid input, determine the state that the system should be in next. Actions can be associated with states to control the behavior of agents. For example, MacLennan (2001) used finite state machines to control agents' signaling behavior. In this study, the tables encoding state transitions were genetically inherited by agents and were allowed to vary and hence evolve under selection pressure for coordinated behavior.

Finite state machines were also used by Saunders and Pollack (1996) in a simulation of the tracking behavior of ants. In this case, they were used to model the default motion of individual ants, which could be overridden when agents detected the simulated equivalent of chemical signals indicating the direction of food sources.

In Noble's (2000) simulations, agents possessed rules governing their signaling behavior during competitive encounters with one another. The agents had access to various 'perceptual' inputs reflecting their current fighting ability, energy level and location within the fighting arena as well as the previous movement of the

---

[1] See also Komarova and Nowak (this volume).

opponent, distance from opponent, previous signal emitted by the opponent, and a number of other factors. The rules were used to relate these inputs to the actions of moving forwards and backwards by varying amounts and producing auditory signals of varying strengths. Moving close to an opponent was interpreted as a physical attack while moving outside of the fighting arena was interpreted as submission. By winning fights, agents accumulated energy which was used to determine the number of offspring each agent would produce. As in MacLennan (2001), agents inherited the rules of their parents (sometimes imperfectly), enabling fighting strategies to evolve.

Hashimoto's (this volume) agents use rules to represent simple internalized grammars. These are phrase structure rules like those used in Chomsky's early work (Chomsky, 1957). Hashimoto combined these rules with what he calls *word-webs*, which represent relationships between the various entities referred to in a conversation.

Rules have also been used to associate meanings with signals (e.g., Batali, 2000; Kirby, 2001). These associations have also been represented using neural networks (e.g., Batali, 1998; Livingstone and Fyfe, 1999).

Neural networks

Agents are often implemented using neural networks. Due to their importance, some basic concepts[2] will be covered here briefly before looking at how they are applied.

Artificial neural networks are inspired by the natural neural networks that constitute the nervous tissue of animals. They are made up of a number of nodes (also called units or neurons) linked to one another via weighted connections that communicate activation levels between them. Each node has an activation level that is determined either by an external input or as a function of inputs from other nodes. Typically, connection weights can vary in strength and can be either excitatory or inhibitory. Learning can be modeled by tuning these weights.

Like their biological counterparts, artificial neural networks process information in parallel allowing multiple constraints to interact simultaneously (McClelland, Rumelhart and Hinton, 1986). Neural networks also enable representations to be distributed in the sense that there is no one-to-one mapping between concepts and nodes (Hinton, McClelland and Rumelhart, 1986). These properties make them less vulnerable to noise and localized lesions than other kinds of representations like rules. A rule-based system will usually rely completely on the accuracy of input and the integrity of the rule-base to produce anything resembling an appropriate output. For a typical neural network, noise or damage will cause output to be less accurate, but still approximately correct. In other words, the quality of solutions degrades gracefully.

To demonstrate this, consider how Kirby's (2001) model deals with noisy input. In that model, a pressure for agents to use shorter signal strings is created by introducing noise that has the effect of corrupting longer signals more often than shorter ones, but this is a fact about rule-based systems which is not generally true

---

[2] For a much more detailed introduction to neural networks, see Rumelhart and McClelland (1986) and McClelland and Rumelhart (1986).

of neural networks. Consider the following pair of incomplete signals: 'p#t' and 'el#ph##t'. It is not possible to identify the first example because the information available is not sufficient to narrow it down to a single word. It could correspond to any of *pat*, *pet*, *pit*, *pot* or *put*. Being a longer word, English speakers can narrow the second example down to a specific word without any trouble presumably as a result of the simultaneous interaction of multiple constraints. Rule-based systems generally do not degrade gracefully so cannot make use of partially specified data. As a consequence, they are just as likely to fail when a small amount of noise is present as they are when a large amount is present.

The capacity to function despite noisy input makes neural networks good for categorical perception, and it is this ability that Harnad (1990) argues is a basic prerequisite for symbolic communication. Without it we wouldn't be able to distinguish between different symbols or between the different objects and concepts they label.

There are many different neural network architectures, but in simulations of language evolution, the most common variants are multi-layer neural networks and recurrent neural networks   (see Figure 2.1). These can be trained to associate patterns of input and output by providing examples of known input-output pairings.

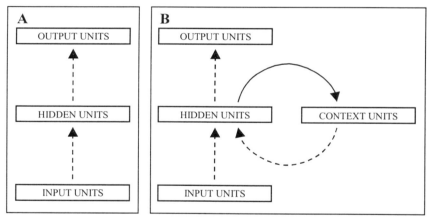

Figure 2.1 (a) A standard multi-layer neural network with one hidden layer, and (b) a recurrent neural network. Dashed arrows indicate sets of trainable connections where each unit in one layer is linked to every unit in the other. The solid arrow in (b) indicates fixed, one-to-one connections responsible for direct copying of activation levels from hidden-layer units to context-layer units.

Multi-layer neural networks, as the name suggests, consist of a number of layers of units each of which feeds its activations on to the next. The input layer receives its activation values from the researcher or from some other system external to the network. This input is fed through the network to generate a pattern of activation in output layer units. For instance, Arbib (this volume) uses a neural network to control the movement of a simulated arm and hand. His network takes as input a representation of a simulated object's shape and location as well as the initial position of the arm. As output, the network produces a representation of a movement trajectory that would allow the hand to grasp the object. Some other examples of input-output mappings are given in Table 2.1.

Table 2.1 Some examples of neural network inputs and outputs from studies of the evolution of language and communication.

Study	Inputs	Outputs
Batali (1998)	Takes one character at each time step.	A vector representation of signal meaning.
Cangelosi and Parisi (1998)	Some units dedicated for communication signals, and others for the perception of food quality and direction.	Some units representing motor responses in the simulated environment, and others for generating a communication signal.
Christiansen and Devlin (1997)	Takes the lexical category of the current word in a training text.	Predictions of the category of the next word in the sequence.
Hazlehurst and Hutchins (1998)	Units representing a shared visual field, the location of the agent's own focus of attention as well as interlocutor's, and some to receive communication signals.	Some units representing a shift in the focus of attention, and others for generating a communication signal.
Saunders and Pollack (1996)	Presence or absence of food and n inputs for 'chemical' signals.	Units indicating whether to follow default movements or the direction of one of n 'chemical' signals, and others for generating communication signals.

The interpretation of inputs and outputs is usually pre-specified by the researcher, but this is not the case for units in hidden layers and for these no clear interpretation will be possible in most cases, even after learning, due to the distributed nature of the representations. The advantage of this is that the researcher does not have to make unnecessary assumptions about internal representations. A disadvantage is that it can make it more difficult to explain how a model does what it does and hence more difficult to extrapolate from it to the real world.

The most commonly-used algorithm for learning in multi-layer neural networks is back-propagation (Rumelhart, Hinton and Williams, 1986). In simple terms, it involves comparing actual output activation levels to target activation levels (often called teaching inputs) to calculate an error value for each output node. The error of a node is then propagated backwards through the network to every node in the previous layer (from which it has an input) and is apportioned according to the relative contribution each made to that error. These errors are then used to modify the weights of connections slightly.

The neurological plausibility of back-propagation is doubtful for a number of reasons. Firstly, it is not clear that real synapses can transmit error backwards. Secondly, we rarely have the opportunity to quantify the errors we make by comparing our output behaviors with correct target behaviors provided by a teacher. Thirdly, unlike learning in the real world, back-propagation typically requires exposure to a vast number of examples.

There are other ways learning can be modeled in a neural network. For instance, connection strengths can be evolved using a genetic algorithm (Montana and Davies, 1999), which may or may not be more neurologically plausible. For a selectionist account of learning at the neurological level see Edelman (1987).

Recurrent neural networks (Elman, 1990) are a variation on standard multi-layer networks, that can learn temporal sequences. In terms of architecture, the only difference between these two types is that recurrent neural networks have a set of *context* units which store the previous activation levels of the hidden units and feed them back as inputs to the hidden layer at the next time step (see Figure 2.1b).

A significant failing of standard multi-layer nets is that the number of inputs that can be presented to the network is fixed. If inputs could be presented to a network sequentially then it would be possible to process data such as words and sentences that can vary in length. This kind of sequential processing is possible using recurrent neural networks.

Elman (1990) trained a network of this kind on grammatical sequences of text (generated using a context-free grammar) using the next word in the sequence as the target output at each time step. After training, the text was presented to the network again and the activation levels in its hidden layer were compared at each time step. The hidden representations for each word clustered (in terms of similarity) into established word classes and subclasses and, for each, the network was able to estimate the approximate likelihood that the next word was a member.

Batali (1994) has applied recurrent neural networks to studies of the Baldwin effect and more recently to the evolution of compositional syntax (Batali, 1998; 2000). Christiansen and Devlin (1997) have also applied recurrent neural networks to the evolution of consistent branching in linguistic structures. For a discussion of the problems associated with representations of syntactic structure in neural networks see Fodor and Pylyshyn (1988).

Modeling interaction and knowledge acquisition

The representation of agents is not the only relevant variable. Agent interactions also shape the linguistic phenomena that emerge. During these interactions, the linguistic proficiency of agents can be evaluated or improved through various structured communication tasks or *language games*. This section concerns the modeling of such interactions and the various ways in which agents come to possess linguistic knowledge.

Learning

Rule generalization

In Kirby's (2001) model, the population of each generation consists of only one individual who imparts knowledge about its language on to the next. Language learning involves finding mappings between signals and structured meanings. The teacher provides a limited number of signal-meaning pairs that the 'child' must use to reconstruct the whole language. The task is simplified somewhat by allowing the child direct access to the meanings associated with each of the signals produced by the teacher.

The following are examples of the kind of rules Kirby (2001) uses to model an agent's knowledge of meaning-to-signal mappings.

(1) a. $\qquad\qquad S : (a_0, b_0) \rightarrow$ abc
　　b. $\qquad\qquad S : (a_0, b_1) \rightarrow$ abd

The terms a_0, b_0 and b_1 are elements of the meaning to be expressed and the strings on the right of the arrow are the signals that are used to express the specified meaning components. In the first example, the meaning specified by a_0 and b_0 is mapped to the string 'abc'. In the second, a_0 and b_1 are mapped to 'abd'. These rules do not capture the similarities between the two mappings so rather than have a single rule for each, Kirby's agents attempt to generalize. In this process, pairs of non-compositional rules like those in (1) are replaced with equally expressive compositional rules like those in (2) where x and y are variables that stand for elements of meaning.

(2) a. $\qquad\qquad S : (x, y) \rightarrow A : x\ B : y$
　　b. $\qquad\qquad A : a_0 \rightarrow$ ab
　　c. $\qquad\qquad B : b_0 \rightarrow$ c
　　d. $\qquad\qquad B : b_1 \rightarrow$ d

Obverter

Oliphant and Batali (1997) use a different approach they call the *obverter* procedure to model the learning of signal-to-meaning and meaning-to-signal mappings. The essence of the procedure is to set the signal-to-meaning (receive) and meaning-to-signal (send) mappings of an agent with respect to the mappings that already exist within the population to maximize the likelihood of mutual intelligibility. The *send* mappings that are adopted are those that are most likely to be correctly decoded given the average receiving response in the population. Likewise the *receive* mappings that are adopted are those that are most likely to result in the agent correctly decoding the signals of others.

From an initially random (and hence uncoordinated) set of communicative behaviors, Oliphant and Batali (1997) demonstrated that if new agents learn the language of their parents using the obverter procedure, an optimally coordinated system is guaranteed to emerge.

As in Kirby's model, the obverter procedure requires that an agent access information about the internal states of other agents. In its strongest form, the obverter procedure requires that agents have access to the internal mappings of *every* other agent in the population. However, Oliphant and Batali (1997) have shown that this strong version can be approximated by sampling the send and receive mappings from a subset of the population.

Imitation

Oliphant and Batali (1997) contrast the obverter procedure with learning by imitation (e.g., de Boer, 1997; Livingstone and Fyfe, 1999) and argue that the latter is not guaranteed to increase communicative accuracy.

De Boer (1997) has modeled the self-organization of vowel systems in populations of interacting agents that possess human-like perceptual limitations. Vowel sounds are communicated to each other during imitation games via representations of their formant[3] characteristics. The vowels that emerge from the iteration of these games are maximally distinct perceptually and resemble counterparts in natural languages.

Livingstone and Fyfe (1999) model meaning-signal mappings using neural networks that allow activation to propagate in both directions between a meaning layer and a signal layer. Learning is achieved by training an agent's network to agree with the mappings of an 'adult' agent. To adjust the weights of agent networks, Livingstone and Fyfe (1999) use the back-propagation algorithm (discussed below).

Self-understanding

Oliphant and Batali (1997) also contrast the obverter procedure with learning that involves agents structuring their send behavior with respect to their own receive mappings – essentially sending signals that they themselves can decode.[4] In a model like this, an agent does not need to be able to inspect the internal states of others to learn appropriate send behaviors, but the model still does not help to avoid the problem of having to access internal states to learn its receive mappings. Batali (1998) implemented a model like this using a recurrent neural network. His network takes a linguistic signal as input that is represented as a sequence of characters presented one at a time. The outputs of the network represent the meaning of the signal once the complete sequence of characters has been fed through. To determine the sequence of characters to send, the network selects, at each time step, the character that, if input to its own network, would guide its output representation closest to the target meaning. This process is repeated until the desired meaning is generated on the output units or until the sequence of characters has reached a maximum length.

---

[3] Formants are bands of energy in the frequency spectra of vowel sounds. Differences in formant frequency relationships correspond to differences in the perception of vowel quality. For an accessible introduction to phonetics see Ladefoged (1993).
[4] This learning procedure is sometimes called *obverter* as well (e.g., Kirby and Hurford, this volume).

Evolution

Genetic algorithms

The evolution of agents can be modeled using a genetic algorithm where their representations are encoded as a population of strings analogous to chromosomes. These are reproduced to a greater or lesser extent according to a fitness function calculated with respect to how well each satisfies the constraints of a given problem. John Holland (1975) first developed and formalized this approach in the 1960s. In his original formulation he used strings of ones and zeros to represent chromosomes, but encodings involving larger alphabets (more than just ones and zeros), real numbers and tree-like structures are now also used. There has been a proliferation of terminology to describe such variants, but classes that are widely recognized include evolutionary strategies (Schwefel and Rudolph, 1995), genetic programming (Koza, 1990) and evolutionary programming (Fogel, Owens and Walsh, 1966).

The fitness function that is used to select the best solutions is analogous to the environment of a species in that the solutions that proliferate are those that best satisfy its constraints. The fitness function is not in any sense a description of a target solution; it is instead a specification of constraints that a solution must satisfy. The actual form that any solution assumes is no more directed than in the biological domain except in examples used in textbooks where fitness functions are defined in terms of the similarity to a target form, but these are useful only for exposition and are obviously without any practical value (if the target solution is already known, there is no need for it to be evolved). Perhaps as a result of such expository simplifications, misunderstandings abound about the directedness of evolution in genetic algorithms and they have attracted criticism on this basis (e.g., Berwick, 1996).

Genetic algorithms can be viewed at different levels of description as optimizing, search or learning algorithms. They are optimizing algorithms because they are used to maximize the fitness of solutions. They are search algorithms because optimization is a kind of search – a search for solutions with (near) optimal properties. Evolution also resembles operant conditioning insofar as it proceeds by penalizing mistakes and rewarding successes – a correspondence recognized by Skinner (1966, 1981) – so genetic algorithms can also be used to model learning (by optimizing connection weights in a neural network for instance). Computational models that involve both evolution and learning (e.g., Cangelosi, 1999) are often constructed using different optimization algorithms for the two aspects (typically a genetic algorithm for the former and back-propagation[5] for the latter), but there is nothing necessary about this.

As in the biological example, genetic algorithms involve selection, recombination and mutation. The selection procedure scores each string within the population according to what is called the objective function, which it uses to determine its fitness relative to others. Strings are then duplicated with a

---

[5] Back-propagation is sometimes referred to as a kind of 'gradient descent' learning - an allusion to an error landscape (not unlike a fitness landscape) on which learning is viewed as the descent of a solution (i.e., the set of weights in a network) to progressively lower points (representing lower error) on the landscape.

probability dependent on their fitness to produce a new population. The recombination (or crossover) procedure mates pairs of strings by randomly choosing a crossover point and producing two offspring each sharing features of both parents (as in Figure 2.2). The usefulness of the crossover procedure will depend on the genotype encoding and the problem domain. As in biology, crossover is not strictly necessary for evolution, but organisms that reproduce this way tend to evolve more rapidly than those that reproduce asexually.

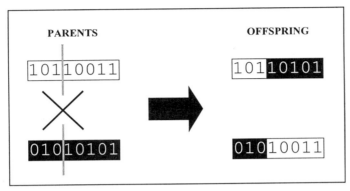

Figure 2.2 The crossover procedure. Two parents are combined to produce two offspring.

The mutation procedure acts with a very low probability to alter features randomly within the genotype. As in biology, mutations are usually deleterious but occasionally produce better solutions than can be generated through crossover alone. To determine a setting for the mutation rate, the need for variation in the population (preventing premature convergence) must be balanced with the need to prevent degrading the population by introducing too many mutants.

In the genetic algorithm literature, individual digits within a genotype encoding are called genes or features. For many problems to which genetic algorithms are applied, this use of gene conflicts with biological usage where genes are (often by definition) regarded as the unit of selection[6] (i.e., the thing that is replicating). Dawkins (1989:28) for instance, defines a gene as "…any portion of chromosomal material that potentially lasts for enough generations to serve as a unit of natural selection". He argues that a section of chromosome counts as a gene if it leads to the organism being selected over others for possessing certain traits and if it is unlikely to be disrupted by crossover or mutation and thus likely to be passed on to its offspring. A section of chromosome has a higher likelihood of being split up by the crossover procedure if it is very long. Consequently, genes tend to be contiguous or closely-packed segments that are relatively short. This definition

---

[6] That genes are the unit of selection is the consensus view in biology, yet many researchers will (in the same breath) equate genes with *cistrons* (a length of chromosome coding for exactly one protein molecule). Dawkins (1989) shows that the one will not always correspond to the other by illuminating cases where segments of chromosome act as a unit of selection, but are longer than an individual cistron.

corresponds to the concept of a building block in the genetic algorithm literature (e.g., Goldberg, 1989) while the biological counterpart of a digit in the genotype encoding is probably a codon or an individual nucleotide.

In models of language evolution, genetic algorithms have been used to evolve weights in populations of neural networks in the ecological simulations of Cangelosi and Parisi (1998), described below, and in Batali's (1994) study of the Baldwin effect where he used them to evolve weights in recurrent neural networks.

Game theory

The emergence of cooperative behaviors like those involved in communication are difficult to explain in adaptive terms because they seem to require reciprocal altruism between interlocutors, and altruists are vulnerable to exploitation by freeloaders (those who take advantage of the kindness of others without contributing in return). Freeloaders will always be more successful than their altruistic counterparts in a population, so the proportion of freeloaders will increase until there are no more altruists to exploit. Other strategies fair better against freeloaders, particularly those that involve remembering individuals who have exploited them in the past so as to prevent being exploited by them again. Complex interactions like these can be understood in terms of a branch of mathematics known as *game theory*. Briefly, a game is a competition between decision-making strategies where the size of the reward obtained during a given round depends not only on the decisions of the given strategy, but also on the decisions of its competitors. These concepts have been applied to evolutionary dynamics most notably by Maynard Smith (1982) where the concept of a reward corresponds to selective fitness. Relating this to the evolution of language, Noble, Di Paolo and Bullock (this volume) simulate the interactions between strategies in an attempt to understand the conditions that could give rise to honest communication systems.

Synthetic ethology

Instead of using an analytical fitness function in a genetic algorithm, fitness can be calculated by simulating an environment with which evolving agents interact. An approach like this was developed using neural network representations for agents by Parisi, Cecconi and Nolfi (1990). This kind of approach (sometimes called *synthetic ethology*) has also been applied to modeling the evolution of language by MacLennan and Burghardt (1994) and more recently by Cangelosi and Parisi (1998) and MacLennan (2001).

Cangelosi and Parisi (1998) evolved linguistic agents in a simulated environment containing edible and poisonous mushrooms that agents could eat and observe and about which they could communicate. The agents were essentially standard multi-layer neural networks with some inputs dedicated to perceptual qualities of the environment and some outputs dedicated to the agent's motor response. The agents evolved to perceive qualities that helped them to distinguish between edible and poisonous mushrooms and thereby produce the appropriate behaviors of approaching and avoiding respectively. In addition to the perceptual inputs and motor outputs, each agent possessed some inputs and outputs dedicated to linguistic signaling, but the form and meaning of signals was not pre-specified, allowing for a communication system to emerge spontaneously and with respect to the other categorization tasks that the agents were performing. The symbols that the agents used were therefore grounded since they derived their meaning from interaction with an environment (Harnad, 1990). In this case, symbols were grounded, not to the real world, but to a simulated one. Nevertheless, this kind of model allows the researcher to avoid dictating the interpretation of linguistic signals, which is enough to observe certain kinds of self-organization in an emerging communication system.

The synthetic ethology approach has been used to model both biological and cultural aspects of the evolution of language, usually by using a genetic algorithm for the evolution aspect and learning via back-propagation for cultural transmission during which parents teach offspring labels for different perceptual inputs, but genetic algorithms and back-propagation are both optimizing algorithms and so either could be used for learning.

Robotic approaches
As in synthetic ethology, a major motivation of robotic approaches is in the modeling of communication systems that have symbols grounded in an environment with which agents interact. However, what makes robotic approaches different is that robots have a material embodiment and communication systems grounded in a physical environment (albeit a controlled laboratory setting) to which they have access via cameras and other sensors.

The robotic approach has been applied to various aspects of language. Steels and Vogt (1997) have used it to simulate the emergence of grounded symbols and Steels (1998; this volume) has used it to simulate the beginnings of syntax.

In Steels and Vogt (1997), robots engage in a language game in pairs. One takes a speaking role and the other, a listening role. The game involves naming objects in their environment and as the simulation proceeds, the robots come to share similar perceptual and lexical distinctions as they adopt representations that progressively lead to better performance in games.

Simulation in Context

The remainder of this chapter places simulation methods in the context of some of the major issues within language evolution research, namely the innateness debate and the extent to which properties of language are adaptive.

Innateness

A central concern in the language evolution literature is to understand the extent to which processes governing linguistic competence and performance are innately specified, and simulations have been used to inform this debate in a number of ways. In some cases, they are being used to demonstrate that certain aspects of syntax are learnable without requiring a model to have prior (i.e., innately specified) knowledge of them (e.g., Batali, 1994, 1998; Elman, 1990). This kind of approach is only informative if certain conditions are met because to say that something is learnable does not in itself help to distinguish between the relevant hypotheses. After all, if a simulated agent could learn to build a nest, it would not demonstrate that birds in the natural world are not hatched with this ability. The learnability argument can only be falsified by demonstrating learning under the conditions in which language is actually acquired (i.e., when constraints such as *on the basis of the limited data available to the child* and *without negative feedback* are placed on the learning procedure). Biological evolution has operated without these constraints. Negative feedback, for instance, is provided in the form of the environment culling genes for certain kinds of traits from the gene pool. So demonstrating that a language can be learned using an unconstrained learning procedure is not unlike demonstrating that a language can be evolved and it doesn't help to inform the debate.

Evolution is more likely to converge on good design features if related designs have sufficient plasticity to enable them to emulate these features (this is the Baldwin effect). Phenotypic plasticity can result from the ability to learn, but is not restricted to learning. Muscle development, callus formation and skin tanning are also examples where exposure to certain kinds of environmental stimuli trigger changes in the phenotype. Phenotypic plasticity in a population has the effect of smoothing out the fitness landscape so that the fitness of closely-related individuals will tend to be more correlated. This means that a peak in the fitness landscape will have a broader base making it easier to find by sampling the space of variations. The perils of trial and error and of missed opportunities will mean that individuals with innate knowledge of some useful skill will always have an advantage over those that can only acquire the skill through learning. Given this, one should expect adaptive abilities that are learnable in one generation to become easier to learn in subsequent generations (see Batali (1994) for a model incorporating this idea). Following this argument to its logical conclusion, we might expect natural selection to continue to shape neural structures or biases to facilitate the learning of adaptive abilities until such a point that these neural dispositions effectively embody the ability or until the cost involved in terms of growth and maintenance of these neural structures exactly offsets the gain in fitness that they confer. There appears to be a continuum here over which evolution might have traversed from the ability to learn, to biased learning, to innate knowledge.

Bates, Elman, Johnson, Karmiloff-Smith, Parisi and Plunkett (1998) have suggested one reason why this kind of adaptive pathway might be implausible. They have expressed their incredulity about the possibility that linguistic and other knowledge could be genetically encoded by saying "it is difficult to understand how 10^{14} synaptic connections in the human brain could be controlled by a genome with approximately 10^6 genes."[7] This appears to be a surprising failure of imagination for these authors given the frequent use of expressions such as 'infinite use from finite means' in the linguistics literature. To understand *how* information about 10^{14} connections could be compressed to this extent one could further appeal to the general finding from chaos theory that complexity (even apparent randomness) can arise from the interplay of a few simple components (e.g.,

---

[7] Recent news from the human genome project indicates that the true number of genes is much less than even this. The current estimate is in the order of 3.5×10^5 which is around a third of the Bates *et al.* (1998) estimate.

Kauffman, 1995). So why should we expect the complexity of the human genetic endowment to be of the same order as the structures they code for? As Dawkins (1991) has stressed, DNA is not a blueprint for the mature organism. A more appropriate analogy, he says, is that of a recipe.

> A recipe in a cookery book is not, in any sense, a blueprint for the cake that will finally emerge from the oven ... a recipe is not a scale-model, not a description of a finished cake, not in any sense a point-for-point representation. It is a set of *instructions* which, if obeyed in the right order, will result in a cake. (Dawkins, 1991:295)

If DNA is not a blueprint, we should not expect the complexity of its instructions to be of the same order as the complexity of the resulting form, but it isn't obvious that the neural structures embodying linguistic knowledge are complex anyway. While the complexity of language *data* has been cited as evidence against its learnability – the generalizations required being too deep to uncover by general-purpose learning mechanisms and the data available to the child too fragmentary – linguists who offer such arguments do not generally believe that the *mental structures* constituting linguistic knowledge are themselves complex.[8] The trend in generative linguistics, especially over the past twenty years, has been to unify the features of universal grammar under fewer and fewer principles. Perhaps as a consequence of this success, the computational system of human language is assumed to be extremely elegant. For instance, Uriagereka (1998) compares its elegance to the growth functions that give rise to patterns in peacock feathers and flower corollas.

Adaptive benefit

The extent to which properties of language are adaptive is another major area of contention. Uriagereka (1998)[9], argues that grammar may have evolved not as an adaptation, but as a particular kind of exaptation[10] or perhaps as a spandrel.[11] Part of the reason he offers for this view is that universal grammar seems to be too elegant to have been crafted like this (principles don't appear to follow from multiple determinants and they don't appear to exhibit the messiness that is characteristic of adapted structures). Another reason involves personal incredulity about functional explanations of any linguistic constraint that disallows

---

[8] Fodor (1998) makes a similar point in response to a recent version of Pinker's argument (Pinker, 1997) that complex functionality necessitates an adaptative explanation.
[9] This view is typically attributed to Chomsky although he has never put a clear version of it in print.
[10] Gould's (1991) label for what Darwin called *preadaptation*. In Gould's terms an exaptation is "a feature, now useful to an organism, that did not arise as an adaptation for its present role, but was subsequently coopted for its current function." (p.43).
[11] Gould and Lewontin's (1979) label for a feature that exists simply because something like it must be present. This is in contrast to adaptations which exist because they are functional in terms of reproductive success. Pinker and Bloom (1990) offer the redness of blood as an example.

semantically interpretable sentences. Take the following example from Uriagereka (1998:65):

(3)   *What have you discovered the fact that English is?

Uriagereka (1998:50) expresses his incredulity thus: "What's the evolutionary advantage of having [a constraint], if it *disallows* the communication (in those terms) of a perfectly fine thought?"

Firstly, 'in those terms' is an important qualification here because this thought certainly *can* be expressed by other means. Secondly, even if this constraint is maladaptive in some ways then this would not be unusual for an adaptation. Cziko (1995:182) makes this point in relation to the anatomy of the vocal tract:

> ...unlike mammals that maintain separate pathways for breathing and feeding, thus enabling them to breathe and drink at the same time, adult humans are at a much higher risk for having food enter their respiratory systems; indeed, many thousands die each year from choking ... The risk of choking to which we are exposed results from our larynx being located quite low in the throat. This low position permits us to use the large cavity above the larynx formed by the throat and mouth (supralaryngal tract) as a sound filter ... We thus see an interesting trade-off in the evolution of the throat and mouth, with safety and efficiency in eating and breathing sacrificed to a significant extent for the sake of speaking.

Some universals may be difficult to explain in terms of gradual adaptations, but others seem quite adaptive by contrast. Pinker and Bloom (1990) suggest some adaptive explanations for case systems, subject-verb agreement and many other universals. Genes that facilitate their possessor making use of a given linguistic feature might replicate more successfully as a consequence of something as arbitrary as the conventionality of the feature within the linguistic community. The linguistic system is part of the environment that selects the genes of language speakers.

Deacon (1997) points out that the reverse may also be true – a linguistic system could be regarded as evolving to fit a niche with the language learner being part of the environment that selects its features. "Language operations that can be learned quickly and easily by children will tend to get passed on to the next generation more effectively and more intact than those that are difficult to learn" (Deacon, 1997:110). Deacon also argues that the limitation on the amount of data that a language learner can use to reconstruct the language of its community provides another kind of selection pressure which may make certain features of language look maladaptive or 'quirky' if viewed in terms of the benefit to language speakers. Chomsky and colleagues reject adaptationist explanations precisely because they are incredulous about functional explanations of linguistic properties for which the language-learner is benefactor, but "if, as linguists often point out, grammars appear illogical and quirky in their design, it may only be because we are comparing them to inappropriate models and judging their design according to functional criteria that are less critical than we think" (Deacon, 1997:110f). The perspective afforded by Deacon's language-adaptive view could explain some of this quirkiness while remaining an adaptive explanation of sorts.

Kirby's (2001) model appears to lend itself well to this kind of analysis. This model involves the transmission of linguistic knowledge from generation to generation. The structure of the agents themselves does not evolve, but the communication systems that they use do. This is achieved by iterated learning in which successive generations of agents are required to learn language from their parents. The model demonstrates that, given a limited number of opportunities for agents to reconstruct the languages of their parents, the communication systems that evolve exhibit the appearance of design for minimizing errors in transmission from generation to generation. This manifests itself in communication systems that exhibit regular compositionality (as opposed to systems in which each structured meaning is expressed via a unique symbol) and the reason for this is that fewer rules are needed to express all the available meaning combinations. Since the description of a compositional language is more parsimonious than that of non-compositional ones, it requires less data to learn and so is more likely to be transmitted intact under conditions in which the linguistic data available to the language learner is limited. Independent evidence for this view has been provided in simulations by Batali (2000) and in a purely mathematical model by Nowak, Komarova and Niyogi (2001).

Kirby also demonstrated that when there is a pressure for shorter strings as well, the incidence of short irregular forms like those observed for commonly used words (e.g., irregular past tense forms such as *went* in English) becomes predictably stable. Kirby introduces a pressure for shorter strings by introducing noise that has the effect of corrupting longer signals more often than shorter ones, but there are plausibly a number of other factors such as time constraints or least-effort principles that could provide pressure for shorter strings.[12]

The idea that languages themselves evolve rather than (or concurrently with) their speakers is not a new one[13], but many have ruled out the possibility. Among them Uriagereka (1998:33ff), but he does so as a result of a misunderstanding of the unit of selection in linguistic evolution. He summarizes his argument as follows.

> Is there any meaning to the claim that English is fitter for survival on the American plains than the language of the Navajo? If English dominates the continent, it does so because of the strongest army in the world, whose finest attribute isn't precisely its verbal brilliance. (Uriagereka, 1998:34)

He argues that even when language change occurs in the absence of such sociological factors, there is no sense in which languages that disappear are less well-adapted than those that remain. That much language change appears to be cyclical gives weight to this view. At the very least, it suggests that not all language change that occurs in modern times is adaptive.

It is easy to reach premature conclusions about adaptive explanations if one isn't adhering to the logic of gene-centrism, and Uriagereka is probably right to

---

[12] See Kirby and Hurford (this volume) for an analysis in terms of least-effort.
[13] Deacon (1997) reviews the history of this idea.

reject an account of language evolution in which the unit of selection is a language for the same reason that modern biologists reject the idea of selection at the level of the species or the individual. It is much easier to understand how evolution could occur when considering linguistic features as competing rather than whole languages.

Another apparent problem has been highlighted at a more fundamental level concerning the adaptive advantage of speaking in the first place (e.g., Ackley and Littmann, 1994; Batali, unpublished; Cangelosi and Parisi, 1998). The benefit to the hearer is taken to be obvious if communication is about the exchange of information, but "[w]hat is the advantage of producing the signal to the individual that produces it? Why should an individual that produces the appropriate signals live longer and have more offspring than other individuals that fail to do so?" (Cangelosi and Parisi, 1998:85). Again, this paradox gets its potency from a confusion over the unit of selection, and framing the question this way suggests a problem that simply does not exist. Under a modern, gene-centric view of evolution, it is the reproductive success of genes and not individuals that is relevant, and the former does not always entail the latter. There may be genuine questions that need to be answered about the reproductive advantage to a speaker's genes, but the wealth of literature on kin selection and evolutionarily stable strategies suggest likely explanations that make this seem much less paradoxical.[14] Briefly, if an individual's behavior increases the likelihood of close kin being replicated, and if that behavior has a genetic basis, then it might still be selected in the gene pool even if it reduces the reproductive success of the specific individual in question. This is because the kin that benefited are likely to possess these genes as well. For example, genes that promote parental care (such as attending to a baby's cries) increase the likelihood of their own replication even though they are at the expense of the parent as an individual, because the parental care genes are likely to be inherited in the crying baby. Incidentally, for an infant, crying appears to be an instance where communication does benefit the 'speaker' (hunger and other unpleasantness go away with some reliability), so at least in this case, explaining the reproductive advantage to the signaler's *genes* is trivial.[15]

Some aspects of language might not demand an adaptive explanation. For instance, conventionality can arise without any selection pressure at all. A feature that is neutral with respect to fitness can become conventionalized across a population simply because, given time, every individual will come to have a common ancestor from whom the feature is ultimately inherited (genetically or culturally).

Some linguistic universals might not be adaptive while others are. Of those that are, some may be adaptive in terms of the benefit to language users while others may be adaptive to the linguistic systems themselves. We should expect the interactions between biological and cultural evolution to be complex. A biologically evolved language acquisition device would be a central feature of the environment to which a culturally evolving language is exposed and as such would

---

[14] See Dawkins (1989) for an accessible introduction that is also a primary text in this field.

[15] The listener isn't typically the one who benefits from a speaker's boasting or persuading either.

be the primary determinant in the selection of linguistic features. At the same time, stable features of a linguistic system might be assimilated into the genome of its speakers. Computational models of these processes might help us to understand the dynamics of such co-evolutionary processes.

Of course, demonstrating the emergence of syntax in a broad sense is not the same as demonstrating the emergence of the particular kind of syntax that linguists call universal grammar, and while the properties of universal grammar are still being debated, it will remain something of a moving target. Nevertheless, computational models serve to demonstrate the consequences of a model's assumptions and what is possible *in principle*.

Conclusion

Although we cannot go back in time to observe the emergence of language in our species directly, we have observed the emergence of modern languages in communities formed when speakers from a mix of languages are brought together in exceptional circumstances (Bickerton, 1990). What is observed initially is that people speak using an impoverished kind of language known as a pidgin, which lacks many of the properties we usually associate with natural languages such as recursive structure, movement and overt morphology. In time, pidgins develop into creoles, which are full-fledged languages with structures characteristic of English or any other natural language. Given this, it is surprising that the transition from pidgin to creole usually occurs in only a single generation – despite children never being exposed to structures that are as rich as those they actually acquire. Also surprising is that creoles formed independently in distant regions of the world seem to share many fundamental properties such as a basic SVO word-order (Uriagereka, 1998).

Computational models of the evolution of language must be reconciled with evidence like this as well as that obtained from a variety of other means. From the paleontological record, it is possible to learn about the environmental conditions in which our ancestors lived including the kinds of social requirements that might have necessitated language and, by indirect means (like looking at skull shape), it is possible to make at least some inferences about brain developments that occurred in parallel. Comparisons between this data and our present knowledge of the neural anatomy of humans and our primate relatives allow us to make inferences about which parts of the brain are necessary for language. Such evidence is problematic for theories that are difficult to relate to such changes. Linguistic data concerning language universals and psycholinguistic data about the way in which language is acquired in childhood also constrain the form that any theory of language origins can take.

As in any scientific discipline, a theory about the origin of language is stronger if it is falsifiable by virtue of the predictions it makes and, in this case, predictions can be made about what should be found in the fossil record, the nature of language acquisition, the features of modern languages and the differences that should be evident in the neural anatomy of human and non-human species. In addition, the evolutionary narrative that a theory proposes should be plausible, and simulating the process computationally can provide an essential test of this. Subtleties arising from co-evolutionary arms-races, kin selection, the Baldwin effect and other processes can be counterintuitive and this is where simulation is most useful. Of particular value are the unexpected features that emerge from a model during simulation that match evidence obtained by other means.

Arguments from personal incredulity are an ideal target for computational modeling because a model need only demonstrate a possibility for them to be weakened, whether the possibility presented is *the* correct explanation for the given feature or not. Fortunately, for computational modelers, this field is filled with arguments of this kind. Bates *et al.* (1998) are incredulous about the possibility that linguistic knowledge is encoded in the genome. Chomsky (1972) is incredulous about the possibility that natural selection can provide a meaningful explanation of the emergence of universal grammar. Uriagereka (1998) is incredulous about the possibility that language change is adaptive. There are probably others. Of course, in each case, a failure to demonstrate a possibility will give weight to the given position.

Aknowledgements

This work was supported by the UK Engineering and Physical Science Research Council (grant: GR/N01118).

References

Ackley DH, Littman ML (1994) Altruism in the evolution of communication. In: Brooks R, Maes P (eds) *Proceedings of the Fourth Artificial Life Workshop*. MIT Press, Cambridge MA

Batali J (1994) Innate biases and critical periods: combining evolution and learning in the acquisition of syntax. In: Brooks R, Maes P (eds) *Proceedings of the Fourth Artificial Life Workshop*. MIT Press, Cambridge MA

Batali J (1998) Computational simulations of the emergence of grammar. In: Hurford J, Knight C, Studdert-Kennedy M (eds) *Approaches to the evolution of human language: Social and cognitive basis*. Cambridge University Press, Cambridge UK

Batali J (2000) The negotiation and acquisition of recursive grammars as a result of competition among exemplars. In: Briscoe EJ (ed) *Linguistic evolution through language acquisition: Formal and computational models*. Cambridge University Press, Cambridge UK

Bates E, Elman J, Johnson M, Karmiloff-Smith A, Parisi D, Plunkett K (1998) Innateness and emergentism. In: Bechtel W, Graham G (eds) *A companion to cognitive science*. Basil Blackwell, Oxford, pp 590-601

Berwick RC (1996) Art imitates life? *Boston Review,* 21(6)

Bickerton D (1990) *Language and species*. University of Chicago Press, Chicago

Briscoe T (2000) Grammatical acquisition: Inductive bias and coevolution of language and the language acquisition device. *Language,* 76: 245-296

Cangelosi A (1999) Modeling the evolution of communication: From stimulus associations to grounded symbolic associations. In: Floreano D, Nicoud JD, Mondada F (eds)

Proceedings of ECAL99 the Fifth European Conference on Artificial Life(Lecture Notes in Artificial Intelligence) Springer-Verlag, Berlin

Cangelosi A, Harnad S (in press) The adaptive advantage of symbolic theft over sensorimotor toil: Grounding language in perceptual categories. *Evolution of Communication*

Cangelosi A, Parisi D (1998) The emergence of a 'language' in an evolving population of neural networks. *Connection Science*, 10: 83-97

Chomsky N (1957) *Syntactic structures.* Mouton, Den Haag

Chomsky N (1972) *Language and mind: Enlarged edition.* Harcourt Brace Jovanovich, New York

Christiansen MH, Devlin JT (1997) Recursive inconsistencies are hard to learn: A connectionist perspective on universal word order correlations. In: *Proceedings of the 19th Annual Cognitive Science Society Conference.* Lawrence Erlbaum Associates, Mahwah, NJ, pp 113-118

Cziko G (1995) *Without miracles: Universal selection theory and the second Darwinian revolution.* MIT Press, Cambridge MA

Dawkins R (1989) *The selfish gene (2nd ed.).* Oxford University Press, Oxford

Dawkins R (1991) *The blind watchmaker.* Penguin Books, London

Deacon TW (1997) *The symbolic species: The coevolution of language and human brain.* Penguin, London

de Boer B (1997) Generating vowel systems in a population of agents. Presented at the *Fourth European Conference on Artificial Life*, ECAL 97, Brighton, UK

Di Paolo EA (2000) Ecological symmetry breaking can favour the evolution of altruism in an action-response game. *Journal of Theoretical Biology*, 203: 135-152

Edelman GM (1987) *Neural Darwinism: The theory of neuronal group selection.* Basic Books, New York

Elman JL (1990) Finding structure in time. *Cognitive Science*, 14: 179-211

Fodor JA (1998) The trouble with psychological Darwinism. *London Review of Books*, 20(2)

Fodor JA, Pylyshyn ZW (1988) Connectionism and cognitive architecture: A critical analysis, *Cognition*, 28: 3-71

Fogel LJ, Owens AJ, Walsh MJ (1966) *Artificial intelligence through simulated evolution.* Wiley, New York

Goldberg DE (1989) *Genetic algorithms in search, optimization, and machine learning.* Addison-Wesley, New York

Gould SJ (1991) Exaptation: A crucial tool for an evolutionary psychology. *Journal of Social Issues*, 47: 43-65

Gould SJ, Lewontin RC (1979) The spandrels of San Marco and the Panglossian paradigm: A critique of the adaptationist programme. *Proceedings of the Royal Society of London*, 205: 581-598

Harnad S (1990) The symbol grounding problem. *Physica D*, 42: 335-346

Hazlehurst B, Hutchins E (1998) The emergence of propositions from the co-ordination of talk and action in a shared world. *Language and Cognitive Processes*, 13: 373-424

Hinton GE, McClelland JL, Rumelhart DE (1986) Distributed representations. In: Rumelhart DE, McClelland JL (1986)

Holland JH (1975) *Adaptation in natural and artificial systems.* University of Michigan Press, Ann Arbor MI

Hutchins E, Hazlehurst B (1995) How to invent a lexicon: The development of shared symbols in interaction. In: Gilbert N, Conte R (eds) *Artificial Societies: The computer simulation of social life.* UCL Press

Kauffman SA (1995) *At home in the universe: The search for the laws of self-organization and complexity.* Oxford University Press, Oxford

Kirby S (1999) Syntax out of learning: The cultural evolution of structured communication in a population of induction algorithms In: Floreano D, Nicoud JD, Mondada F (eds) *Proceedings of ECAL99 the Fifth European Conference on Artificial Life(Lecture Notes in Artificial Intelligence)* Springer-Verlag, Berlin

Kirby S (2001) Spontaneous evolution of linguistic structure: An iterated learning model of the emergence of regularity and irregularity. *IEEE Transactions on Evolutionary Computation and Cognitive Science*, 5(2): 102-110 (Special issue on Evolutionary Computation and Cognitive Science)

Kirby S, Hurford JR (1997) Learning, culture and evolution in the origin of linguistic constraints In: Husband P, Harvey I (eds) *Proceedings of the Fourth European Conference on Artificial Life.* MIT Press, Cambridge MA

Komarova NL, Niyogi P, Nowak MA (2001) The evolutionary dynamics of grammar acquisition. *Journal of Theoretical Biology*, 209: 43-59

Koza JR (1990) Genetic programming: A paradigm for genetically breeding populations of computer programs to solve problems. Technical Report STAN-CS-90-1314, Stanford University

Ladefoged P (1993) *A course in phonetics (3rd ed.).* Harcourt Brace Jovanovich, Fort Worth

Livingstone D, Fyfe C (1999) Dialect in learned communication. In: Dautenhahn K, Nehaniv (eds) *AISB '99 Symposium on Imitation in Animals and Artifacts*, Edinburgh, The Society for the Study of Artificial Intelligence and Simulation of Behaviour

MacLennan BJ (2001) The emergence of communication through synthetic evolution. In: Patel M, Honavar V (eds) *Advances in Evolutionary Synthesis of Neural Systems.* MIT Press, Cambridge MA

MacLennan BJ, Burghardt GM (1994) Synthetic Ethology and the evolution of cooperative communication. *Adaptive Behavior*, 2: 151-188

Maynard Smith J (1982) *Evolution and the theory of games.* Cambridge University Press, Cambridge UK

McClelland JL, Rumelhart DE (eds) (1986) *Parallel distributed processing: Explorations in the microstructure of cognition (Vol 2: Psychological and biological models).* MIT Press, Cambridge MA

McClelland JL, Rumelhart DE, Hinton GE (1986) The appeal of parallel distributed processing. In: Rumelhart DE, McClelland JL (1986)

Montana DJ, Davies LD (1989) Training feedforward networks using genetic algorithms. In: *Proceedings of the International Conference on Genetic Algorithms* Morgan Kaufmann

Noble J (2000) Talk is cheap: Evolved strategies for communication and action in asymmetrical animal contests. In: Meyer J-A, Berthoz A, Floreano D, Roitblat H, Wilson SW (eds) *From Animals to Animats 6: Proceedings of the Sixth International Conference on Simulation of Adaptive Behavior.* MIT Press, Cambridge MA

Nowak MA, Komarova NL, Niyogi P (2001) Evolution of universal grammar. *Science*, 291: 114-118

Oliphant M, Batali J (1997) Learning and the emergence of coordinated communication. *Center for Research on Language Newsletter*, 11(1)

Parisi D, Cecconi F, Nolfi S (1990) Econets: Neural networks that learn in an environment. *Network*, 1: 149-168

Pinker S (1997) *How the mind works.* WW Norton and Company, New York

Pinker S, Bloom P (1990) Natural language and natural selection. *Behavioral and Brain Sciences*, 13: 707-784

Rumelhart DE, Hinton GE, Williams RJ (1986) Learning internal representations by error propagation. In: Rumelhart DE, McClelland JL (1986)

Rumelhart DE, McClelland JL (eds) (1986) *Parallel distributed processing: Explorations in the microstructure of cognition (Vol 1: Foundations)*. MIT Press, Cambridge MA

Saunders GM, Pollack JB (1996) The evolution of communication schemes over continuous channels. In Maes P, Mataric M, Meyer J-A, Pollack JB, Wilson SW (eds) *From Animals to Animats 4: Proceedings of the Fourth International Conference on Simulation of Adaptive Behavior*. MIT Press, Cambridge MA

Schwefel H-P, Rudolph G (1995) Contemporary evolution strategies In: Mor'an F, Moreno A, Merelo JJ, and Chac'on P (eds) *Advances in Artificial Life: Proceedings of the Third International Conference on Artificial Life*. Springer-Verlag, Berlin

Skinner BF (1966) The phylogeny and ontogeny of behavior. *Science*, 153: 1205- 1213

Skinner BF (1981) Selection by consequences. *Science*, 213: 501-504

Steels L (1998) The origin of syntax in visually grounded robotic agents. *Artificial Intelligence*, 103: 133-156

Steels L, Vogt P (1997) Grounding adaptive language games in robotic agents. In: Husband P, Harvey I (eds) *Proceedings of the Fourth European Conference on Artificial Life*. MIT Press, Cambridge MA, pp 474-482

Uriagereka J (1998) *Rhyme and reason: An introduction to minimalist syntax*. MIT Press, Cambridge, MA

Part II

EVOLUTION OF SIGNALING SYSTEMS

Chapter 3

Adaptive Factors in the Evolution of Signaling Systems

Jason Noble, Ezequiel A. Di Paolo and Seth Bullock

Introduction

Many of the chapters in this book will approach human language with an eye to its unique features, such as recursive syntax, or a large learned lexicon. We propose to take a wider view, seeing human language as one among many animal communication systems, and focusing on the selective pressures affecting the origin and maintenance of such systems. The possibility that human language arose from animal communication through a process of evolutionary change demands that we attend to the conceptual problems at the heart of our current understanding of animal signaling. In doing so we may throw light upon not only the origins of human language, but also its character.

The chief theoretical problem that comes to light when we look at the evolution of communication is accounting for the amount of honesty that is apparently involved (Johnstone, 1997; Noble, 2000a). Let us specify a hypothetical communicative scenario, such as a warning call to alert other animals about an approaching predator, or a display to advertise one's suitability as a mate. We can then construct a game-theoretic model, which allows us to consider the advantages and disadvantages of communicative and non-communicative behavior in our scenario. There usually turns out to be a tempting payoff for cheats, liars, bluffers, or free-riders, which means that communication should not be evolutionarily stable. It can therefore be difficult to use the model to explain the apparent prevalence of real-world communication in the situation we are modeling.

The problem is sometimes solved by constructing a more subtle game-theoretic model; for example, we might take into account the effects of kin selection, and find that communication will be selected for, as long as signaler and receiver are from the same family group. However, if they are to remain tractable, game-theoretic and other mathematical models can only be made more elaborate up to a point. If we suspect that communication is occurring in a certain natural context, and yet the best game-theoretic model we can construct tells us that communication should not be stable, what are our options?

Moving to individual-based computational modeling lets us test the idea that stable communication may emerge from low-level details of space, time, and interactions between organisms that cannot be captured in a conventional game-theoretic model. We are most interested in evolutionary simulation models (Di Paolo, Noble and Bullock, 2000), which involve the explicit modeling of individual organisms interacting in a shared environment. Evolution is incorporated in the sense that the more successful organisms (where success is defined by a criterion analogous to energy accumulation) will have a greater likelihood of passing on their genetic material to the next generation. Variation is introduced through mutation, i.e., the occasional random alteration of the transmitted genetic information.

The current chapter looks at simulations of the evolution of communication in the ecological domains of feeding, sexual choice, and contests over resources. We hope to demonstrate the power of individual-based evolutionary simulation modeling to explore more subtle hypotheses about signal evolution than is typically possible using conventional methods.

The Role of Ecological Feedback

There is an increasing recognition of the non-trivial effects that many ecological factors can have on evolution in general, and on the evolution of social behavior such as communication in particular. For instance, it had been suggested that the effects of limited dispersal could enhance the local coefficient of relatedness between interacting individuals (Hamilton, 1964), thus facilitating the evolution of cooperative behavior. However, contrary to this initial intuition, the effect of increased relatedness due to local interactions may be overcome by the effects of increased local competition when the scale of dispersal and the scale of interactions coincide (Taylor, 1992a, 1992b; Wilson, Pollock, and Dugatkin, 1992), a result that does not contradict the theory of kin selection if relatedness is properly calculated (Queller, 1994). This cancellation of positive and negative effects on fitness is challenged when the scales of dispersal and density regulation do not coincide (Kelly, 1992; 1994). It has also been shown that altruism may evolve in such viscous populations if organisms are modeled as discrete entities and the associated stochasticity is taken into account (Goodnight, 1992; van Baalen and Rand, 1998; Krakauer and Pagel, 1995; Nakamaru, Matsuda, and Iwasa, 1997). These cases can be regarded as evidence that ecological dynamics – which can include the effects of spatial situatedness, distribution of resources, mating strategies, and the activity of other species – can transform a simple evolutionary problem into a complex and non-intuitive one.

So far, the best way to approach such problems has proven to be a combination of traditional analytical models and individual-based computer simulations, in which factors such as space, discreteness of individuals, and noise can be included naturally. As an example of particularly non-intuitive ecological effects, we may consider the model introduced by Di Paolo (2000), in which the evolution of altruism in an action-response game is studied via a series of analytical and simulation approaches. Action-response games (e.g., Hurd, 1995) are fairly general models of social interaction which include signaling behavior as a special case. The evolution of altruism in such games can, for instance, be equated to the evolution of honest signaling systems.

An action-response game

As in other cases, the game proposed by Di Paolo (2000) starts by assuming a situation of conflict of interest between two actors. The game affords various interpretations but a useful one from the communication point of view is the situation in which an animal (first player) has found a source of food and must decide (by choosing between two possible actions) whether to attract the attention of another conspecific (second player) to this source or to distract it. The second player has a choice of two actions: approaching the first player or ignoring it. Approaching the source of food results in both players sharing the payoff in equal measure, while if the first player manages to distract the attention of the second, it will have access to a larger than half share of the energy contained in it. The degree of conflict c represents the energy proportion that the first player gets in this situation. If $c = 0.5$ there is no conflict from the perspective of the first player; conflict is mild for values of c slightly greater than 0.5 and more significant as c approaches 1.

The word 'coordination' is used to describe the outcome of interactions that lead to the joint exploitation of the source. Without losing generality, it is possible to suppose that this happens in half of the four possibilities that the 2-signal 2-response scheme affords. Signals and responses can be either of types 'O' or 'E' (originally for 'odd' and 'even'), and the outcome of the interaction is denoted by a signal-response pair: OE or EO (in which cases coordination occurs), or OO or EE (in which cases the first player does not share the food). The signal and response given by each individual player can be described by one of the above four strategies, which are genetically determined. During its lifetime, a given individual will play sometimes the first and sometimes the second role. The cooperation/conflict relationships between the four strategies are described in Figure 3.1. Each arrow is interpreted as connecting the initiator and responder strategies of those interactions that result in coordination. Thus a player using strategy 'OE' will behave altruistically only towards players using either 'OE' or 'EE' and this is indicated by the two arrows starting at the 'OE' node. An initiator playing 'EE', in contrast, will not behave altruistically towards players using 'OE' since there is no arrow from 'EE' to 'OE', although it will behave altruistically towards individuals playing 'OO' or 'EO'. In a randomly constituted population, the proportion of cooperative coordination will be 50%. In order to say that cooperative interactions have evolved, the proportion should rise above this value.

Notice the cyclic structure of part of the resulting graph indicating a kind of Rock-Paper-Scissors situation which, at first sight, suggests that no single strategy may become dominant because it will always be invaded by its 'neighbor' strategy in the graph. A detailed game-theoretic analysis of this game for the case of infinite, random mating and random playing populations (Di Paolo, 2000) leads precisely to this conclusion for all values of c. The constitution of the population will oscillate by following the straight arrows in Figure 3.1. There are no evolutionarily stable strategies in this case.

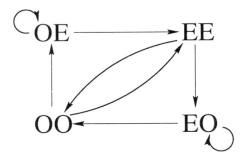

Figure 3.1 Relations of cooperation/conflict between the four strategies in the game, see text for details.

Broken symmetries

The above result changes for the case of a finite population and the introduction of noise to the evolutionary dynamics. In this case there is a single stable point attractor in which half of the population play strategy OO while the other half play strategy EE. However, the proportion of cooperative interactions at this equilibrium is only 50%, i.e., equal to the baseline level of cooperation. Thus cooperative coordination cannot evolve under these conditions.

Both these models remain quite abstract and further assumptions could be relaxed. For instance, the population could be considered to be distributed in space so that interactions, as well as reproductive events, are local. The distribution of energy in the food sources could also be described by a local variable, so that a kind of ecological coupling would be introduced resulting in differences in quality between local environments depending on how those environments are exploited. A continuous-time model of this situation, based on partial integro-differential equations, leads to the conclusion that players will tend to aggregate into discrete clusters even if they are uniformly distributed initially, but within those clusters the different strategies will oscillate as in the game-theoretic model with cooperative interactions once again at baseline levels.

This is as far as the purely analytical approach can go. If more assumptions are to be relaxed, such as treating individual players as discrete entities instead of 'densities' in the distribution of strategies, an evolutionary simulation modeling scenario must be contemplated. In such an individual-based model each player accumulates energy by interacting with others and thus drawing energy from food sources. This amount of energy must not only cover the energy survival costs (same for all individuals) but must be enough for the player to eventually reproduce. Noise is present at different stages in the model, e.g., in the asynchronous updating scheme, in the choice of co-participants in the interaction, in the choice of food sources which may differ in their energy value, etc. (There is, however, no noise in the production of signals and responses.)

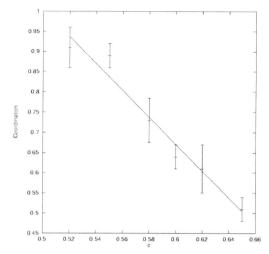

Figure 3.2 Average value for the level of cooperative coordination for different values of c. Each point is the average of five simulation runs. The line represents a linear regression, (correlation coefficient: -0.982). Error bars indicate standard deviation.

The simulation model can be explored both in the spatial and non-spatial cases. In the latter, the result are again oscillations. However, in the spatial case, the level of cooperative interactions depends on the degree of conflict c. For positive, but small, conflict (c slightly greater than 0.5) the population shows a high and stable level of games resulting in cooperative coordination. This level decreases linearly with c until it reaches the baseline level for $c \approx 0.65$ (Figure 3.2).

This result is explained by looking at the spatial patterns that form spontaneously in the population and how they break many of the in-built symmetries of the abstract situation by allowing for reciprocal interactions between the evolving population and its environment.

As with the continuous spatial model, stable clusters can also be observed as a consequence of a dynamical equilibrium between two tendencies: the tendency for the population to concentrate in a small region so as to maximize the chances of finding a partner to interact with, and the tendency to move away from dense regions to places where local resources are exploited less frequently. Once a cluster is formed there is an equilibrium in the rate of energy consumption per unit of space. This equilibrium would seem to establish a degree of spatial 'neutrality' in the sense that spatial position does not matter for the rates of energy intake and offspring production. Players at the center interact more often than those at the periphery but they do so for poorer resources. If the rates of net energy intake were different, the cluster would not be at equilibrium. Such a homogeneous rate of energy consumption (and reproduction) is indeed what is observed in the simulations. However, it is not true that spatial position is neutral in evolutionary terms.

If a player is born from a parent near the periphery of the cluster there is a high chance that it will be placed 'outside' the cluster in the sense that it will have a very small number of neighbors. Those players will tend to die before they reproduce. In fact, the chances of originating a lasting genealogy of players diminish as the originating position moves from the center to the border of a cluster. This is a geometrical consequence of the stochastic and local character of the process of offspring allocation. Given this, we would predict that the position of a cluster's ancestors would tend to be concentrated near the center of the cluster as one travels backwards in time, and this is what is observed. Thus, it is reasonable to conclude that a player's position within a cluster, although not under genetic control, plays a role in determining its fitness.

The above phenomenon is a case of symmetry breaking of the spatial homogeneity. Other symmetries are also broken by the center/periphery structure such as the frequencies with which individuals play each role; with central individuals playing the role of responder more often.

By analyzing the evolutionary stability of different strategies in view of these conditions it is possible to show that for low values of (positive) conflict – even though a cluster composed of altruistic strategists (OE or EO) can indeed be invaded locally by non-altruistic strategies – the environmental conditions, in terms of available energy and rates for role assignment, are such that a very small increase in the local density of invaders renders them unviable in the central region. Invasions will occur locally but will be followed by the local disappearance of the invaders, leaving a gap at the center of the cluster which either allows the altruists to re-invade or causes the cluster to split into two smaller ones (see Figure 3.3). This effect is harsher near the center of the cluster which is where most lineages originate.

It is important to point out that altruism, in this case in the form of honest signaling, is favored by a combination of discreteness and ecological organization. Neither of these factors is sufficient on its own, as shown by the results of the continuous spatial model and the non-spatial individual-based model. The rupture of spatial homogeneity is essential for altruism to be favored in the case of low conflict. But some of the ensuing broken symmetries occur only as a consequence of the discreteness of the players, for example, the dependence of the genealogy length on spatial position within a cluster. Discreteness also plays a role in the local extinction that may occur when non-altruistic players begin to invade the center of an altruistic cluster. If sufficiently fine-grained density values were permitted within the model, such local extinctions would not occur. Rather, invading strategies would be allowed to take very small, but non-zero, density values corresponding to less than 1 individual in the region of interest. Because of their reduced energy consumption, these 'infinitesimal' individuals would be able to subsist in the unfavorable environment until eventually local energy would have been replenished and they would begin to increase in density.

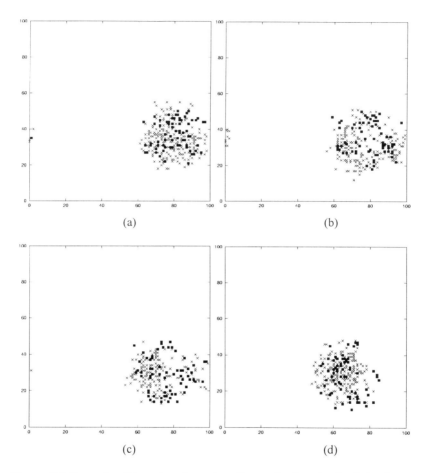

(a)   (b)

(c)   (d)

Figure 3.3 Example of invasion dynamics of a predominantly altruistic cluster. Each frame shows the cluster constitution at intervals of 3000 time steps (about half an average individual lifetime); $c = 0.55$. Altruists are marked with xs and invaders with squares. The first frame (top left) shows a high concentration of invaders in the central region. The following frame shows a density gap in the same central region due to the local disappearance of invaders. The bottom-left frame shows an expanded gap leading to a separation into two clusters, and only one of these survives in this case as shown in the last frame.

Even though there is no significant interaction between clusters, and, consequently, interpretations of these results in terms of group selection are not possible at this level, there is a sense in which such an interpretation could be reasonable for units of selection *within* a cluster (van Baalen and Rand, 1998). The local invasions of altruistic players, followed by local extinction, can indeed be interpreted in terms of the viability of two different groups in a specific local environment that one of them sets and the other cannot change fast enough to adapt to. On the flip-side of this interpretation, and also following van Baalen and Rand (1998), we could equally say that an appropriately defined coefficient of relatedness, taking into consideration density-dependent effects, would bring this result within the domain of Hamilton's rule. This is also a viable interpretation of the results, even though a simple estimation of purely genetic coefficients of relatedness (following Queller and Goodnight, 1989; Queller, 1994) was inconclusive in this respect. Finally, it would also be possible to construe these results as a consequence of reciprocal altruism (Trivers, 1971) although there would be little or no difference between this and the kin-selective interpretation since there is no segregation into different species in this model. However, a constant fact in all these possible interpretations remains the two-way coupling between selection and ecological dynamics and the resulting broken ecological symmetries due to the activity of the players. The importance of such couplings has been long noted (Lewontin, 1983) and has been recently highlighted under the label of 'niche-construction' (Laland, Odling-Smee, and Feldman, 2000; but see also Bullock and Noble, 2000).

Evolutionary Simulation Modeling and the Handicap Principle

Since Israeli ornithologist Amotz Zahavi first presented his theory a quarter of a century ago (Zahavi, 1975, 1977), the *handicap principle* has been the subject of energetic debate within the evolutionary biology literature. Briefly, Zahavi suggested that extravagant displays such as unwieldy, colorful tail-feathers, or protracted bouts of exhausting bellowing, which are used by creatures throughout the natural world to advertise mate quality, fighting prowess, etc., may not be costly by accident, but because it is only through their extravagance that their trustworthiness is guaranteed.

Zahavi's insight was to suggest that the costs incurred in producing such displays might enforce honesty amongst signalers if these costs were somehow connected to the quality being advertised such that they favored those signalers with more of whatever was being advertised (the best fighters, the highest quality mates, etc.). For example, an honest advertisement of a predator's ability to efficiently catch prey might be the extent to which the predator deliberately wastes food items. Wasting food is always costly, but it is more costly if you are unlikely to be able to get any more. Since poor predators cannot afford to waste hard-won prey items, a system in which predators demonstrate their ability through wasting as many food items as they can afford to cannot be invaded by cheats who exaggerate their ability, since the increased costs that this would entail ensure that bluffing is simply not worth their while.

This notion of waste as a signal of quality is reminiscent of the concept of "conspicuous consumption" discussed by Thorstein Veblen (1899), a turn-of-the-century sociologist. Veblen noted that members of the "leisure class" persistently and protractedly overindulged themselves. They left expensive food uneaten, rarely wore their opulent clothes, and spent much of their time and money pursuing costly pastimes for no purpose other than their own entertainment. He suggested that this seemingly senseless hedonism was also a way in which the members of the leisure class demonstrated their class membership. That is, the purchase of prohibitively expensive goods and services could be understood as an indicator of the procurer's wealth. This index of societal status was an effective one because those of lower status could not afford to make the "advertisements" of which wealthier individuals were capable. Indeed, at the lower extreme of the scale, the funds of the poorest individuals were more than accounted for by the demands of simply staying alive, leaving no extra money to "waste" upon "unproductive consumption".

Initially, Zahavi's theory suffered considerable skepticism. Evolution by natural selection was understood to produce *efficient* systems – the opposite of the scenarios Zahavi described. Why would evolution favor wasteful exhibitionism? More specifically, if a peahen were to choose a mate on the basis of an advertisement, surely the advertisement (which her male offspring would be likely to inherit) would not be chosen for its ability to seriously handicap its owner, increasing the likelihood that her offspring would die before themselves winning mates? If this was the price of honesty, surely it would make more sense to choose a mate at random and spare one's offspring the handicap? But despite these worries, and only intermittent empirical and theoretical support over the next decade, the handicap principle achieved increasing notoriety.

The rise of the handicap principle in the face of almost continuous criticism (e.g., Davis and O'Donald, 1976; Maynard Smith, 1976, 1978, 1985; Kirkpatrick, 1986) is perhaps attributable to two factors. The first is that Zahavi's theory filled a theoretical vacuum left by the collapse of group-selectionist accounts of signaling. Prior to the reassessment of group selection in the mid-1960s (Hamilton, 1964), the evolutionary function of signaling behaviors could be explained in terms of the benefits that they conferred upon a signaling community as a whole. Mating displays, aggressive posturing, informative dances, begging cries, warning coloration, and danger signals, if honest, enable the efficient distribution of resources (food, sex, shelter, etc.). This efficiency derives from the flow of useful information between the members of an honest signaling system – each member gains much of their information from other members, without having to collect it individually. Contrast a beehive, foraging as a unit on the basis of shared information, with the less efficient behavior of the same bees obstinately foraging solo, or the difference between settling contests by honest aggressive displays of strength and settling the same disputes through fighting.

However, although the increased efficiency afforded by honest signaling is of benefit to those groups that employ such signals, *individuals* within these groups often stand to gain by freeloading, bluffing, cheating, lying, double-crossing, exaggerating, misleading, or crying wolf. For the individual, then, honesty is not always the best policy. With dishonesty comes mistrust, and eventually the collapse of an honest signaling system, undermined by deceit. But although the selfish actions of individuals were expected to compromise the stability of natural signaling systems, such systems appeared to be the frequent products of evolution. Signaling systems are everywhere in nature. If signaling systems are evolutionarily fragile, why are they so ubiquitous? Zahavi's handicap principle at least offered an explanation, even if it appeared counter-intuitive.

The second factor in Zahavi's favor was the rise of game-theoretic modeling in behavioral ecology (Maynard Smith, 1982). From 1985 onwards, a series of successful, game-theoretic models (Enquist, 1985, being the first, and Grafen, 1990, being the foremost amongst these) demonstrated the soundness of the handicap principle's central tenets, succeeding where population genetic models had previously failed (see Maynard Smith, 1985, for a review). As evolutionary game theory benefited from its success in dealing with ideas which had proven hard to explore using alternative modeling approaches, the handicap principle gained credibility. Although the handicap principle does not yet enjoy the status that Zahavi believes it deserves (having thus far failed to eclipse Darwin's theory of sexual selection), both the vocabulary and explanatory perspective associated with it have attained a central position within current evolutionary thinking.

However, while handicap thinking spreads within biology and beyond (e.g., Miller, 2000), the theoretical biology community face several unanswered questions. In this section, we will try to demonstrate how simulation models of the kind already introduced can help to answer these questions, and reveal new problems that have been neglected up until now. Three issues will be raised in the next sections, before an evolutionary simulation model with which to address them is introduced.

Balancing the handicap books

While the costs of signaling have clearly been the focus of work on the handicap principle, certain important aspects of these costs remain unclear in Zahavi's writing. Crucially, Zahavi's verbal arguments offer little clue as to the way in which handicap costs are perhaps balanced by the benefits to the signaler of achieving whatever goal the signal is intended to bring about. For each individual signaler, must handicap displays reduce their fitness (through loss of time and energy, increased risk of predation, etc.) to a greater extent than these signals on average increase it (through gaining copulations, victories, food, shelter, etc.)?

Zahavi sometimes appears to consider the net costs involved in signaling, when, for instance, he asserts that "it is reasonable to expect a population in its optimal fitness to benefit from a handicap" (Zahavi, 1977, p. 604). At least, then, at the population level, the costs of bearing handicaps are assumed to be more than compensated for by the associated benefits. At the level of the individual, matters are not as clear-cut, "so long as the [signaler] ... does not deviate to grow its handicap larger than it can afford, the handicap [may persist] as a marker of honest advertisement" (ibid.), i.e., handicap costs are limited in some way, but how? Compounding this vagueness, when describing natural examples, Zahavi rarely discusses the benefits obtained from signaling, and the manner in which these benefits balance the costs.

Furthermore, Zahavi's terminology is not easy to reconcile with a notion of the handicap principle couched in terms of net costs. For example, as Hurd (1995) and Getty (1998a, 1998b) point out, if the costs involved in signaling must be balanced by consonant benefits, then in what sense are these costs a 'handicap'? However, if these costs are not so balanced, what is the value of signaling? Although the exaggerated costs incurred by a bluffer might be characterized as a handicap, since these costs would be larger than the bluffer could afford, this is not the sense in which Zahavi proposed the term. For Zahavi, honest signalers suffer a handicap – they must do so in order to demonstrate their honesty.

Not surprisingly, this confusion has led authors to multiple interpretations of the handicap principle. Wiley (1983), for example, characterizes Zahavi's (1975) paper as claiming that "signals should evolve to become a *net* handicap to signalers" (p. 176, our emphasis). In contrast, Adams and Mesterton-Gibbons (1995) reach the opposite conclusion, stating that their model *differs* from the handicap principle in that "the net benefit for a given advertisement may not increase monotonically with the signaler's strength" (p. 406), implying that typical handicap thinking proposes that signalers gain a net *benefit* from signaling. Later we will use an evolutionary simulation model to explore what the costs and benefits are for signalers that are involved in a handicap signaling system.

Need vs. quality?

A second, separate but related issue concerns the conflicting roles of signaling costs and signaling benefits in stabilizing handicap signaling systems. Can, as Zahavi implies, honesty only be ensured by (gross) signaling costs varying such that some signalers stand to lose less from signaling than others and are thus able to signal more? Or might honesty also be maintained by (gross) signaler benefits varying such that some signalers stand to gain more from signaling than others and are thus able to signal more? Johnstone (1997) has usefully divided handicap models into these two kinds. The first attempts to account for the evolutionary stability of the honest advertisement of *quality* as the result of the manner in which the gross *costs* of signaling vary with quality (e.g., Grafen, 1990; Hurd, 1995). The second kind attempts to account for the evolutionary stability of the honest advertisement of *need* as the result of the manner in which gross signaler *benefits* vary with need (e.g., Godfray, 1991; Maynard Smith, 1991).

The latter kind of model includes that used by Godfray (1991) to demonstrate the evolutionary stability of a strategy in which nestlings honestly advertise their hunger (need) by varying the strength of their begging calls. Godfray showed that such a strategy is evolutionarily stable if the costs of begging are the same for all chicks, but the value of any particular parental resource to a begging chick increases with the chick's hunger. In such situations, hungry chicks beg more than satiated chicks because the resources are worth more to them.

The former kind of model includes Grafen's (1990) treatment of a similar scenario, in which a very different stable begging equilibrium was derived. If we assume that the parent wishes to feed the *highest quality* chicks rather than the most needy, Grafen (1990) showed that we can expect chick begging to be an honest indicator of quality if the value of parental resources are the same for all chicks, but the cost of any particular begging display is greater for the lesser quality chicks. In such situations, high quality chicks beg more than lower quality chicks because the signals are more affordable to them.

Are these two scenarios distinct, though complementary, classes of handicap signaling, or two extreme cases from a wider range of possible signaling systems?

The attainability of honesty

Until now, we have been concerned with arguments from theoretical biology concerning whether signaling systems can be evolutionarily *stable*. Since evolution has been continuing for billions of years, theoretical biologists assume that the systems we see around us are stable; if not they would most likely have been replaced by some other system that was stable. Since signaling systems are so prevalent and so widespread, it is hard to imagine that each is unstable – in a state of evolutionary flux, poised at the brink of collapse (although some have pursued this idea, Dawkins and Krebs, 1978; Krebs and Dawkins, 1984). However, there is nothing to prevent an evolving system from admitting of several different evolutionarily stable situations. In fact, it is becoming clear that many if not most interesting evolutionary systems feature *multiple equilibria* of this kind. Evolutionary simulation models are well-suited for addressing this issue. Whereas existing formal modeling paradigms (game-theoretic models and population-genetic models, for example) are able to disentangle the contributions of the various ideas and theories discussed in the previous two sections, evolutionary simulation models are ideally positioned to deal with matters of evolutionary change, modeling as they do the manner in which populations subjected to evolutionary pressures change over evolutionary time. Amongst other things, such models allow us to explore questions of *equilibrium* selection – which of a number of possible equilibrium states will an evolving system reach from some initial ancestral condition? Here we will compare two different conclusions that may be drawn from the empirical observation that fuelled Zahavi's initial papers introducing the handicap principle, and use an evolutionary simulation model to decide between them.

Observation: Many natural signaling systems appear to feature "extravagant" signals.

Conclusion 1: Costliness is necessary to ensure the stability of honest signaling systems.

Conclusion 2: Costliness is necessary to ensure the attainability of honest signaling systems.

The first conclusion has been widely explored in the theoretical biology literature. While it has been shown that signal costs are able to stabilize signaling systems, it is unclear whether these costs are "extravagant" or "handicaps" in the sense implied by Zahavi's papers. The second conclusion has been largely

Plymouth University
The Charles Seale-Hayne Library

Customer ID: *****36528

Title: Simulating the evolution of
language /
ID: 9004998267
Due: 25/07/2015 23:59:00 BST

Total items: 1
13 Jun 2015
Checked out: 7
Overdue: 0
Hold requests: 0

Due date of borrowed items subject
to change. Please check your
University email regularly for recall
notices to avoid incurring a fine.

Thank you for using the
3M SelfCheck™ System.

unexplored (although Yachi, 1995, has attempted to characterize the conditions under which handicap signaling might evolve).

An evolutionary simulation model

To recap, the evolutionary simulation model presented here was designed to explore three issues: (i) what is the character of handicap signaling at equilibrium, and how does this character vary with the model's parameters? (ii) what conditions must be met in order for handicap signaling to be evolutionarily *stable* given that both signaling costs and signaler benefits vary with signaler state? (iii) are handicap signaling equilibria *attainable* from appropriate initial conditions, and how does this attainability vary with the model's parameters? For present purposes, a brief description of the model will be given. Full details of the model can be found in Bullock (1997; 1998).

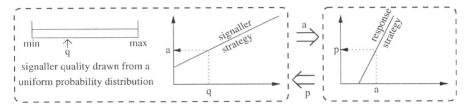

Figure 3.4 A simple signal-response game in which a signaler is allocated an internal state (q, sometimes referred to as quality) at random, and produces an advert (a) with magnitude determined by an inherited signaling strategy (in this case a linear mapping). Adverts may not have negative sign. This advert is passed to the receiver, who produces a response (p) determined by an inherited response strategy (in this case also a linear mapping). Responses are truncated to lie within the range $[q_{min}, q_{max}]$. Receiver fitness is calculated as $\frac{1}{1+|p-q|}$, increasing with the accuracy with which the response matches the signaler's state. Signaler fitness is calculated as $pq^R - aq^S$. See text for explanation.

For each evolutionary run, a population of signalers and a population of receivers coevolved for 1000 generations. Fitness scores were determined by pairing up signalers with receivers and allowing them to play a simple signal-response game (see Figure 3.4). Receiver fitness was awarded proportional to the accuracy with which receiver response, p, matched the internal state of the signaler, q. Signaler fitness was calculated in a slightly more complicated fashion as the benefit of obtaining a response, pq^R, minus the cost of signaling, aq^S, where R and S are model parameters fixed for the course of an evolutionary run. The interests of signalers and receivers conflict, since signalers always benefit from as large a response as possible, whereas receivers benefit from matching their response to a signaler's internal state.

In order to address the three issues raised above, the evolutionary dynamics of this scenario were explored under a range of different cost benefit parameters and from a variety of initial conditions. Crucially, we need to manipulate both (i) the manner in which signaler quality influences the negative effect of signal cost on signaler fitness and (ii) the manner in which signaler quality influences the positive effect of receiver response on signaler fitness. The signaler fitness function was designed such that these two manipulations could be achieved by varying two model parameters, S and R, respectively (see Figure 3.5 Left).

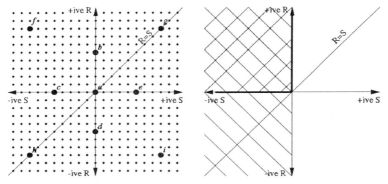

Figure 3.5 *Left* – The model's R-S parameter space was extensively sampled. Nine of the 21 x 21 scenarios explored are labeled. *(a)* $R = 0$, $S = 0$: signaler state does not affect the cost of making an advert, nor the benefit of gaining a response. *(b)* +ive R, $S = 0$: the value of a response increases with signaler quality, the cost of advertising is independent of signaler quality. *(c)* $R = 0$, -ive S: the value of a response is independent of signaler quality, whereas the cost of advertising decreases with increased signaler quality. *(d)* –ive R, $S = 0$: as *(b)*, but the value of responses decreases with increasing signaler quality. *(e)* $R = 0$, -ive S: as with *(a)*, the effect of signaler quality on signal cost is balanced by its effect on the value of responses. (f) +ive R, -ive S: responses are more valuable to high quality signalers, who also pay less for any given advert. *(g)* and *(h)*: as with *(a)*, the effect of signaler quality on signal cost is balanced by the effect on the value of responses. (i) –ive R, +ive S: not only is signaling more costly for high quality signalers, but they also gain less benefit from a receiver response. *Right* – Previous models' predictions of parameters values that support honest handicap signaling equilibria: Zahavi (1975, 1977) diagonal hatching; Grafen (1990) cross-hatching; Godfray (1991) and Maynard Smith (1991) bold vertical line; and Hurd (1995) bold horizontal line. The current model predicts honest signaling equilibria will exist in the part of parameter space lying above the line $R = S$.

The right-hand panel of Figure 3.5 depicts areas of the model's parameter space that previous models have predicted will support evolutionarily stable honest signaling. For runs in which $S < 0$ (diagonal hatching), the costs of advertising decrease with signaler quality – this is the condition predicted to guarantee honesty by Zahavi's handicap principle (1975, 1977). Several models have supported Zahavi, in suggesting that portions of this area of the parameter space admit of honest signaling equilibria. However, analysis of the model presented here (Bullock, 1998) suggests that honest signaling will only be stable for scenarios in which $R > S$. While this finding is not incommensurate with those of previous models, it contradicts Zahavi's basic premise that the manner in which signaling cost varies with signaler state (i.e., the value of S in this model) is all that determines whether handicap signaling is stable or not.

Evolutionary runs were carried out from three kinds of initial condition. (i) Honest: initially signalers played $a = q$, while receivers played $p = a$, (ii) Random: initially signaler and receiver strategies were determined at random, (iii) Mute-Deaf: initially signalers played $a = 0$, while receivers played $p = 0$. After 1000 generations, each run was terminated and the state of the evolved populations examined. In this way the evolutionary simulation model was used to explore the evolutionary dynamics of a range of scenarios, and, for each scenario, to assess whether stable handicap signaling equilibria could be achieved from a variety of initial conditions.

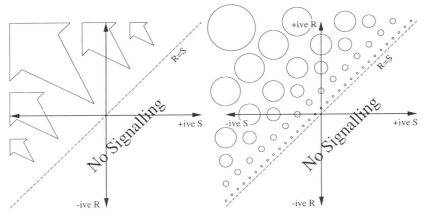

Figure 3.6 The evolutionary simulation model reveals that signaling equilibria exist for scenarios in which $R > S$, but not otherwise. *Left* – The range of signaling exhibited at equilibrium, and the magnitude of the largest adverts, increases exponentially as R outstrips S. *Right* – The size of the basins of attraction for signaling equilibria also increase as R outstrips S.

The simulation results (see Figure 3.6) supported the analytical results in that no signaling behavior was observed for scenarios in which $R \leq S$, whereas signaling equilibria were observed for all scenarios in which $R > S$. In addition, where signaling equilibria were discovered, both the character of the signaling at these equilibria and their attainability, varied with the relationship between R and S.

For scenarios in which R is only slightly larger than S, stable signaling systems exhibited a relatively small range of signals, with the largest signals exhibited themselves being relatively small. These signaling equilibria were also associated with relatively small basins of attraction, which ensured that evolution did not tend to reach them from Random or Mute-Deaf initial conditions.

In contrast, for scenarios in which $R \gg S$, signaling equilibria exhibited a very wide range of signal sizes, with the largest signals being orders of magnitude more massive than the smallest. Furthermore, the basins of attraction for these equilibria were also much larger (and hence more easily attainable from Random or Mute-

Deaf initial conditions) than those discovered for scenarios in which R is only slightly larger than S.

Discussion of the model

These findings have several implications for our understanding of the handicap principle and how it affects the character of natural signaling systems. First, rather than conclude from the existence of seemingly extravagant natural signaling systems that extravagance is necessary in order that such systems remain evolutionarily *stable*, we might now surmise that these observations are due to the relative ease with which such signaling systems are *attained* by evolving populations of signalers and receivers. The simulation model above suggests that although a wide range of stable signaling systems are possible, some featuring relatively restrained signals, while some feature larger and perhaps seemingly extravagant displays, it is only the latter that are easily achieved from non-signaling ancestral scenarios. These results suggest that we may find examples of subtle signaling systems which nevertheless are evolutionarily stable under conditions in which the net cost of signaling decreases only slowly with signaler quality (i.e., R is only slightly larger than S). Such systems might be those in which either the resource being signaled for is itself of limited (and relatively constant) value, or in which the cost of advertising does not vary to a great extent with the property being advertised. For instance, the "I-see-you" signals made by a small bird to a stalking cat might be one such system. Despite involving subtle signals that do not appear costly to the casual observer, such a scenario would still be a handicap system, stabilized by the net costs of signaling.

A second implication of the model is that it is net costs that must be considered when dealing with handicap signaling systems. Contra Zahavi, signaling equilibria were sometimes exhibited by the model under conditions in which making signals was *more* costly for signalers of high quality (S > 0). Similarly, and again contra Zahavi, signaling equilibria *fail* to exist for some model scenarios which meet Zahavi's handicap criterion, i.e., in which signal costs decrease with signaler quality. Only once consideration is given to the *balance* between the manner in which signal costs vary with quality, and the manner in which signaler benefits also vary with quality, can the distribution of honest signaling equilibria across the model's parameter space be understood.

Signaling in Contests

Animal contests – disputes over resources such as food, territory or mates – are good examples of interactions in which the interests of the participants seem to be maximally opposed. This is particularly true of struggles over the control of an indivisible item: one's gain is necessarily another's loss. Nevertheless, animals contesting the possession of a resource are often observed to settle the dispute by exchanging signals or threat displays rather than engaging in an all-out fight. For example, mantis shrimps *Gonodactylus bredini* contest the ownership of small cavities in their coral reef habitat. These contests sometimes result in physical combat, but often an opponent is deterred by a claw-spreading threat display (Adams and Caldwell, 1990). Red deer stags *Cervus elaphus* compete for control of groups of females, but unless two stags are closely matched in strength, the weaker will usually retreat after a roaring contest and/or a parallel walk display (Clutton-Brock, Albon, Gibson and Guinness, 1979).

What is happening in these cases? Are the competing animals likely to be exchanging honest signals, informing each other of their fighting ability or their intention to attack? (And if not, what is the function of their aggressive displays?) Intuitively, settling contests by signaling makes sense. We can see that an all-out fight is usually a bad idea: fighting is energetically expensive, and there is always a risk of injury or death. The early ethologists suggested that threat displays were honest signals of aggressive intent that benefited the species by preventing costly fights, but the naive group-selectionist overtones of this idea mean that it is no longer taken seriously. Moreover, standard game-theoretic predictions (Maynard Smith, 1982) suggest that in contest situations, it will not be evolutionarily stable for animals to exchange signals of strength or aggressive intent because would-be honest signalers will always be less fit than bluffers. According to this perspective, there is no room in the arena of animal contests for the cooperative exchange of arbitrary signals; the aggressive displays observed in nature are either unfakeable because of physical constraints, or are the uninformative result of a manipulative arms race (Krebs and Dawkins, 1984).

On the other hand, some theorists have argued that, in effect, competing animals share enough of a common interest in avoiding serious injury that honest signaling can be evolutionarily stable. Enquist (1985) presents a game-theoretic model in which contestants are either strong or weak, and cost-free, binary signals are exchanged before the decision to fight or flee is taken. Enquist concludes that, under certain conditions,[1] the honest signaling of fighting ability, referred to as strategy *S*, will be evolutionarily stable. Fights will occur only between evenly-matched opponents, and weak animals will defer to signals denoting strength.

Enquist's conclusion is driven by the assumption that weak animals cannot afford to risk confronting stronger opponents and must be honest about their shortcomings (and in this sense Enquist's model can be considered a handicap signaling model). However, Caryl (1987) notes that in real contests weak animals may be able to bluff (i.e., signal that they are strong) and then rapidly retreat if challenged. Even if weak bluffers are briefly attacked as they flee, the expected cost of such attacks may well be lower than the cost of an extended fight with another weak animal; this state of affairs would invalidate Enquist's result.

We will now look at Enquist's argument in the light of an evolutionary simulation model of contests over an indivisible resource (Noble, 2000b). The aim of using a simulation is to avoid oversimplification. In particular, time will be modeled in an approximately continuous fashion: in Enquist's model there are only

---

[1] $0.5v - c > v - d$, where v is the value of the resource, c is the cost of an escalated fight between two equally matched opponents, and d is the cost to a weak animal of being attacked by a strong one.

two time steps – an exchange of signals followed by a choice of actions – and thus the model may fail to capture critical aspects of real-time interactions. In a more realistic model of animal combat, is it true that weak animals have so much to lose by bluffing that selection will favor the honest signaling of fighting ability?

The model

The simulation will be described only briefly; full details are given in Noble (2000b). The contests commence with two players facing each other in a one-dimensional arena (Figure 3.7). Each player has a fighting ability and an accumulated energy score – neither of these properties can be perceived by the opponent. The pair are assumed to be competing for possession of a food resource. At each time step, a player can move forwards or backwards by up to a metre, and can produce an auditory signal of variable intensity.

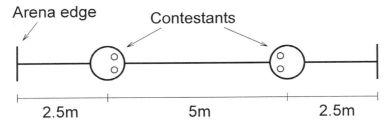

Figure 3.7 Starting positions of the two competing players. The arena is 10m wide; one player starts 2.5m from the left edge, the other 2.5m from the right edge.

Contests can end due to one player fleeing, one player seriously injuring the other, or because a time limit has been reached. If one player moves backwards far enough to leave the arena, the other obtains the food resource and the contest is over. If both players leave the arena simultaneously, neither obtains the food item. If the two players are within 50cm of each other, any forward movement counts as an attack (and thus attacks may be one-sided affairs). During every time step that one player attacks another, it may succeed in inflicting a serious injury with a probability proportional to its fighting ability. If a serious injury occurs, the contest ends at once and the injured player forfeits the resource. The players have access to 11 perceptual inputs, on which they can base their decisions about movement and possible signaling. Briefly, each player has access to privileged information about its own state, such as its fighting ability and its energy level, and can observe the signals and movements of the other player. Players are also aware of their position in the arena.

The perceptual inputs are translated into outputs (movement and signaling) using a production system. A player's production system specifies its contest strategy; the system is also the genotype that will be passed on if the player succeeds in reproducing. Production systems include up to six rules; a typical rule might be "If (own energy level < opponent's signal) and (time elapsed < 72 time steps) then advance 32cm and signal 85%." At each time step in a contest, a player's production system is given fresh inputs, and the movement and signaling outputs are provided by the first rule to have all of its conditions met. In case no rules fire, the genotype includes default movement and signaling values.

The simulation is organized into days. Each day, every player in the population plays out a contest with a random opponent. The result is that each player participates in at least one contest per day, and expects on average to participate in two. After all the contests have been resolved, reproduction takes place: any players with more than a threshold level of energy are randomly paired up and allowed to reproduce sexually. Each pair produces a single offspring.

The energy budget for the simulation has been set up such that the biggest cost, by far, is due to being seriously injured. This reflects the findings of Riechert (1982) that, in spider contests, the long-term fitness costs of serious injury – and, of course, death – are orders of magnitude greater than other costs such as energetic expenditure associated with threat displays. The average cost of being seriously injured is approximately double the value of the contested resource. Furthermore, the cost to a weak player of an extended fight with a strong player is high enough that Enquist's conditions for stable communication of fighting ability are met.

Simulation results

Genotypes in the initial population were generated randomly. Each simulation run continued for the equivalent of about 7500 generations. Ten evolutionary runs were performed, each with a different random seed value. Contests lasted, on average, 19.4 time steps. This indicates that the players tended to engage each other in some way, as immediately fleeing the arena would take only three time steps.

Contests were resolved 66.0% of the time through one or both players fleeing, and 24.2% of the time through one player inflicting a serious injury on the other. The remaining 9.8% of contests reached the maximum time limit.

After 7500 generations, the median strategy was recorded for each population. A common theme across all 10 strategies was an aggressive default movement, combined with at least one rule spelling out conditions under which the player would retreat. Figure 3.8 illustrates this by showing one of the simpler strategies in full. A player following this strategy will start the contest with the default move of advancing one metre: the initial distance to the opponent is always five metres, and so the rule will not be satisfied. If two competing players are both following this strategy, after two time steps they will each have moved forward two metres, and they will be only a metre apart. At this point, the rule may fire. If one player is relatively weak (i.e., its fighting ability is less than 53% of the maximum value) then it will retreat 93cm, as for this player the distance to the edge will be 4.5m. If the second player is strong, it will pursue the first, ensuring that the weaker contestant eventually flees from the arena, although the stronger one will never get quite close enough to attack. Two strong players will clash head on; neither will ever retreat. The behavior of two weak players is more interesting: they face each other one metre apart, and then each takes a 93cm step backwards. They then move forward one metre again. Next they will advance yet again and fight, as they will be exactly 4.57m from the edge of the arena and thus the rule will not fire.

What can we make of this strategy? Does it involve the honest signaling of fighting ability? The first point to notice is that users of this strategy pay no attention whatsoever to their opponent's signals. There is no significance in the fact that a "louder" signal is given when advancing than when retreating, because in a population of players all playing this strategy, no-one will be listening. However, there is some indication that players may be signaling, or at least giving away information, through their movements. When weaker players reach the moment of truth, one metre from their opponent, they reveal their low fighting ability by retreating. The interpretation of this result will be considered presently.

```
If  Own fighting ability < 53%
    Distance to opponent < 1.70m
    Distance to edge < 4.57m
    then retreat 93cm and signal 47%

Otherwise advance 1m and signal 80%
```

Figure 3.8 The median strategy evolved in run 9. The default movement is maximally aggressive and the single rule specifies conditions for retreating.

Space precludes a detailed analysis of all 10 of the evolved strategies. However, if we look at the sensory inputs the players actually used in their decision making, we find that the most popular were the distance to the opponent, one's own fighting ability, the distance to the edge of the arena, and the change in the distance to the opponent. The sensory inputs associated with the opponent's signaling activity were attended to only infrequently.

The results presented so far suggest that the exchange of honest signals of fighting ability via the signaling channel is not favored by selection. However, the 10 simulation runs each began with a randomly generated set of initial strategies. It is possible that stable signaling strategies exist, but that their basins of attraction in genotype space are not large enough for the strategies to emerge given random initial conditions. We therefore looked at what happens when an analogue of Enquist's signaling strategy S is programmed into the initial population.

```
If  Own fighting ability < 40%
    Opponent's signal > 50%
    then retreat 1m and signal 0%

If  Own fighting ability < 40%
    then advance 1m and signal 0%

Otherwise advance 1m and signal 100%
```

Figure 3.9 An analogue of Enquist's (1985) strategy S, expressed in the framework of the players' production system. The default strategy is an aggressive advance and a loud signal. The first rule specifies that weaker players will retreat from a loud signal, and the second, that they will advance without signaling if they hear no signal.

Figure 3.9 shows the way in which strategy *S* was implemented as a two-rule production system. The cutoff point between weak and strong was set at 40% as this was the approximate mean fighting ability implemented in the 10 runs described above. For stronger players, the chosen action will always fall through to the default behavior of aggressively advancing while making a loud signal. For weaker players, rule one or rule two will always fire. This means that weaker players will announce their status by always signaling with zero intensity. If a weak player detects a signal (i.e., a strong opponent) it will retreat, but if there is no signal it will advance to fight its presumably weak opponent.

The evolutionary stability of strategy *S* was investigated by conducting another 10 runs, with players in the initial populations set to play strategy *S*. These simulations can therefore show us whether or not a population of strategy *S* players is resistant to invasion by mutant strategies. Looking at the proportion of the time that various sensory inputs were used to make decisions in the evolved players, it became clear that strategy *S* was not able to resist the invasion of alternative strategies: for example, the "Distance to opponent" input was used most often, despite not being present in the initial population. Inspection of the median strategies showed them to be very much like those that evolved in the basic model, with any signaling behavior on the part of the opponent being largely ignored.

Why is *S* not stable against invasion? Enquist (1985) shows that it is an ESS under conditions that might appear to be satisfied here: why the inconsistency? Enquist's argument for the evolutionary stability of *S* rests on the idea that weak contestants must honestly signal their weakness because they cannot afford the risks of being injured by a stronger contestant. The results discussed so far present a different picture, in which weak players do not signal their weakness at all, and only give away information about their state by retreating at the last possible moment. It may be that, in the current model, weak players can afford to behave in this way because the condition $d > \frac{1}{2} v + c$ – identified by Caryl (1987) as unrealistic – is not met. That is, the model lacks a mechanism that would maintain a high value of d (where d is the cost to a weak contestant of facing up to a stronger one). Consider the pattern of behavior outlined for the strategy shown in Figure 3.8. Clearly, if weak players can bluff it out against stronger opponents, up to a point, and then retreat without being harmed, then d is not particularly high.

Discussion

Enquist's (1985) model suggests that weak contestants have so much to lose by bluffing that selection will favor the honest signaling of fighting ability. The simulation reported here shows that this claim is very much dependent on Enquist's idiosyncratic way of modeling animal combat. Given more realistic signaling and movement over an extended period of time, reliable signaling of fighting ability did not evolve. This result held, whether the members of the initial population were allocated random strategies, or were programmed to play an analogue of Enquist's strategy *S*. Results in the latter condition show that strategy *S* is not an ESS in the current model, which must detract from Enquist's claims of generality. These findings support and extend Caryl's (1987) claim that Enquist's model of animal combat is implausible.

Although disagreeing with his conclusions, we can sympathize with Enquist's motivation. Field observations of behavior in animal contests sometimes do seem to contradict the game-theoretic conclusion that talk is cheap (e.g., Hansen, 1986; Dabelsteen and Pedersen, 1990). There really is a need for explanation in such cases: either the appearance of signaling is an illusion, or our models are leaving something out. But unfortunately Enquist (1985) settled on some questionable assumptions in his attempt to explain apparent honesty in contests.

The available signaling channel was not used by the players, but there was evidence that they were gaining information about fighting ability based on observations of each other's movements. Does this count as communication? A poker analogy may be useful: if you are bluffing with a terrible hand, the other players do not know whether your cards are strong or weak. If someone calls your bluff, by seeing your bet and then raising again, you will probably fold. By doing so, you have given the other players information about your strength (i.e., they now know that you had a poor hand). However, the reason you folded was not to provide information to others, but because it was the best way to minimize your expected losses at that point. Similarly, the weak players using the strategy shown in Figure 3.8 are giving away information about their weakness when they back off from immediate confrontation at time step three. But their choice at this point is to retreat or to start fighting against an opponent that may well be stronger than they are. The expected costs of entering such a fight are higher than the costs of retreating, so the player retreats. Information is conveyed to the opponent by this behavior, but it is not the function of the retreat to be informative.

Overall Conclusion

Conventional game-theoretic models in biology abstract away from the individual organism and incorporate radical simplifying assumptions such as random mating in homogeneous populations, the absence of spatial distribution, and the lack of significant ecological feedback. Evolutionary simulation models are able to highlight the importance of many of these assumptions through exploring their contribution to a model's evolutionary dynamics. Study of the evolution of communication and language is just one domain of enquiry that is crucially concerned with interactions between individuals mediated by an environment. It is hoped that the individual-based evolutionary simulation models presented here demonstrate the methodological value of taking a comparative modeling approach to problems of this kind.

Finally, it is worth stressing that simulation results are no substitute for empirical evidence. If a simulation establishes the plausibility of a hypothesis, this is not the same as establishing its truth. The claim here is only that simulation methods can demonstrate the logical coherence (or indeed incoherence) of a particular model, and that they may suggest new hypotheses for empirical investigation (see Di Paolo *et al.*, 2000, for a more complete treatment of these issues).

References

Adams ES, Caldwell RL (1990) Deceptive communication in asymmetric fights of the stomatopod crustacean Gonodactylus bredini. *Animal Behaviour*, 39: 706-716

Adams ES, Mesterton-Gibbons M (1995) The cost of threat displays and the stability of deceptive communication. *Journal of Theoretical Biology*, 175: 405- 421

Bullock S (1997) *Evolutionary simulation models: On their character and application to problems concerning the evolution of natural signaling systems*. PhD thesis, School of Cognitive and Computing Sciences, University of Sussex Brighton UK

Bullock S (1998) A continuous evolutionary simulation model of the attainability of honest signaling equilibria. In: Adami C, Belew R, Kitano H, Taylor C (eds) *Artificial Life VI*. MIT Press, Cambridge MA, pp 339-348

Bullock S, Noble J (2000) Evolutionary simulation modeling clarifies interactions between parallel adaptive processes. *Behavioral and Brain Sciences*, 21: 150-151. Commentary

Caryl PG (1987) Acquisition of information in contests: The gulf between theory and biology. Paper presented at the *ESS Workshop on Animal Conflicts*, Sheffield UK

Clutton-Brock T, Albon SD, Gibson RM, Guinness FE (1979) The logical stag: adaptive aspects of fighting in red deer (*Cervus elaphus L*). *Animal Behaviour*, 27: 211-225

Dabelsteen T, Pedersen SB (1990) Song and information about aggressive responses of blackbirds Turdus merula: Evidence from interactive playback experiments with territory owners. *Animal Behaviour*, 40: 1158-1168

Davis JWF, O'Donald P (1976) Sexual selection for a handicap: A critical analysis of Zahavi's model. *Journal of Theoretical Biology*, 57: 345-354

Dawkins R, Krebs JR (1978) Animal signals: Information or manipulation? In: Krebs JR, Davies NB (eds) *Behavioural ecology: An evolutionary approach*. Blackwell, Oxford, pp 282-309

Di Paolo EA (2000) Ecological symmetry breaking can favour the evolution of altruism in an action-response game. *Journal of Theoretical Biology*, 203: 135-152

Di Paolo EA, Noble J, Bullock S (2000) Simulation models as opaque thought experiments In: Bedau MA McCaskill JS Packard NH Rasmussen S (eds) *Artificial Life VII: Proceedings of the Seventh International Conference on Artificial Life*. MIT Press Cambridge MA, pp 497-506

Enquist M (1985) Communication during aggressive interactions with particular reference to variation in choice of behaviour. *Animal Behaviour*, 33: 1152-1161

Getty T (1998a) Handicap signaling: When fecundity and viability do not add up. *Animal Behaviour*, 56: 127-130

Getty T (1998b) Reliable signaling need not be a handicap – commentary. *Animal Behaviour*, 56: 253-255

Godfray HCJ (1991) Signaling of need by offspring to their parents. *Nature*, 352: 328-330

Goodnight KF (1992) The effect of stochastic variation on kin selection in a budding-viscous population. *American Naturalist*, 140: 1028-1040

Grafen A (1990) Biological signals as handicaps. *Journal of Theoretical Biology*, 144: 517-546

Hamilton WD (1964) The genetical evolution of social behaviour I and II. *Journal of Theoretical Biology*, 7: 1-16; 17-32

Hansen AJ (1986) Fighting behaviour in bald eagles: A test of game theory. *Ecology*, 67: 787-797

Hurd PL (1995) Communication in discrete action-response games. *Journal of Theoretical Biology*, 174: 217-222

Johnstone RA (1997) The evolution of animal signals. In Krebs JR, Davies NB (eds) *Behavioural ecology: An evolutionary approach (fourth edition)*. Blackwell, Oxford, pp 155-178

Kelly JK (1992) Restricted migrations and the evolution of altruism. *Evolution*, 46: 1492-1495

Kelly JK (1994) The effects of scale dependent processes on kin selection: Mating and density regulation. *Theoretical Population Biology*, 46: 32-57

Kirkpatrick M (1986) The handicap mechanism of sexual selection does not work. *American Naturalist*, 127: 222-240

Krakauer DC, Pagel M (1995) Spatial structure and the evolution of honest cost-free signaling. *Proceedings of the Royal Society of London Series B*, 260: 365-372

Krebs JR, Dawkins R (1984) Animal signals: Mind reading and manipulation. In: Krebs JR, Davies NB (eds) *Behavioural ecology: An evolutionary approach (2nd edition)*, Blackwell, Oxford, pp 380-402

Laland KN, Odling-Smee J, Feldman MW (2000) Niche construction, biological evolution, and cultural change. *Behavioral and Brain Sciences*, 21: 131-146

Lewontin RC (1983) Gene, organism, and environment. In: Bentall D (ed) *Evolution from molecules to men*. Cambridge University Press.

Maynard Smith J (1976) Sexual selection and the handicap principle. *Journal of Theoretical Biology*, 57: 239-242

Maynard Smith J (1978) The handicap principle - a comment. *Journal of Theoretical Biology*, 70: 251-252

Maynard Smith J (1982) *Evolution and the theory of dames*. Cambridge University Press Cambridge

Maynard Smith J (1985) Mini review: Sexual selection handicaps and true fitness. *Journal of Theoretical Biology*, 115: 1-8

Maynard Smith J (1991) Honest signaling: The Philip Sidney game. *Animal Behaviour*, 42: 1034-1035

Miller GF (2000) *The Mating Game: How sexual choice shaped the evolution of human nature*. William Heinemann, London

Nakamaru M, Matsuda H, Iwasa Y (1997) The evolution of cooperation in a lattice-structured population. *Journal of Theoretical Biology*, 184: 65-81

Noble J (2000a) Cooperation competition and the evolution of prelinguistic communication In Knight C, Studdert-Kennedy M, Hurford J (Eds) *The emergence of language*, Cambridge University Press, pp 40-61

Noble J (2000b) Talk is cheap: Evolved strategies for communication and action in asymmetrical animal contests. In: Meyer J-A, Berthoz A, Floreano D, Roitblat H, Wilson SW (eds) *From Animals to Animats 6: Proceedings of the Sixth International Conference on Simulation of Adaptive Behavior*. MIT Press, Cambridge MA, pp 481-490

Queller DC (1994) Genetic relatedness in viscous populations. *Evolutionary Ecology*, 8: 70-73

Queller DC, Goodnight KF (1989) Estimation of genetic relatedness using genetic markers. *Evolution*, 43 258-275

Riechert SE (1982) Spider interaction strategies: Communication vs coercion. In: Witt PN, Rovner J (eds) *Spider communication: Mechanisms and ecological significance*. Princeton University Press, Princeton NJ, pp 281-315

Taylor PD (1992a) Altruism in viscous populations - an inclusive fitness approach. *Evolutionary Ecology*, 6: 352-356

Taylor PD (1992b) Inclusive fitness in a homogeneous environment. *Proceedings of the Royal Society of London Series B*, 249: 299-302

Trivers RL (1971) The evolution of reciprocal altruism. *Quarterly Review of Biology*, 46: 35-57

van Baalen M, Rand DA (1998) The unit of selection in viscous populations and the evolution of altruism. *Journal of Theoretical Biology*, 193: 631-648

Veblen T (1899) The theory of the leisure class. In: Lerner M (ed) *The portable Veblen*. Viking Press, New York, pp 53-214. Collection published 1948

Wiley RH (1983) The evolution of communication: Information and manipulation. In: Halliday TR, Slater PJB (eds) *Communication* (*vol 2 of Animal Behaviour*). Blackwell, Oxford, pp 156-189

Wilson D, Pollock GB, Dugatkin L (1992) Can altruism evolve in purely viscous populations? *Evolutionary Ecology*, 6: 331-341

Yachi S (1995) How can honest signaling evolve? The role of handicap principle. *Proceedings of the Royal Society of London Series B*, 262: 283-288

Zahavi A (1975) Mate selection - a selection for a handicap. *Journal of Theoretical Biology*, 53: 205-214

Zahavi A (1977) The cost of honesty (further remarks on the handicap principle). *Journal of Theoretical Biology*, 67: 603-605

Chapter 4

Evolving Sound Systems

Bart de Boer

Human languages use an amazing variety of subtly different speech sounds to convey meaning. With the exception of sign languages that are used and developed by communities of deaf people, all human languages use sound as the primary signal. The sounds, or more accurately the differences between sounds, that humans use for distinguishing meanings can be very subtle. Two different sounds that would be perceived as identical by a speaker of one language might make an important distinction in meaning in another. For example, in the Bahing language of East Nepal, the word /m• r• / means "monkey", while the word /m• r• / means "man". Speakers of neighboring and European languages alike are generally not able to perceive this distinction, an unlimited source of fun to the Bahing people.

In the UCLA Phonological Segment Inventory Database (UPSID), a database that now contains 451 languages (Maddieson, 1984; Maddieson and Precoda 1990) 921 different speech sounds occur. The language with the largest inventory of speech sounds in the database is !Xu⊕ (Snyman, 1970, 1975), a Khoisan language of Southwest Africa with 141 sounds, while the languages with the smallest inventories are Rotokas (Firchow and Firchow, 1969) an East-Papuan language and Mura-Pirahã (Sheldon, 1974; Everett, 1982) a South-American language, both with only 11 sounds. According to Maddieson (1984) usually languages have between 20 and 37 sounds in their repertoires. However, these repertoires are not chosen randomly. Some sounds occur much more often in the languages of the world than others. Lindblom and Maddieson (1988) have found that languages tend to use a set of basic articulations first. Such basic articulations are simple articulations that involve only one articulatory gesture and minimal displacements of the articulators. When the repertoire becomes larger, languages tend to use what Lindblom and Maddieson call 'elaborate' articulations, which involve larger displacements and simultaneous actions of multiple articulators. Finally, when a language's repertoire becomes even larger, 'complex' articulations will be used. These consist of combinations of the two previous types.

There are other patterns to be found in the sound repertoires as well. Examples of such patterns are symmetries. In consonant systems, for example, if a language has a voiced sound at a certain place of articulation, it is very likely to have a voiceless sound at the same place of articulation. Comparable symmetries are found in vowel systems.

Regularities are not just found in the repertoires of sound systems, but also in the way sounds are combined into words and syllables. It is possible to make a hierarchy of sounds with respect to whether they tend to occur close to or far from the nucleus of a syllable. This hierarchy is called the *sonority hierarchy* (Vennemann, 1988). Some sounds, such as vowels, are very sonorous and tend to occur at the nucleus of a syllable, while others, such as plosive consonants (p, b, t etc.) are little sonorous and tend to occur at the periphery of a syllable. Whenever sounds occur in sequence, it turns out that they almost always increase in sonority towards the nucleus of a syllable. For this reason, "play" is a possible word in English, while "*lpay" is not.

Phenomena that occur in many languages are often called *universals*. Although the term universal implies validity for all languages, there are very few non-trivial phenomena that occur in all known human languages. For this reason the term universal is often used for phenomena that occur in a (large) majority of human languages. All parts of language: syntax, morphology, semantics, phonology, can have their own universals. This paper will concentrate on universals that have to do with sound systems.

Universals might be explained in different ways. The first possible explanation would be that all languages are historically related. Although there is still some controversy over the exact evolution of *Homo sapiens*, it is most likely that modern humans came from Africa some 200000–300000 years ago. Genetic diversity within the species *Homo sapiens* is so small that it is very likely that at one time in its early history the species must have consisted of only a few thousand individuals. It is not unlikely that all these individuals spoke dialects of the same language. However, given the speed with which languages change, and given the amount of time during which different groups of humans have been isolated from each other, it is highly unlikely that any trace of the original relationship between all human languages remains. Tentative reconstructions of "proto-world" (Ruhlen, 1994) although enthusiastically embraced by the popular press, should be regarded with the utmost scepsis. Another reason why deep historical relations between human languages alone cannot explain universals is that there are also universals of language change (e.g., Labov, 1994). Quite different languages seem to change along similar paths.

A second possible explanation is that language universals are a reflection of innate human capacities for language. Such an innate capacity does not only have to be in the form of a "universal grammar" as investigated by some researchers, but could also consist of more general cognitive mechanisms that are used for using and learning language. The innate capacity for language is also determined by physical and physiological factors, such as the shape of the vocal tract, accurate control over breathing and the way the ear processes sound. Innate factors obviously play a role in determining universals of human language. However, the problem of innate factors as explanation for language universals is that they themselves have to be explained as the result of evolution, or possibly as exaptations of pre-existing body parts and cognitive mechanisms.

This leads to the third possible explanation of language universals: that they are functional optimizations for communication over a noisy channel. Human language seems to be optimized for communication in a number of respects. The frequency with which different vowels occur in human language can for example be explained by the optimization of acoustic distinctiveness. If one optimizes a system with a fixed number of vowels so that the average distance between them is maximized, systems that occur frequently in human languages tend to appear. Now such functional optimization could be a result of the interactions between the speakers, listeners and learners of a language or the result of an evolutionary process. Also, the preference for languages that are functionally optimal over languages that are not could, over a long period of time, influence the mechanisms that are used for learning language through a process that is called the Baldwin effect (Baldwin, 1896).

Possibilities for Modeling

The role of innate properties versus the role of functional optimization and the way by which the different human adaptations to speech have evolved can be investigated with computer models. Traditionally, linguists prefer to solve theoretical disputes with linguistic data and physical, cognitive or philosophical arguments. However, language origins and evolution can hardly be investigated by looking at modern languages, and the complexity of theories of evolution of populations is such that their behavior cannot be predicted by simple philosophical argument. For this reason computer models are used more and more to test and create hypotheses. The study of speech has a long tradition of using computer and other electronic equipment. Due to the fact that speech works with objectively measurable and recordable signals, it can be manipulated relatively easily. From the 1950s onwards important discoveries were made by manipulating recorded signals and synthesizing artificial ones. Another advantage of the fact that speech signals can be measured in a relatively objective way, is that predictions of models can be easily compared with observations of real language data.

Different aspects of the evolution of speech can be investigated with computer models. One can try to reconstruct the evolution of the human vocal tract, one can use computer simulations to find out what factors (such as articulatory ease, acoustic distinctiveness etc.) have played a role in evolution, but one can also use computer models to investigate how much of speech is learnt and how much of it is innate.

Different approaches to modeling speech

One interesting and important way in which computer models have been used to study the evolution of speech (and language indirectly) is by reconstructing the vocal tract of fossil hominids, most notable Neanderthals (Lieberman and Crelin 1971). These vocal tract models can then be manipulated and excited with an artificially generated glottal pulse. By studying the resonances of the model, the range of possible vowel sounds that could be made by the hominid under study can be estimated. Although this technique comes closest to actually being able to listen to our hominid ancestors, the technique is not quite uncontroversial, mostly because important parts of the vocal tract (tongue, pharynx, larynx) do not fossilize very well. Interesting and exciting as these results may be, they do not quite model the origins and evolution of speech (they only reconstruct one stage of the evolution from fossil data) so they fall somewhat outside the scope of this chapter.

Apart from such direct modeling techniques, roughly three computational paradigms have been used for investigating the evolution of speech. The first paradigm is that of straightforward optimization of sound systems on the basis of different criteria (e.g., Liljencrants and Lindblom 1972; Lindblom *et al.*, 1984, Schwartz *et al.*, 1997b). The paradigm is illustrated in Figure 4.1. The figures are added for illustration, but also to be able to compare the different paradigms at a glance.

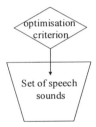

Figure 4.1 Schematic view of optimization.

The optimization criteria include factors such as acoustic distinctiveness, acoustic stability, articulatory ease or learnability. Through optimizing different (combinations of) criteria and checking whether the sound systems that are predicted conform to what is found in human languages, one can find out what criteria are important for the formation of human sound systems.

Optimization is probably the technique that is least controversial in its applications, as its dynamics are relatively simple: there is an optimization criterion and it results in sound systems that look like human sound systems or not. Discussion is possible on the implementation of the optimization criteria or on the interpretation of the sound systems that are found, but the optimization process itself is not controversial. The relative simplicity of optimization is also a disadvantage. It can be applied only to relatively simple problems. As soon as multiple optimization criteria interact, the optimization process becomes more difficult and decisions have to be made about which solutions to investigate. Also, the relative importance of the different criteria and the way they interact might be controversial. However, optimization is a good technique for checking which criteria play a role in determining the sound systems that are found in human languages. How these criteria have become important and how the optimization process takes place in human language use and learning can then be investigated with different techniques.

The second paradigm is that of genetic algorithms (GAs). The genetic algorithm is a technique that is based on the way evolution works in nature. The algorithm has a population of potential solutions, all of which are coded as artificial genes (usually in the form of bit strings). These genes are converted into possible solutions to the problem at hand (sound systems in the case of evolution of speech, e.g., Redford *et al.*, 1998; in press) and are evaluated with a fitness function. This fitness function is a function that gives a high value for good solutions and a low value for bad solutions. Just as in nature, solutions with a high fitness are allowed to create offspring, while bad solutions are removed from the population. The genes of the offspring are created by combining the genes of the parent solutions. Usually combination methods inspired by nature, such as mutation and crossover, are used. It is clear that for the proper functioning of a genetic algorithm the right fitness function as well as the right coding of solutions in genes are essential. The GA is illustrated in Figure 4.2.

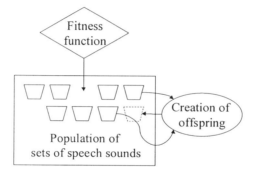

Figure 4.2 Schematic view of a genetic algorithm.

Basically, GAs also optimize on the basis of an optimization criterion (the fitness function), but they are much more flexible and robust than straightforward optimization algorithms. They can therefore be used to model more complex optimization problems and even problems in which the optimization criterion changes over time. Also, GAs work with a population of solutions, instead of with a single one. This is more realistic in the case of language, as language is typically used in a group of individuals rather than by a single individual. Finally, genetic algorithms are modeled after Darwinian evolution, and are as such ideally suited for modeling real evolution.

Their resemblance to real biological evolution is possibly the biggest advantage of genetic algorithms for research into the evolution of speech. But modelers who enthusiastically embraces genetic algorithms as their paradigm of choice should be aware that there are a large number of design decisions to be made in building a GA for investigating the evolution of speech. Decisions have to be made what to encode as genes and how to implement the fitness function. Also, it is very important to not confuse biological evolution of the human faculty for speech and cultural evolution of human languages. Although historical relations between languages and historical change of languages are often expressed in terms similar to those of biological evolution and although there are definite and valid similarities between the processes of biological evolution and language change, one should not confuse the two processes in one's model. The two processes are clearly distinct and operate on totally different time scales. They do influence each other, but this influence happens because the properties of a learned system (the language) influence the fitness of individuals that have to learn it, and is an interesting subject of investigation in itself.

The third paradigm is inspired by game theory and Wittgenstein's (1967) ideas on language games. Language games as a paradigm for modeling of evolution of language were first used by Steels (1995, 1997). In this research the notion of a game is not very well defined, but language games have a number of properties in common. There usually is a population of agents that each have certain linguistic knowledge and that can interact with each other. The rules of the game determine how the interactions are structured and what information is exchanged. The agents can update their linguistic knowledge on the basis of the interactions they have taken part in. Usually all agents follow the same strategy for updating their knowledge. The language game paradigm is illustrated in Figure 4.3.

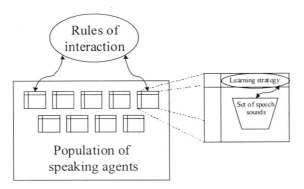

Figure 4.3 Schematic view of a language game.

Language games are a useful model of linguistic interactions between humans. The rules of the game and the strategy for updating an agent's knowledge can be varied to create different types of games for investigating different parts of language. Of course, one has to make simplifications while using language games. In real human language, different parts of language influence each other and interactions between language users can be highly complicated and dependent on extra-linguistic context. In this respect, the language game model is not different from other computational models of the study of language, but it is necessary to keep in mind what simplifications one has made and how these might influence the outcome of the games.

Strictly speaking, language games cannot be used to study the evolution of language, as the agents do not change over time. However, language games can be used to investigate to what extent properties of language can be explained as the result of interactions between agents and to investigate what must be programmed into the agent (i.e., what must be innate) so that it can learn a certain aspect of language. Such aspects as have to be pre-programmed will have to be explained by evolutionary models, such as genetic algorithms.

As both the genetic algorithm and the language game paradigm work with a population of agents, it is obvious that the two can be combined (Glotin's and Berrah's work, to be discussed later, come close to implementing such a system). However, the sound systems would not be coded into genes, but the properties of individual agents. In such a system it could be investigated, for example, how different learning techniques can evolve, or whether it is possible to reconstruct the evolution of the human vocal tract on the basis that it enables speakers to produce a wider range of possible speech sounds. The combination of the language game and the GA is illustrated in Figure 4.4.

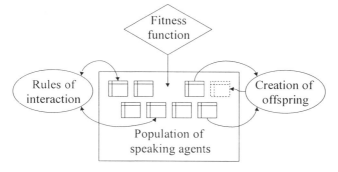

Figure 4.4 Combination of GA and language game.

The combination of these two techniques makes it possible to investigate the interactions between biological evolution and cultural evolution, without running the risk of confusing time scales or genetically and culturally transmitted information, as mentioned above. Although the paradigm of language games with evolving agents is the one that comes closest to human reality, there are still a number of problems. All problems with respect to how agents are coded into genes, and how the fitness function is implemented also occur here, as well as the problems with respect to the simplification of interactions that were mentioned with the language games. Another important problem is that the combination of the two highly complex mechanisms might result in behavior that is hard to explain. It may no longer be possible to determine which mechanism caused which part of the complete behavior, or to reconstruct how the system came up with the solution that was found.

Another problem with systems that have to simulate many iterated operations with populations of agents, or sets of speech sounds, is that their running time can become prohibitively long. For example, the most realistic speech synthesizers that exist take approximately 1000 times as long to calculate a speech signal than the actual duration of the signal. It is not possible to simulate a realistic number of interactions in a population of any size with such a model. It is therefore essential that the right simplifications be found and that reasonably realistic, but fast models of the speech phenomena under study be used. An important part of modeling the evolution of speech (and perhaps of any cognitive phenomenon) is therefore the trade-off between speed and realism.

Modeling different aspects of speech

Not only are there different possible approaches to the problem of modeling speech, there are also different aspects of speech that can be modeled. Here again, there is a trade-off between accuracy and speed. As speech sounds pronounced in sequence influence each other, and as this influence is of great importance to understand language change, it would be desirable to have a model that is as complete as possible. That is, a model that is able to produce a sequence of consonants and vowels as well as an intonation contour. However, there are a number of problems with modeling such complex utterances. The first problem, that was already mentioned in the previous section, has to do with the lack of speed of complex articulatory models. But this is not the only problem. Another problem is that actually very little is known about how sounds in sequence are produced, perceived and processed.

Linguists generally make descriptions of human languages in terms of *phonemes*, the sounds that are able to make distinctions in meaning. An example is the distinction between English /r/ and /l/ which have many minimal pairs (words that differ only in one sound, and that have different meanings) such as 'rate' and 'late'. However, in a language such as Japanese, this particular distinction is not used, there are no minimal pairs with [r] and [l], and so in that language [r] and [l] are said to be *allophones* of one phoneme /l/.

Although phonemes have great descriptive value, it is not quite clear what their role is in storage and processing of speech sounds. It is quite possible that processing of speech is done on different levels of complexity, both on a level higher and lower than that of the phoneme. This is because when people pronounce words, they do not produce a string of nicely distinguishable phonemes. Instead, they produce a sequence of speech gestures that influence each other mutually, so that different phonemes overlap and become indistinguishable. Little is known about how this process works in articulation, and even less is known about how the speech signal is converted into strings of phonemes and words by the listener. Any model that works with complex utterances therefore has to make assumptions about how these processes work. But such assumptions reduce the realism that was sought by using more complex speech signals.

A final problem with modeling complex utterances with the computer is that inevitably time sequences have to be learnt. This is actually an area of machine learning that is very hard, and for which very few general-purpose algorithms are available.

For the time being, all attempts at modeling have tried to tackle only a subset of the possible speech sounds and the possible speech universals. Successful models have been made of models and simple (abstract) syllables, while work is in progress on tone systems and intonation.

A Short History of Modeling

Probably the first attempt at making a computer model to explain universals of speech sounds was made by Liljencrants and Lindblom (1972). This model performed an optimization of randomly initialized vowel systems with a fixed number of vowels. The optimization used a function that was based on the potential energy of repelling magnets or electrically charged particles with equal polarity (this potential energy is higher whenever such particles are closer together). By shifting the individual vowels in the system, this energy function was minimized. Liljencrants and Lindblom found that vowel systems that were optimized in this way showed remarkable similarities with vowel systems found in human languages, although there were some discrepancies. Later re-implementations using modified distance functions (e.g., Vallée, 1994; Schwartz *et al.*, 1997b) have succeeded in making progressively better approximations of human vowel systems.

Subsequently, Lindblom *et al.* (1984) tried to use an optimizing model for explaining phonemic (that is combinatorial) coding of syllables. The syllables consisted of a simple consonant followed by a vowel. Although the systems that emerged were phonemically coded, their model has not had the success of the model for vowels, because there are many more parameters in it and it is much more difficult to replicate the results.

Only in the mid-1990s did work on explaining sound systems with computer models get a new impulse with systems that were based on populations of sound systems and agents. The first to make an agent-based implementation to investigate the emergence of vowel systems was Glotin (Glotin, 1995; Glotin and Laboissière, 1996; Berrah *et al.* 1996) of the Institut de Communication Parlée (ICP) in Grenoble, the same institute where Schwartz *et al.* (1997b) do their research. He made a model in which a population of talking agents tries to develop a shared repertoire of (a fixed number) vowels. His agents have both an acoustic and an articulatory representation of the vowels, and adapt their vowel systems on the basis of their interactions. The agents are also subject to a genetic algorithm, which is (according to Glotin, personal communication) not meant to be a model of actual biological evolution of the agents, but rather of the way sound systems are transferred from parents to children. This is a weak point of the research, as the influence of the genetic algorithm and the interactions between the agents are difficult to separate. Another problem with the model was that it was computationally intensive, and that therefore only a few simulations with small populations and small numbers of vowels could be run. In a way, this work was ahead of the computing power of the time.

This model has been the basis of a number of subsequent research efforts, however (Berrah, 1998; de Boer, 1997; 2000; de Boer and Vogt, 1999). Berrah's work was a direct continuation of Glotin's research. Berrah's model is a simplification of Glotin's model, in that the agents no longer have an articulatory representation of the sounds they use, only an acoustic one. This reduces the computational load considerably and allows more experiments with larger populations and larger numbers of vowels to be run. Berrah extends Glotin's model by investigating what he calls the "Maximum Use of Available Features". By allowing the agents to use an extra feature (which could be length, nasalization etc. in human languages, but which he models as an extra abstract dimension of the acoustic space) he shows that this is only used whenever the number of vowels in the agents' repertoires exceeds a certain threshold. His simulations also contain a genetic component, which makes it sometimes hard to tell when a particular phenomenon is due to interactions between the agents and when it is due to the actions of the genetic algorithm.

de Boer's work has concentrated on predicting vowel systems from interactions in a population. The agents have both an articulatory as well as an acoustic representation of their vowels, but use a much simpler articulatory model than Glotin's model. Also, the agents do not evolve, although experiments have been done with changing populations (de Boer and Vogt, 1999). They interact through language games (in this experiment called imitation games) only. It has been shown that vowel systems of human languages, and the relative frequencies with which they occur can be predicted quite well with this model.

More recently research has started to investigate syllable systems with genetic algorithms and population models relating in a similar way to the optimizing simulation used by Lindblom *et al.* (1984), as Glotin's, Berrah's and de Boer's relates to Liljencrants' and Lindblom's (1972) model. Redford *et al.* (1998, in press) have made a model that is based on a genetic algorithm. The population consists of words, which in turn consist of a closed set of phonemes. Redford *et al.* use a number of rules that determine how hard it is to produce and perceive different combinations and sequences of phonemes. On the basis of this a fitness for all the words in the population is calculated and selection and recombination take place. They try out different combinations of rules and investigate which rules are most important to predict syllables that are like those found in human languages.

Other work on predicting properties of more complex utterances is underway, but still largely unpublished. Pierre-Yves Oudeyer of the Sony computer science laboratory in Paris, France is working on predicting repertoires of syllables using more realistic signals. Emmanuelle Perrone of the Institut des Sciences de l'Homme is also working on predicting consonant-vowel syllables in the framework of imitation games. Eduardo Miranda of the Sony computer science laboratory in Paris, France is working on modeling intonation contours, while professor William Wang of the electronic engineering department of the City University of Hong Kong and co-workers Mieko Ogura and Jinyun Ke are working on modeling tone systems within the framework of genetic algorithms.

A Case Study

In order to illustrate the ideas outlined above, a case study will now be presented, based on de Boer's model of the emergence of vowel systems. At every point, the description will discuss the design decisions that have been made. Full details will not be presented, as these can be found in de Boer (1997, 1999; de Boer and Vogt, 2000). The fact that a genetic component is lacking in this system makes it somewhat different from most computational modeling of the origins of sound systems. However, this work provides a good example of using computer models to actually predict and investigate linguistic phenomena that can be checked directly with real language data. Other work is best studied in the original sources. As a genetic component is a very important factor in modeling evolution and origins of language, the possibilities of integrating this model with a genetic algorithm will be discussed, although so far this has not been implemented.

Vowels were chosen as the subject of research for two reasons. First of all, they are the easiest speech sounds to model. Typically, a vowel signal is constant over time and both its articulatory and acoustic characteristics can be described by very few parameters: in this model three real numbers for articulation and four real numbers for the acoustic signal. Secondly, vowels are the speech signals for which most is known about their distribution over the languages of the world. This makes it relatively easy to compare results of simulations with what we know about real human languages. Easy and objective comparison with human language data makes simulations much more convincing for a linguistic audience.

It was decided to investigate the process of change in vowel systems from a cultural perspective rather than an evolutionary perspective because, though vowel systems of human languages change over time, they continue to show the same near-universal characteristics. There are exceptional vowel systems, however, that do not conform to the universals. Therefore, it would seem unlikely that a strong innate constraint determines their shape. Rather, as Steels proposed (1995) in the context of vocabulary, self-organization in a population might be the force that causes human vowel systems to show universal tendencies. Of course, genetic evolution has also played an important role in shaping the vocal tract, but this might then be thought of as a process that is driven by cultural evolution.

This is an illustration of the different time scales on which cultural evolution and genetic evolution operate. Both self-organization in a population and cultural evolution operate on a very brief time scale. New words can spread in a language in less than ten years, and the grammatical foundations of an entire language can be dramatically changed in a few thousand years. Genetic evolution operates on a much slower time scale, and it is therefore to be expected that self-organization in the population and cultural evolution change the pressures on genetic evolution. Because of this impact, it is important to understand the dynamics of cultural evolution and self-organization.

Therefore, it was decided to leave out any genetic component in the first implementations of the model and rather to work with a population of agents playing language games. This also makes it easier to analyze the behavior of the system and to determine what phenomena are caused by which processes. Of course, genetic evolution of the agents can be introduced as well, and suggestions will be made as to where this could be done.

The agents that make up the population were designed to be as simple as possible while preserving the crucial characteristics necessary for investigating the characteristics of human vowel systems. They were equipped with a simple articulatory synthesizer that was based on measurements of vowel parameters taken from Vallée (1994). This synthesizer takes as input the three articulatory parameters necessary to describe a simple vowel: position, height and rounding (Ladefoged and Maddieson, 1996) and outputs the first four formant frequencies. These represent the center frequencies of the four most important peaks of the vowel's acoustic spectrum. The agents' perception uses a distance function that is calculated in the space that has as dimensions the first and the so-called effective second formant. The effective second formant is the weighted sum of the three highest formants and represents the perceptual phenomenon that multiple peaks in the higher part of the spectrum can be replaced by one single peak and still be perceived as the same. The particular calculation used is adapted from Mantakas *et al.* (1986).

The agents store vowels in terms of both acoustic and articulatory prototypes. There is a one-to-one association between the two types of prototypes. Prototypes are centers of categories. Whenever a signal is perceived, the distance to all the acoustic prototypes is calculated and the one that is closest is considered to be the one that is recognized. In the case of production, an articulatory prototype is chosen and the corresponding acoustic signal is produced, but noise is added to this by shifting the formant frequencies somewhat. During the process of learning a repertoire of vowels, prototypes can be added, deleted or shifted in order to match the vowels of other agents in the population more closely. For doing this, agents can only base themselves on the behavior of other agents; they cannot look at the other agents' vowel repertoires directly. Storing phonemes in terms of prototypes seems to be cognitively plausible. It has been observed that different types of speech signals are perceived in terms of prototypes (e.g., Cooper *et al.*, 1952; Frieda *et al.*, 1999) and also that other linguistic and cognitive concepts are stored and processed in terms of prototypes as well (e.g., Lakoff, 1987).

In a model of this kind, the interactions between the agents are as important as the architecture of the agents themselves. In human language, linguistic interactions do not just consist of an exchange of linguistic symbols. There is always a context, both in the form of a linguistic context and the situation in which the conversation is taking place. This situation has a physical aspect, i.e., the environment in which the conversation is taking place, but it also has a social aspect and a pragmatical aspect (and possibly other dimensions as well). All these aspects influence the linguistic exchange. It is clear that modeling a complete linguistic exchange is extremely difficult.

However, when one is only interested in the sounds of language, one can in principle ignore everything that has to do with meaning. Instead, one can use interactions that are based on imitation. In imitation, the same constraints on sound systems apply as in real linguistic interactions. For imitation to be successful, sounds have to be easily distinguishable, as well as easy to produce, just as they should be in a complete communication system. For this reason, the interactions between the agents in the system under study consisted of agents trying to imitate each other. In analogy with the term language game, these interactions will be called *imitation games*.

In an imitation game two agents are picked from the population at random. One of these agents is assigned the role of *initiator* of the imitation game, the other is assigned the role of *imitator*. Although the roles of the agents in an imitation game are not symmetrical, all agents in the population have equal probability of playing both roles. Although it is the case that in human learning of sound systems the roles of infants and adults are not symmetrical, it was decided not to implement this in the model. First of all, it would have introduced more parameters and more arbitrary design decisions and secondly, the aim of the research was not so much to model the way sound systems are acquired, but to investigate whether universal tendencies of vowel systems can be explained as self-organization in a population of language users.

The initiator of the imitation game chooses a random vowel from its repertoire, and produces it, while adding a small amount of noise. The imitator perceives this sound, finds the acoustic prototype of the vowel from its repertoire that is closest to it and produces the corresponding articulation, again adding noise. The initiator then perceives this signal, finds its closest vowel, and checks whether this is the same as the one it originally produced. If it is the same, it gives a "non-verbal feedback" to the imitator that the imitation was successful, while if it was not the same, it gives feedback that it was a failure. These steps include the main aspects of a linguistic utterance using a sound: production under constraints and with error, analysis in terms of a finite set of categories, and grounding of these categories outside the agent using non-linguistic cues. Although it is true that infants do not receive direct feedback about the quality of the sounds they produce, there must be a mechanism to provide a connection between meanings in the outside world and the sounds an infant perceives, otherwise an infant would not be able to learn which sounds in its language can distinguish meaning and which sounds cannot. The feedback in the case of human infants learning language is probably derived from the extra-linguistic context in which the utterance takes place, or by the ability to achieve a goal with a given utterance or not.

In reaction to the feedback, and based on the success of the vowel in previous games, the imitator can shift the vowel it used or add a new vowel. Both agents also keep track of how many times the vowel was used and how many times it was used successfully. Also, both agents regularly throw away vowels that have been tested a few times and have been found to be unsuccessful most of the time, and merge vowels that are too close together. Finally, a random vowel can be inserted with low probability, in order to make sure that the agents' repertoires become as large as possible. Merges, shifts and splits of vowels also occur in human languages, but these are different processes. They occur not just in a single speaker but in the language as a whole and are often caused or accompanied by dialectal variation, allophonic variation or different realizations in different speech styles. As such they are not directly comparable to the changes that occur in the agents in the simulation. It is likely, however, that such individual actions, which are basically part of the learning process, ultimately can cause changes in the language as a whole.

The agents start out with an empty repertoire and are in principle able to produce all basic vowels. This means that the system is not biased towards any language in particular, and that the results of the simulations can therefore be assumed to say something about human language in general.

Running the simulations results in the emergence of realistic vowel systems. A representative example is given in Figure 4.5. The figure consists of five frames, each representing a stage in the development of the vowel system. In each frame, the effective second formant and the first formant of all the acoustic prototypes of all the agents in the population are projected. The effective second formant is projected on the horizontal axis and the first formant is projected on the vertical axis. The usual directions of the axes are reversed, so that the vowels are projected in the way phoneticians usually project vowel systems, with [i] in the upper left corner, [u] in the upper right corner and [a] below. Note that not every point in the square can be reached by the agents' articulations. The available acoustic space is roughly triangular with the tip at the bottom of the graph.

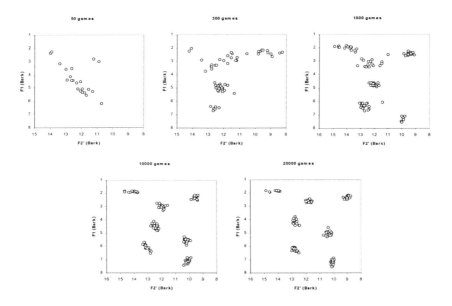

Figure 4.5 Emergence of a realistic vowel system.

The first frame shows the situation after 50 games. The agents start out empty, and as there has only been little time for agents to interact with each other, the most important process so far is random insertion of new vowels by agents that initiated an imitation game and direct imitation of these vowels by the agents that played the role of imitator in an imitation game. The vowels are therefore quite widely dispersed through the available acoustic space, but they do not cluster very much. During subsequent imitation games, the agents' vowels gradually move together. Also, due to the random insertion of new vowels, other clusters emerge, but not all agents have prototypes that correspond to all clusters. This situation is illustrated by the second frame of Figure 4.5, taken after 300 games. When the interactions continue, the clusters tend to stabilize and contract, and become dispersed over the available acoustic space. This becomes apparent after about 1000 imitation games (frame 3) and is almost finished after 10000 imitation games (frame 5). After 10000 imitation games, the clusters have become compact, and the available acoustic space is almost completely covered. However, the dispersion of the clusters over the available space is perhaps not quite optimal, yet. The dispersion gradually becomes better, until it is quite natural after 20000 imitation games (frame 6). The vowel system that emerges is natural, and could be found in a human language. It is not completely static, though. Vowel prototypes can move, so that the actual phonetic realizations of the vowels might change a little over time. Also, in rare cases, clusters may approach each other and be merged, or, if there is room, a new cluster might emerge.

Although a realistic vowel system emerges from the simulation illustrated in Figure 4.5, this does not establish that the simulation always results in realistic vowel systems emerging. In order to investigate this, many runs of the system need to be done, and the results compared with what is known about human languages. For one thing it is possible to define a measure of the dispersion of the vowels in the population of agents. It has been found that vowels in human languages tend to be dispersed more than in randomly created systems, and are actually quite close to being optimally dispersed (Liljencrants and Lindblom, 1972). It turns out that emerged systems, too, are almost optimally dispersed over the available acoustic space.

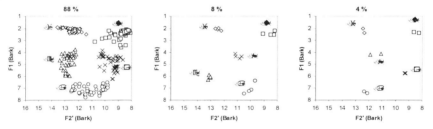

Figure 4.6 Classification of emerged five-vowel systems. The symbols used for the different clusters are meaningless, but were chosen such that every cluster has its own symbol and, conversely, different vowels of an agent are depicted by different symbols.

But it is also possible to compare emerged vowel systems with human ones directly. This can be done by running the simulation many times, then classifying the emerged vowel systems and comparing this classification with the classification one can make of human vowel systems. This is illustrated in Figure 4.6. Here five-vowel systems that emerged from the simulation for one setting of the parameters are classified in three different types. The symmetrical type occurs in 88% of the cases, the type with more front vowels than back vowels (and one central vowel) occurs in 8% of the cases, while the type with more back than front vowels occurs in 4% of the cases. This compares very well with the percentages found by Schwartz *et al.* (1997a): 87% the first type, 4% the second type and 2% the third type (these percentages do not add up to 100%, as they also found types that did not emerge in our simulations). Although the match between merged systems and real human language data is particularly good, excellent matches were also found for systems of six and seven vowels. For systems of four, eight and nine vowels, matches were good, but not as good. For three-vowel systems, the right types were predicted, but the so-called "vertical" three-vowel system, which is quite rare in human languages, occurred relatively frequently. However, the study has shown that the universal tendencies of human vowel systems can be explained as the result of self-organization under constraints of perception and production. Compared to previous models, fewer parameters (such as vowel system size) have to be determined beforehand. Also, the system does not just generate the most frequent human vowel systems, but less frequent ones, too, in approximately the right numbers.

The model could be augmented with a genetic algorithm that works on the agents in the population in several ways. One way is to let the learning parameters (such as the probability of adding random new vowels, the minimal quality required for keeping vowels or the distance criterion for merging vowels) of the agents change over time in a genetically determined way, and select for the agents that imitate the best. In this way, parameters that have to be tuned by hand in the present model could be set in a more objective way. Another way is to let the agents' production or perception evolve over time. Production would be particularly interesting as it seems that the human vocal tract is specially adapted to language. One could imagine a population of agents that start with a uniform tube with only a few control parameters, which is evaluated on how well they can imitate each other and how many different sounds they can distinguish. It would be interesting to investigate whether a vocal tract that is similar to that found in humans evolves.

Conclusion and Future Work

It has been demonstrated by different researchers that the evolution of human speech sounds can be investigated successfully with computer models. Different aspects of speech, such as vowel systems, syllables, tone systems and intonation have been investigated, or are being investigated. The approaches taken have consisted of either pure optimization, the use of genetic algorithms, the use of a population of language-using agents or a combination of these. The most realistic would be a system consisting of a population of agents that learn speech from each other, but that are also subject to genetic evolution. However, such a system would have many parameters and many points on which a (more or less) arbitrary design decision would have to be made. Also, it might turn out to be difficult to analyze the behavior of such a system. For the time being most systems either concentrate on population dynamics or on evolution, but in the future the two will definitely have to be combined.

In future work, too, more complex utterances have to be tackled. So far vowels in isolation and simple consonant-vowel syllables have been the main subjects of investigation. But for more insight into language change and evolution, longer combinations of arbitrary sounds have to be studied. For this, more realistic and more computation-intensive models will be needed. However, computing power available to the average researcher has increased so much in recent years that such models have now become computationally feasible. It will still be necessary, though, to find appropriate simplifications in order to make realistic, but tractable models.

Also, for the study of more complex sounds, machine learning algorithms are needed that are able to learn temporal sequences and that are able to extract patterns from such sequences. This is an area of research that is still very open in the machine learning community. An interesting aspect is that the ability to learn sequences and to find patterns in them is also a necessary prerequisite for learning syntax and grammar. Perhaps an interesting exchange of ideas and models between the investigation of the origins of syntax and the origins of speech is possible.

Speech is the aspect of language that is most concrete. It is therefore easiest to make an objective comparison between real linguistic data and the outcomes of a computer model in research into the evolution of speech. Also, paleontology data can only tell us something about our ancestor's capacity for speech, never about other aspects of language. Speech is therefore ideal for investigating and modeling the evolution of language. So far, we have only scratched the surface.

Acknowledgements

The work that is described in this chapter was performed at the artificial intelligence lab of the Vrije Universiteit Brussel in Brussels Belgium. It was written at the Center for Mind, Brain and Learning at the University of Washington in Seattle. I would especially like to thank all the researchers whose ongoing and unpublished work is mentioned.

References

Baldwin JM (1896) A new factor in evolution. *The American Naturalist*, 30: 441-451, 536-553. Reprinted In: Belew RK, Mitchell M (eds) *Adaptive individuals in evolving populations: Models and algorithms (SFI Studies in the Sciences of Complexity, Vol XXVI)*, Addison Wesley, Reading MA, 1996

Berrah AR (1998) *Évolution artificielle d'une société d'agents de parole: Un modèle pour l'émergence du code phonétique.* Thèse de l'Institut National Polytechnique de Grenoble, Spécialité Sciences Cognitives

Berrah AR, Glotin H, Laboissière R, Bessière P, Boë L-J (1996) From form to formation of phonetic structures: An evolutionary computing perspective. In Fogarty T, Venturini G (eds) *ICML '96 Workshop on Evolutionary Computing and Machine Learning*, pp 23-29

Cooper FS, Delattre PC, Liberman AM, Borst JM, Gerstman LJ (1952) Some experiments on the perception of synthetic speech sounds. *Journal of the Acoustical Society of America*, 24: 597-606. Reprinted in: Fry DB (ed) *Acoustic phonetics*, Cambridge University Press, pp 258-272

de Boer B (1997) Generating vowels in a population of agents. In: Husbands P, Harvey I (eds) *Proceedings of the Fourth European Conference on Artificial Life*, MIT Press, Cambridge MA, pp 503-510

de Boer B (2000) Emergence of vowel systems through self-organisation *AI Communications*, 13: 27-39

de Boer B, Vogt P (1999) Emergence of speech sounds in a changing population. In: Floreano D, Nicoud J-D, Mondada F (eds) *Proceedings of the Fifth European Conference* on *Artificial Life, ECAL 99 (Lecture Notes in Artificial Intelligence, Volume 1674)*, Springer-Verlag, Berlin, pp 664-673

Everett DL (1982) Phonetic rarities in Piraha. *Journal of the International Phonetic Association*, 12: 94-96

Firchow I, Firchow J. (1969) An abbreviated phoneme inventory. *Anthropological Linguistics*, 11: 271-276

Frieda EM, Walley AC, Flege JE, Sloane ME (1999) Adults' perception of native and nonnative vowels: Implications for the perceptual magnet effect. *Perception Psychophysics*, 61: 561-577

Glotin H (1995) *La vie artificielle d'une société de robots parlants: émergence et changement du code phonétique.* DEA sciences cognitives - Institut National Polytechnique de Grenoble

Glotin H, Laboissière R (1996) Emergence du code phonétique dans une societe de robots parlants. *Actes de la Conférence de Rochebrune 1996: du Collectif au social*, Ecole Nationale Supérieure des Télécommunications, Paris

Labov W (1994) *Principles of linguistic change.* Blackwell, Oxford

Ladefoged P, Maddieson I (1996) *The sounds of the world's languages.* Blackwell, Oxford

Lakoff G (1987) *Women, fire, and dangerous things: what categories reveal about the mind.* Chicago University Press, Chicago

Lieberman P, Crelin ES (1971) On the speech of Neanderthal man. *Linguistic Inquiry*, 2: 203-222

Liljencrants L, Lindblom B (1972) Numerical simulations of vowel quality systems: The role of perceptual contrast. *Language*, 48: 839-862

Lindblom B, MacNeilage P, Studdert-Kennedy M (1984) Self-organizing processes and the explanation of language universals. In: Butterworth B, Comrie B, Dahl Ö (eds) *Explanations for language universals*, Walter de Gruyter, Berlin, pp 181-203

Lindblom B, Maddieson I (1988), Phonetic universals in consonant systems. In: Larry M, Hyman C, Li N (eds) *Language, speech and mind.* Routledge, London, pp 62-78

Maddieson I (1984) *Patterns of sounds.* Cambridge University Press

Maddieson I, Precoda K (1990) Updating UPSID In: *UCLA Working Papers in Phonetics*, 74, pp 104–111

Mantakas M, Schwartz JL, Escudier P (1986), Modèle de prédiction du 'deuxiéme formant effectif' F2 - application à l'étude de la labialité des voyelles avant du français. In: *Proceesings of the 15th journées d'étude sur la parole Société Française d'Acoustique*, pp 157-161

Redford MA, Chun Chi Chen, Miikkulainen R (1998) Modeling the emergence of syllable systems. In *Proceedings of the 20th Annual Meeting of the Cognitive Science Society (COGSCI-98, Madison, WI).* Lawrence Erlbaum Associates, Hillsdale NJ, pp 882-886

Redford MA, Chun Chi Chen, Miikkulainen R (in press) Constrained emergence of universals and variation in syllable systems, *Language and Speech*

Ruhlen M (1994) *The origin of language: Tracing the evolution of the mother tongue.* Wiley, New York

Schwartz JL, Boë L-J, Vallée N, Abry C (1997a), Major trends in vowel system inventories. *Journal of Phonetics*, 25: 233–253

Schwartz JL, Boë L-J, Vallée N, Abry C (1997b), The dispersion-focalization theory of vowel systems. *Journal of Phonetics*, 25: 255-286

Sheldon SN (1974) Some morphophonemic and tone rules in Mura-Pirahã. *International Journal of American Linguistics*, 40: 279-82

Snyman JW (1970) *An introduction to the !Xuⱸ (!Kung) language.* Balkema, Cape Town

Snyman JW (1975) *Zu/ohasi: fonologie woordeboek.* Balkema, Cape Town

Steels L (1995) A self-organizing spatial vocabulary. *Artificial Life*, 2: 319-332

Steels L (1997) The synthetic modelling of language origins, *Evolution of Communication*, 1: 1-34

Vallée N (1994) *Systèmes vocaliques: de la typologie aux prédictions*, Thèse préparée au sein de l'Institut de la Communication Parlée (Grenoble-URA CNRS no 368)

Vennemann T (1988) *Preference laws for syllable structure.* Mouton de Gruyter, Berlin

Wittgenstein L (1967) *Philosophische untersuchungen.* Suhrkamp, Frankfurt

Chapter 5

The Evolution of Dialect Diversity

Daniel Livingstone

Introduction: From Dialect to Dialectology

Observations on dialect diversity have been recorded for thousands of years, including Old Testament stories and early writings and literature from around the globe. Language diversity remains the subject of much study today – largely in the related fields of socio-linguistics, historical linguistics and dialectology. One key question is *why* is there so much difference in dialects. To some the question was irrelevant as diversity was somehow obviously a natural feature of human language, one not requiring much explanation – it simply was. Currently the question of why diversity should exist to the degree that it does has been taken quite seriously, with strong differences of opinion apparent.

On one side, there are various arguments that linguistic factors alone could not be responsible for language diversity, and hence there must be some specific reason for its emergence and degree. Some additional extra-linguistic factor which causes diversity where otherwise none would exist. The extent of dialect diversity, and the speed of its evolution caused Dunbar to remark that it "is so striking and so universal that it cannot be a simple accident of evolution: it must have a purpose" (Dunbar, 1996: page 158).

Opposing this is the view that the nature of language, how it is transmitted and learned and the role of contact between individuals, provides sufficient means to explain the origins of language diversity. In this chapter we briefly set out some of the arguments for both positions, concentrating on arguments based on the possible requirements of social factors to effect change and diversity.

With plausible and convincing arguments presented on both sides of the divide, it is hoped that the use of simulation-based models may help resolve the debate.

We review a number of existing models, and present an overview of our own approach, including some recent work applying de Boer's model of evolving sound systems (also see Chapter 4) to this problem. Using simulation-based models is not the only possible approach, and we also briefly describe some of the difficulties that may result from the use of more traditional mathematical modeling techniques.

We conclude that simulation models have much potential for the investigation of linguistic diversity and dialect. But first we present a brief overview of some characteristics of linguistic diversity, and a brief literature review.

Dialect and Linguistic Diversity

As stated, the study of dialects and language diversity is of interest to academics in a number of sub-disciplines of linguistics, as well as to academics in other, related, fields. A good starting point for an overview of some of the different approaches taken is the Cambridge Encyclopedia of Language (Crystal, 1987).

Crystal presents an overview of some key characteristics of human language diversity. One important finding of dialectology is that the boundaries between different dialects or languages are not always easy to define. Geographically close dialects may have sufficiently similar grammars or lexicons to allow speakers from each dialect to understand each other quite well, despite differences in the dialects. Geographically distant dialects may not be at all mutually intelligible, despite being at either end of a chain of dialects where every dialect is intelligible to speakers of the neighboring dialects (Figure 5.1). For example, several such dialect continua exist in Europe, blurring the boundaries between the different languages within the Germanic, Scandinavian, Romance and Slavic language groups.

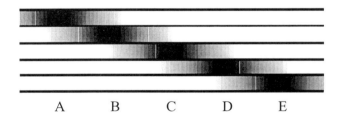

A   B   C   D   E

Figure 5.1 A schematic dialect continuum from dialect A to dialect E, showing some degree of mutual intelligibility between adjacent dialects (after Crystal, 1987: p. 25).

Further, the geographical boundaries between dialects may not be easy to determine. Dialects may differ in their lexical, semantic, morphological or phonological features. Sampling the language use of individuals in some area results in a map, on which boundaries, termed *isoglosses*, may be drawn to show where language use is distinct on either side according to a particular linguistic feature.

It may be expected that these lines will be largely coincident, forming clear dialect boundaries. Often, however, the boundaries are not even nearly coincident, as individuals near a boundary may use differing mixtures of lexical and grammatical items from the surrounding major dialects (Figure 5.2). Only when viewed at a more distant scale is it possible to determine distinct dialect areas – with poorly demarcated boundaries between them.

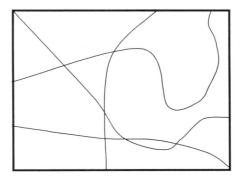

Figure 5.2 Dialect boundaries (after Crystal, 1987: p. 28).

As if this picture were not complicated enough, it is also widely recognized that no two individuals use language in quite the exact same way – in a sense every individual speaks their own particular dialect, or *idiolect*. What is viewed as a dialect is merely some norm derived by sampling many idiolects. This, as we shall see, is highly significant when examining or attempting to explain linguistic diversity.

Sharp linguistic boundaries that do exist often coincide with significant geographical boundaries or with strong cultural boundaries (c.f. Chambers and Tridgill, 1980). Where such boundaries exist, limiting the interactions of individuals across the divide, there may be many coincident isoglosses splitting the sides and a resultant lower degree of mutual intelligibility. The aim of modeling language diversity is not to reproduce these characteristic patterns, but to examine what processes may give rise to them, and before we proceed to build models we must examine the explanations which already exist.

In recent years much of the research has centered on social variations in language use, propelled by the impact of the work of Labov, who detailed significant quantitative differences in language use by individuals according to social groupings (Labov, 1972). This fuelled the growth of the field of socio-linguistics. Within socio-linguistics the predominant view of why such variation exists is that it serves some social function, the benefits provided leading people to learn and use particular variants over others (for example, Chambers, 1995; Trudgill and Cheshire, 1998).

In contrast, in historical linguistics the prevalent view has been that language changes are due to pressures internal to language as a system – most famously with Grimm's 'Sound Laws' (Grimm, 1822). Such a view holds that changes in one part of the system may lead to increased opportunities for further change. The internal structure of a language, rather than the requirements of the speakers, is seen as the major force driving change. In both cases it appears that the favored explanation is heavily influenced by the different forms of evidence and phenomena that each discipline studies. For example, two opposing arguments are presented by Milroy (1993) and Lass (1997), skeletons of which are repeated here. Milroy's argument emphasizes that the best linguistic systems, for optimal communication, would be ones where every speaker used the same language. The differences in dialect are not just learned but actively maintained by speakers, where otherwise it would be expected that the dialects would merge and unify. Therefore, the differences must be of some use to the language users.

Lass' reply is unusual in historical linguistics literature, in its extensive use of evolutionary theory. It also gives a very central role to the idiolect. The language used by a single individual is considered to be a member of a hyper-variable species. With no two idiolects the same, language norms are simply averages of the different language features in use. Linguistic 'junk' and redundancy allows variability without serious loss of communicative ability. Another result of this is that learners are not required to learn perfectly the existing language norms.

It is not easy to resolve this argument using the available linguistic evidence. To prove either theory conclusively, language and dialect evolution must be tested in a population where the precise effects of social influence on language change are well understood. In this chapter, we hope to show that simulation models are useful in bypassing this problem, and for providing reliable supplementary evidence. Simulation models are not the only technique available, however. In the next section we give an overview of other modeling approaches and argue that simulation models provide a more flexible and powerful means for testing theories explaining the emergence of dialects.

Approaches to Modeling Dialect Diversity

Here we very briefly review some of the different modeling approaches that have been used to study the development of linguistic diversity. This is not a thorough critique of the different approaches but, it is hoped, it provides some justification for the use of simulation micro-models in attempts to better understand the evolution of dialects.

Mathematical models could be a better approach than the use of simulation to describe the process of language change and the evolution of dialects. Pagel (2000) presents a mathematical model for examining the rate and pattern of linguistic evolution. The equations presented are useful for describing the degree of diversity that exists or the rate of diversification, but not for explaining what causes diversity in the first place. Pagel's equations are useful in comparing the rate at which languages evolve into other, distinct, languages, and the rate of the linguistic diversification within different language groups. The model developed does not allow for close investigation of particular changes, or of what happens within languages as they develop, presenting a more global picture instead.

This is quite different to the approach of Cavalli-Sforza and Feldman (1978). Their theory of Cultural Evolution, although intended more generally for modeling the evolution of cultural traits, can be applied to the evolution of language. They present a mathematical model that is, in essence, quite simple – a summation of the influences on a single individual determines which traits are acquired by that individual. With the presence of mutation, innovations may occur and over time the cultural traits in the population may evolve – particular traits succeeding in the population at large or otherwise. The possibility that this model might also be used to describe language evolution is emphasized by similarities it has with the description of the action of Social Networks on language (Milroy, 1980), appearing to be a mathematical formulation of the same. To actually use this model requires the use of computer simulation – the very large number of calculations required would make the use of this model infeasible on all but the smallest of populations with the minimum of cultural traits otherwise.

Where models that abstract away the spatial distribution of speakers are unable to capture this important aspect of dialect evolution, mathematical models which seek to represent this are complex and not feasible without the aid of computers. An alternative approach, fully leveraging the power of computers, is to build a simulation model, and a number of such models have been built.

For example, Niyogi and Berwick (1995) use a dynamical systems based model to study language change. Even so, there would be some problems in extending or adapting their approach to study linguistic diversity. By considering the distribution of dialects only as a proportion of the population which uses it, no spatial information is used in the model. The effect of this is that the acquisition of grammars is not reliant on any spatial constraints, and there can be no comparison with the emergence of dialects across geographical or social distances in the evolution of human languages. Briscoe (2000) argues that a micro-simulation model has a number of advantages over Niyogi and Berwick's more mathematical approach, and such simulations with more realistic population models may even produce different results.

In the following sections we review some works that have applied the micro-simulation approach to the evolution of linguistic diversity and closely related problems.

Related Simulation-based Models

The emergence of 'dialects' has been observed in a number of Artificial Life and other micro-simulation models of language and signal evolution. In many cases the existence of dialects is noted but not investigated – research being focused elsewhere, such as the evolution of a signaling ability or the emergence of conventionalized signals (e.g., Werner and Dyer, 1991; Hutchins and Hazlehurst, 1995; Livingstone and Fyfe, 2000).

In some cases even work that is not about language at all may be relevant. Axelrod (1997) presents a model to investigate the dissemination of culture through a spatially arranged population. In the model, neighboring sites may interact if they have at least one cultural trait in common, and as sites interact they share traits and slowly converge. Eventually a stable distribution emerges where a limited number of groups survive, within each group all sites having identical sets of traits, and no traits being shared with sites belonging to neighboring groups.

Viewing language as a cultural trait, this is obviously relevant to the evolution of dialects. The results are at odds with observed phenomenon in human language, however (see Figure 5.2, above, and related discussion).

In other cases the relevance of the results to the question of the origins of linguistic diversity is clearer. For example, Kirby (1998) builds a simulation model to show that Universal Grammar constraints may not be innate constraints at all, but merely the outcome of learning over time leading to a reduced set of surviving grammars. In this work Kirby shows results indicating the existence of geographically distributed dialects of grammars.

Maeda *et al.* (1997) examine the effect of language contact. The results show subsequent language reorganization, but after this dialect diversity is absent from the population – again, not results that contrast with those observed in the real world.

Other related work has looked at the process of language change, apart from the question of dialect. Steels and Kaplan (1998) demonstrate how various linguistic and extra-linguistic errors can lead to continued language change. While natural language errors are somewhat more systematic than the random errors introduced in this model, the model successfully demonstrates the large influence such errors may have on language innovation. A similar model, based on artificial neural networks, is presented by Stoness and Dircks (1999). In this model, it is found that noise is not required to maintain competition between forms. This appears to be due to the networks learning either one of two similar signals for particular internal meanings – an ability not present in the Steels and Kaplan model, where similarities in the lexical forms are ignored by the agents.

There yet remain further works that explicitly explore the origins and nature of linguistic diversity, and these we consider next.

Simulation Micro-models of Dialect Diversity

Arita and Taylor (1996) present what is possibly the first attempt to explain the origin of linguistic diversity using a micro-simulation model. They hold that it is the spatial distribution of individuals that is the key factor in the emergence of dialects. While this is a plausible position, it is not strongly supported by the model, which relies on *genetic* mutation for the emergence of linguistic diversity. Language is inherited, with mutation producing diversity and learning leading to increased convergence. If the spatial distribution of speakers is indeed a factor in the emergence of dialect diversity, then it must be able to work when the only means of language transmission is through learning – as it is for human language

Innate language is again used by Arita and Koyama (1998), in their investigation into the evolutionary dynamics of vocabulary sharing. Mutation rate is again identified as being an important factor in the emergence of diversity in the vocabularies, but without an identified linguistic equivalent. The degree of vocabulary sharing is also related to the availability of resources – rather than vocabularies, it is cooperative strategies that are being evolved here, as evidenced by cases where the evolved communication strategy is not to communicate at all.

A significant contribution to the use of micro-simulation for exploring the emergence of linguistic diversity has been made by Daniel Nettle in a series of papers. Nettle also sees learning as a force for convergence, and argues that additional, socially motivated, factors are required for the emergence of dialect diversity.

First, in Nettle and Dunbar (1997) a model is developed which shows how dialects may be used to indicate group membership, and how such a marker may be used in the evolution of cooperation. Groups of cooperative agents are able to resist invasion from non-cooperative individuals. This is used as the basis for an argument that dialects emerged for this *reason* – something that we argue against later in this chapter. Nettle presents two further models that support his arguments that social status and social functions of dialect differences are prerequisites for the emergence of dialect diversity (Nettle, 1999a; 1999b).

The model presented in Nettle (1999a) arranges language learners into a series of small groups, the language used consisting of a model of a vowel sound system. Learners pass through five life-cycle stages, and all language acquisition occurs during the first stage, where the new language agents learn from the other agents in the same village. Each group contains four individuals at each of the life-cycle stages (twenty in total at any time). After the fifth life stage, the elderly are replaced by a new set of infants. The infants each learn a sound system according to the set of sound systems in use by the existing group members, plus a small amount of noise. After this, all the individuals are 'aged' one stage. No learning occurs after the first stage.

It is seen that, unless the groups are completely isolated from one another, diversity does not emerge. Adding in social status changes the findings significantly. Each individual has a 25% chance of gaining high social status after the first life stage. Learners only learn language from those individuals with high status within the village. Otherwise, for any vowel, the sound learnt is the average formant frequency values used by all of the adults in the population for that vowel, plus a slight perturbation due to noise. With social status included small differences between groups may become magnified over time and it is found that contact between groups no longer eliminates diversity.

The model presented in Nettle (1999b) has many major differences, but retains the same agent life cycle, where agents pass through five life stages, before being replaced by new learners. Again, learning only occurs during the first life stage. Apart from this there are few similarities. Inspired by social impact theory (Latané, 1981), there are no subgroups within this model, all agents existing on a single spatial array. Instead of learning vowels, agents acquire one of two grammars, p or q. In determining which grammar a learner acquires, the impact of all the surrounding grammars is calculated. This forms a sum of all the surrounding grammars, weighted by distance. Then, if the result is in favor of one grammar, that is the grammar acquired. Several factors may be varied in this model, but the general finding is that sustained diversity requires that social status exerts a very large influence upon the acquisition of grammar.

These last two models each have design features which lead directly to these results. In the former, vowels are learnt by an explicit averaging of the vowels in use already in the local group. In the latter, the impact measurement and forced selection of a grammar from one of two distinct grammars – without the possibility of acquisition of elements of different grammars – is a form of thresholding.

Nettle argues in his work that the effects of averaging and thresholding would work to stifle diversity, were it not for the effect of social status, and uses these models as demonstrations. He then uses models in which these are enforced by the language acquisition rules he has built in. It is not proven that under more realistic learning conditions, where language is acquired as the result of many interactions, or where there is a possibility of learning grammars that are different but compatible with surrounding grammars, that averaging or thresholding will prove to be the barrier to diversity that Nettle argues they are.

In the next section we present our own model, which was developed after an earlier model of the co-evolution of language and physiology (Livingstone and Fyfe 1998; 2000). The acquisition of signals occurs over many stochastic interactions between the signal learners, and as we shall see the results are quite unlike those described by Nettle.

Emergence of Dialects in Spatially Organized Populations of Simple Signal Learning Agents

We observed the existence of what appeared to be dialects in some of the results of our earlier model of the co-evolution of physiology and language (Livingstone and Fyfe, 1998, 2000). The model used does not include any apparent social or adaptive benefits which might give rise to such diversity. The emergence of dialects in this work was an unexpected result that was not investigated at the time. That the result was at odds with other work, claiming the need for social function, gave impetus for further investigation. To begin with, we used essentially the same model as our earlier work, removing some features no longer required (principally removing evolution of the agents themselves – as now it is only the evolution of the language used by the agents that we wish to examine) and adding others. However, the implementation of the individual language learner was not modified at all.

The following section details our implementation of language learners, after which we present the population model and results. In a later section we discuss and compare our results against those of Nettle.

A simple language agent

The language used by the agents in our model is a greatly simplified one, the agents themselves capable of only a small repertoire of signals. A very small set of arbitrary 'meanings' exists, common to all agents. The agents produce conventionalized signals to indicate the current meaning, and listener agents attempt to interpret what the original meaning was from the received signal. Each agent participates in many interactions, sometimes as signal producer, sometimes as listener.

An agent is implemented by a fully connected Artificial Neural Network (ANN), with two layers of nodes – three internal state nodes to hold the 'meanings' and three signal nodes (plus a signal bias node).

The internal state is a sparse bipolar vector (+1 or –1 at each node, only one being set to +1 for any one meaning). The signals are non-sparse bipolar vectors (arbitrary vectors with +1 or –1 at each node). This representation allows three possible meanings and 2^3 (= 8) possible distinct signals. This provides some degree of redundancy in the representation of meaning, the significance of which becomes clear when viewing the results of the model.

In signal production, the meaning is presented to the input nodes, and fed forward through the ANN to generate a signal for that meaning. Each output is thresholded to a bipolar value. Signals are interpreted in a similar way – the incoming signal is presented at the signal layer and fed back to generate a meaning vector. A winner-take-all comparison at the internal state layer determines which one of the nodes has the greatest activation, and this node is set to +1, the remaining to –1.

Using a standard ANN learning algorithm to train the signal production of the agents would allow the agents to learn to use the same signals for given meanings, but would not train the agents to use *distinct* signals for each meaning (c.f. Fyfe and Livingstone, 1997). Oliphant (1997) overcomes this problem by using the signal production behavior of the population to train the signal reception behavior of the agents and the signal reception behavior of the population to train the transmission behavior. We make use of a similarly inverted training algorithm for our ANN.

Similarly to Steels' 'naming game' (Steels, 1996), one agent takes a turn as a teacher, another as a student. A meaning is presented to the teacher, which then produces a signal. This signal is presented to the signal layer of the student and fed-back to produce a generated meaning. The difference between the original meaning, x, and the generated meaning, x', is used to update the weights. This is shown in the equation below:

$$\Delta w_{ij} = \eta\left(x_i - x_i'\right)y_j$$

As is shown, learning only occurs when the learner misclassifies the signal.

Spatially organized populations of language learners

As well as using a very simple ANN to represent each agent – each with six nodes, excluding bias – we use a very simple population structure. Agents are placed in a single row. The ends of the row are *not* connected – so with the exception of the agents at the row ends, each agent has two immediate neighbors. Communication between agents – for learning or for evaluating the success rate of the learned signaling schemes – is limited by pre-determined neighborhoods based on distance. An agent may communicate with any other agent in its neighborhood.

As an example consider the following. During the training period, agent A has been selected to be the student in one round. The agent which will act as teacher is selected from the agents within the neighborhood of A. The chance that each of the agents within the neighborhood has of being selected is determined by a normal-distribution curve centered on A. Agent B, immediately next to A, is more likely to be selected than some agent C, located more distantly from A. This creates a population where communication is localized but which does not have any explicit group boundaries.

We also adopt a population aging structure considerably reduced from that used by Nettle (1999a). Agents pass through only two life stages, child and adult, opposed to the five life stages of agents in Nettle's model. Children and adults populate separate rows of identical size. As a child, agents learn only from adults – the position within each row being used in the determination of the neighborhoods. After training, the existing adults are removed, the current child population is aged and a new row of children populated. A variation on this, where the children have an additional short period of signal acquisition where they learn only from their peers, leads to no significant differences in the recorded results.

We have tested this model using a variety of initial and environmental conditions. The principal factors that are varied between different tests are:

- **Neighborhood size**. By varying the standard deviation of the normal curve which is used to determine the likelihood of a neighbors selection, the effective neighborhood size can be varied from one that only includes the immediate neighbors to one that includes the entire population. The default neighborhood includes only a small number of agents to either side of the currently selected agent. Additionally, the neighborhood can be changed to cover the entire population with a uniform distribution.

- **Initial conditions**. The first generation has no previous generation to learn from, and so must be initialized somehow. It is possible to initialize the population such that all agents use the same signaling scheme. Alternatively, and by default, random values may be used. Training of the following generation can then begin (or of the current, if peer-to-peer learning is to be used).

- **Noise**. By default the communication between agents is noise free and the perfect replication of signaling schemes across generations is possible – something that is not possible in the human language learning for several reasons. First, the linguistic evidence presented to children is insufficient to allow perfect replication. It is also possible that errors in language use or comprehension, on the part of learners or the speakers who provide them with evidence, may help drive language change (Steels and Kaplan, 1998), or that individual variations in language use (Chambers, 1995) might have some impact. In the model we use noise to represent these extra forces that may act on language replication from generation to generation. With a small probability, noise may cause individual bits in the signal to be inverted (+1 or −1 or vice versa).

Additional factors that can be varied include the population size, the number of meanings and/or signals used or the learning rate used in the ANN learning rule.

Results

Here we present only an overview of the results and highlight some advantages and disadvantages of the approach used in the model – most of the results have been previously presented in more detail elsewhere (Livingstone and Fyfe, 1999a; 1999b).

Under the default conditions distinct dialects are readily apparent in each generation of the population, and the dialects themselves evolve over generations, gaining and losing speakers eventually being replaced by other dialects. Within the populations dialect continua form whereby chains of mutually intelligible signal schemes span across the population, linking dialects that may, or may not, be mutually intelligible (as shown in Figure 5.1). Between distinct dialects, the point where signal use changes tends to be different for each internal state, forming indistinct boundaries between the dialects (similar to the position presented in Figure 5.2).

In some cases convergence is observed, where the whole population learns the same signaling scheme. Once this occurs (in the cases where this has been observed, it happens only after a very large number of generations has passed – on the order of several hundreds of thousands of generations) signal diversity is unable to return to the population. Similarly, when starting with a converged signaling scheme, diversity never emerges. The addition of a small amount of noise is sufficient to introduce variation into the signaling schemes, which rapidly, within tens or hundreds of generations, creates diversity and distinct dialects across the population. Noise is also able to prevent convergence occurring.

Increasing the neighborhood size tends to reduce the number of distinct dialects. At the limit, where any agent might interact with any other with uniform probability, a single global dialect emerges. Some variation exists inside this dialect, with different sets of signals being used for each meaning. For a given meaning an agent may use any one of the signals from the set, but each agent is able to correctly interpret the signals in use. This result is in accordance with the view that dialect differences flow automatically from any non-uniformity in the interactions of speakers in a community (c.f. Sweet, 1888, para 189).

By using the three bit binary representation for signals, eight different signals were possible. The advantages this provides for visualization of the results (as described below) prevented us from changing this in different runs.

Review and evaluation

Our model has some important advantages, as well as limitations.

The signal representation allows a signal to be plotted as a color pixel, and this makes visualization of the results very easy – not just for comparing the dialects within a generation, but across many hundreds of generations. More important is the ease with which the results can be compared to the patterns of human linguistic diversity, and the similarities with these observations. This increases our confidence in the claim that social motivation, a factor absent in our model, is not a requirement for the evolution of diversity in human language.

A severe limitation of the model is the constrained representations of signals and meanings. Meanings are pre-determined, and limited by the three possible internal states. The signals are similarly restricted, this time to eight possible discrete forms. The ability of agents to interpret two or more signals as having the same meaning is an advantage, however. Similar signals are more likely to be interpreted the same, and this brings a degree of redundancy to the signal schemes, a feature absent in most other models. The importance of this is supported by the weight given to the role of linguistic junk by Lass (1997), who considers redundancy to be a key feature of language in the processes of language change.

However, the results of this model are opposed by results from three different models used by Nettle, even before other models are considered. Additional corroboration for our results is highly desirable, and this is what we seek to gain, as detailed in the remainder of this chapter.

Emergence of Dialects in Spatially Organized Populations of Vowel Learning Agents

Corroboration can be gained in the same manner as used by Nettle – by using alternative models that provide qualitatively similar results. However, rather than develop a new model from scratch, we chose to base our new model on some existing work, de Boer's model of emergent vowel systems (see Chapter 4).

We use an implementation of individual agents based closely on de Boer's original, adding a more structured population model. Initially we use a linear arrangement of agents – similar to the one used in our previous model.

The agents used here have some additional advantages, beyond simply allowing us to re-test our results. The signals exist in a continuous domain, in which there is a very wide range of possible signals. The number of signals an agent learns is not fixed, allowing for much more open-ended evolution of signaling schemes.

The phonological model used is quite realistic, and produces emergent vowel systems very like those that appear in human language. This may aid the 'believability' of our results as the individual sound systems can be closely compared to real sound systems, not just the overall patterns of diversity and change. It is also possible to use the agents in a variety of population models, representing different social conditions, which makes the model potentially extremely flexible. These are not explored here, however.

A disadvantage of using the phonological model in comparison to the previous model is that visualization may not be quite so easy. In particular, a clear view of the dialect used by each speaker, when displaying the vowel systems used by the whole population, may not be achievable. This is not an insurmountable problem, as we shall see next.

de Boer's model of emergent phonology

As in the model used previously, de Boer's agents learn over the course of many repeated interactions, rather than a single update. de Boer's model has been thoroughly documented, enabling an independent re-implementation to be attempted (de Boer, 1997; 2000; de Boer and Vogt, 1999). This was done, and testing determined that it performed qualitatively the same as de Boer's own implementation.

In de Boer's implementation, small populations of twenty vowel learners exist and interact with each other according to a uniform random selection. No form of spatial or social structure is used.

An example of an emergent vowel system from our implementation of the model is shown in Figure 5.3. Superimposed onto this are the approximate positions of the major vowel sounds of the English language.

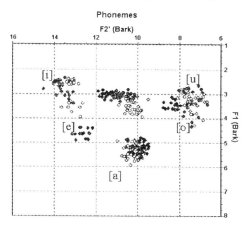

Figure 5.3 An emergent phonology. Clusters appear in areas of the phoneme-space where multiple agents have learned the same vowels. The approximate positions of the major vowels of English have been superimposed on the graph.

Experimental setup

The model is enhanced to include a larger population of agents, spread across a spatial array. One hundred agents are arranged in a single row, the ends of which are not connected, as in our previous model. The algorithm for agent learning is not altered from de Boer's original, other than to enable the neighborhood-based selection of partners. Once an agent has been selected a partner is required for learning. The partner is selected from a position along the line on either side of the original, from within a range of ten agents to either side. Agents near the ends of the line have as a consequence fewer other agents to communicate with. The selection of partners is from a uniform random distribution within the neighborhood.

Table 5.1 sets out the parameter settings for the run detailed in the following results. As before, a number of runs were performed for a variety of parameter settings; some of these are discussed below.

The large number of training rounds is used to provide a good chance of each and every agent receiving somewhere on the order of 200 training examples, or more. The chosen neighborhood selection limit was determined by the size of the population divided by ten. This figure ensures that there is reasonable distance between the far ends of the population. The actual neighborhood size is double this, as agents can communicate with others on either side up to this distance.

Table 5.1 Parameter settings for testing the evolution of dialects with the emergent phonology model

Parameters			
Population	100	Training rounds	25,000
Neighborhood size	10	Noise (%)	15

In de Boer's work the effect of varying noise is well documented. Smaller noise values allow the emergence of vowels systems with more vowels than occur with larger noise values, which tend to have fewer, larger, clusters. The noise value we use is in the mid-range of the noise values investigated by de Boer.

Results
Running the model produces what appears to be a rather messy result, shown in Figure 5.4, compared to the previous (Figure 5.3). This could represent a fairly noisy system with four or perhaps six different vowels, the clusters being indistinct and seemingly overlapping. By breaking the population into groups, and displaying only the different vowels used by agents within the groups, one or two groups at a time, a better picture of the vowel use can be gained.

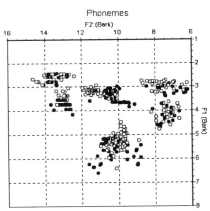

Phonemes

Figure 5.4 The emergent vowel system of the population.

The diagrams in Figure 5.5 show the same emergent vowel system. The population has been split into five arbitrary continuous groups. The first twenty agents of the population form the first group, the next twenty are placed in group two, and so on. Note, these groups exist only for the purpose of displaying the results – and no related boundaries exist during the negotiation of the phonology.

Some of the individual diagrams remain a little unclear – it is not always obvious whether one or other of the clusters represents one or two vowels. Even within a single group, there is some distance, and it is possible that different agents within a group have learned slightly different sets of vowels. The groups themselves have not been chosen with regard to how close the vowel systems of the individuals within the group are – rather, the groups are a completely arbitrary division of the population. As such we should not expect that within groups the vowel systems should be very similar or that there must be distinct differences between adjacent groups.

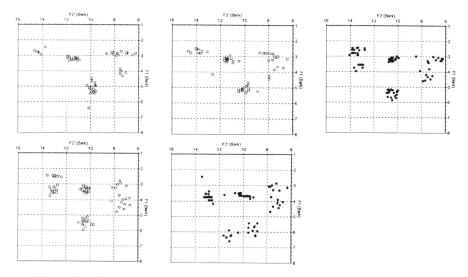

Figure 5.5 The emergent vowel system of the population. Each diagram shows the phonemes used by a different continuous subgroup of the population.

Nevertheless, it would appear that most groups have developed a four or five vowel system, which is largely shared amongst the agents within a group, with gradual shifts between the groups. Figure 5.6 emphasizes the differences that exist across the population. In this figure, the phonemes used by the agents of the first and the last groups are shown together (as white and black dots respectively).

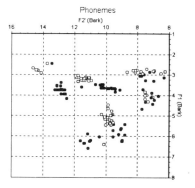

Figure 5.6 The emergent vowel systems of the first and last subgroups of the population.

As with the previous work, a dialect continuum has emerged in the population. Minor changes exist within neighborhoods, allowing successful communication therein. Across the population, more major shifts and differences exist. Despite the significant differences between the two models used to investigate the evolution of dialect diversity, the qualitative result is the same – the negotiation of a communication scheme/phonology within a population, where neighborhoods limit interactions, gives rise to emergent dialects without any requirement for any need or motivation to create the different dialects.

Future Goals and Research

In this chapter we have shown how different micro-simulation models have been used to explore issues in the evolution of linguistic diversity – primarily in the requirement of social motivation for the emergence of dialects.

Yet, more detailed work is required before this issue is settled. Such is the evidence of the extent to which dialect can affect social behavior, that the idea that not just some, but *all*, language change and diversity is socially motivated will not be easily dispelled. The arguments for and against this idea are unlikely to be progressed much further by simply building further models which find similar results to those already obtained – whether for or against the need for socially motivated change. Instead, future work would be better to build on the existing models. A direction for this is suggested by the idea of a language ecology.

Language ecology (Haugen, 1971, Mülhäusler, 1996) considers languages as existing in an environment, the environment being human society. Different societal organizations and population mixes form different ecologies, these being the conditions under which languages evolve. Mufwene (1997) extends this to consider also the particular mix of languages and variants present, and their typological features, adding the structural states of languages to the list of relevant ecological conditions. Artificial language ecologies may be modeled in a micro-simulation by embedding phonological or grammatical language learners in models of different social structures. Modifying the existing language acquisition rules to incorporate social motivation may lead to different simulated outcomes. If particular patterns of language change – e.g., the widespread adoption of one of a number of existing dialects versus creolization, observed in particular societies – can be successfully re-enacted with such models, then the experiments might aid in our understanding of the influence social motivation has had on the evolution of dialects and language diversity.

Conclusions

We hope to have shown, in our own work, that social function is not required for the evolution of linguistic diversity as well as to have indicated what will be a fruitful direction in which to take future model based research. Rather than elaborate on this, we conclude this chapter with a brief consideration of some of the issues important in the development of simulation models of language diversity.

As also noted by Briscoe (2000), implementing language learning by means of a single update, without modeling the individual learning interactions which occur as part of a stochastic process, can lead to results that differ quite significantly. This may lead to quite opposite conclusions, as has been shown by our work versus that of Nettle. Language learning involves many stochastic events, and the learning environment is different for every learner – and it makes sense to capture this in our models. Contact has a significant role to play in cultural evolution of language, not just contact between speakers of different languages but the contact between speakers of different lects of the same language, a role often underestimated. With the noted difficulties in incorporating contact and population distribution in purely mathematical models, micro-simulation is the only feasible approach whereby non-rhetorical models can be developed which include the effect of contact on the evolution of languages.

While it may be appropriate to build models with complex linguistic ecologies, it may be less so to use models in which the appropriate use of dialect brings direct adaptive benefits. As in seen in Arita and Koyama (1998), where dialect diversity becomes subsidiary to the evolution of (non-)cooperative behavior, the introduction of genetic evolution can produce results highly unlikely in the cultural evolution of language. As genetic evolution has little role to play in the evolution of languages, models would be best to eliminate it entirely.

Finally, it is worthwhile to consider what the reason for using a simulation model is in the first place. There is much evidence of language change gathered in linguistic atlases, and a great many studies of local and national dialects the world over. Why do we need artificial evidence?

Simulation allows the development of models which can incorporate different aspects of society as well as different aspects of language and language learning. By testing the models under different conditions the role played by these can be investigated more closely in a model – it is not possible to toggle social influence on and off in real life, like it is in a model. If the real-world evidence is able to support more than one interpretation, then maybe our simulated evidence can provide the balance of evidence required to come down on one side.

References

Arita T, Koyama Y (1998) Evolution of linguistic diversity in a simple communication system. In: Adami C, Belew RK, Kitano H, Taylor CE (eds) *Proceedings of Artificial Life IV*, MIT Press

Arita T, Taylor CE (1996) A simple model for the evolution of communication. In: Fogel LJ, Angeline PJ, Bäck T (eds) *Evolutionary Programming V*, MIT Press

Axelrod R (1997) The dissemination of culture: A model with local convergence and global polarization. *Journal of Conflict Resolution*, 41: 203-226

de Boer B (1997) Generating vowel systems in a population of agents. Presented at the *Fourth European Conference on Artificial Life*, ECAL 97, Brighton, UK de Boer B (2000) Emergence of sound systems through self-organization. In: Knight C, Studdert-Kennedy M, Hurford J (eds), *The evolutionary emergence of language: Social function and the origins of linguistic form.* Cambridge University Press, pp 177-198

de Boer B and P Vogt (1999) Emergence of speech sounds in a changing population. In: Floreano D, Nicoud J-D, Mondada F (eds) *Proceedings of the Fifth European Conference on Artificial Life, ECAL 99 (Lecture Notes in Artificial Intelligence, Volume 1674).* Springer-Verlag, Berlin, pp 664-673

Briscoe EJ (2000) Macro and micro models of linguistic diversity. Presented at *The 3rd International Conference on the Evolution of Language*, Paris, April 3rd-6th

Cavalli-Sforza LL, Feldman MW (1978) Towards a theory of cultural evolution. *Interdisciplinary Review*, 3: 99-107

Chambers JK (1995) *Sociolinguistic theory.* Blackwell

Chambers JK, Trudgill P (1980) *Dialectology.* Cambridge University Press

Crystal D (1987) *The Cambridge encyclopedia of language.* Cambridge University Press

Dunbar RIM (1993) Co-evolution of neocortex size, group size and language in humans. *Brain and Behavioral Sciences*, 16: 681-735

Fyfe C, Livingstone D (1997) Developing a community language. Presented at *The 4th European Conference on Artificial Life, ECAL 97*, Brighton, UK

Grimm J (1822) *Deutsche Grammatik, Part I: Zweite Ausgabe.* Göttingen, Dieterich

Haugen E (1971) The ecology of language. In: Dil AS (ed) *Ecology of language, essays by einar haugen.* Stanford University Press, Stanford CA, pp 325-339

Hutchins E, Hazlehurst B (1995) How to invent a lexicon: the development of shared symbols in interaction. In: Gilbert N, Conte R (eds) *Artificial societies: The computer simulation of social life.* UCL Press

Kirby S (1998) Fitness and the selective adaptation of language. In: Hurford J, Knight C, Studdert-Kennedy M (eds) *Approaches to the evolution of human language.* Cambridge University Press, Cambridge UK, pp 359-383

Labov W (1972) *Sociolinguistic patterns.* University of Pennsylvania Press, Philadelphia

Lass R (1997) *Historical linguistics and language change.* Cambridge University Press, Cambridge UK

Latané B (1981) The psychology of social impact. *American Psychologist*, 36: 343-365

Livingstone D, Fyfe C (1998) A computational model of language-physiology coevolution. Presented at *The 2nd International Conference on the Evolution of Language*, London

Livingstone D, Fyfe C (1999a) Modeling the evolution of linguistic diversity. In: Floreano D, Nicoud J-D, Mondada F (eds) *Proceedings of the Fifth European Conference on Artificial Life, ECAL 99 (Lecture Notes in Artificial Intelligence, Volume 1674).* Springer-Verlag, Berlin

Livingstone D, Fyfe C (1999b) Dialect in learned communication. In: Dautenhahn K, Nehaniv (eds) *AISB '99 Symposium on Imitation in Animals and Artifacts.* The Society for the Study of Artificial Intelligence and Simulation of Behaviour, Edinburgh

Livingstone D, Fyfe C (2000) Modeling language-physiology coevolution. In: Knight C, Studdert-Kennedy M, Hurford J (eds), *The evolutionary emergence of language: Social function and the origins of linguistic form.* Cambridge University Press, pp 199-215

Maeda Y, Sasaki T, et al. (1997) Self-reorganizing of language triggered by "Language Contact". In: *ECAL 97*, Brighton, UK

Milroy L (1980) *Language and social networks.* Basil Blackwell, Oxford

Milroy J (1993) On the social origins of language change. In: Jones C (ed) *Historical linguistics: Problems and perspectives.* Longman, London, New York

Mufwene SS (1997) What research on creole genesis can contribute to historical linguistics In: Schmid MS, Austin JR, Stein D (eds) *Historical linguiustics*. John Benjamins, Amsterdam

Mülhäusler P (1996) *Linguistic ecology: Language change and linguistic imperialism in the pacific region*. Routledge, London

Nettle D (1999a) *Linguistic diversity*, Oxford University Press

Nettle D (1999b) Using social impact theory to simulate language change. *Lingua*, 108: 95-117

Nettle D, Dunbar RIM (1997) Social markers and the evolution of reciprocal exchange. *Current Anthropology*, 38: 93-99

Niyogi P, Berwick RC (1995) A dynamical systems model for language change, AI Memo No 1515 (CBCL Paper No 114), Massachusetts Institute of Technology

Oliphant M (1997) *Formal approaches to innate and learned communication: Laying the foundation for language*. PhD Thesis, University of California, San Diego

Pagel M (2000) The history, rate and pattern of world linguistic evolution. In: Knight C, Studdert-Kennedy M, Hurford J (eds), *The evolutionary emergence of language: Social function and the origins of linguistic form*. Cambridge University Press, pp 391-416

Steels L (1996) Emergent adaptive lexicons. In Maes P, Mataric M, Meyer J-A, Pollack J, Wilson SW (eds) *From Animals to Animats 4: Proceedings of the Fourth International Conference on Simulation of Adaptive Behavior*. MIT Press, Cambridge MA

Steels L, Kaplan F (1998) Stochasticity as a source of innovation in language games. In: Adami C, Belew RK, Kitano H, Taylor CE (eds) *Proceedings of Artificial Life IV*, MIT Press

Stoness SC, Dircks C (1999) Investigating language change: A multi-agent neural-network based simulation. In: Floreano D, Nicoud J-D, Mondada F (eds) *Proceedings of the Fifth European Conference on Artificial Life, ECAL 99 (Lecture Notes in Artificial Intelligence, Volume 1674)*. Springer-Verlag, Berlin

Sweet H (1888) *The history of English sounds*. Clarendon Press, Oxford

Trudgill P, Cheshire J (eds) (1998) *The sociolinguistics reader (Volume 1: Multilingualism and variation)*. Arnold, London

Werner GM, Dyer MG (1991) Evolution of communication in artificial organisms In: Langton CG, Taylor CE, Farmer JD, Rasmussen S (eds) *Artificial Life II*. Addison-Wesley

Part III

EVOLUTION OF SYNTAX

Chapter 6

The Emergence of Linguistic Structure: An Overview of the Iterated Learning Model

Simon Kirby and James R. Hurford

"The most basic principle guiding [language] design is not communicative utility but reproduction – theirs and ours ... Languages are social and cultural entities that have evolved with respect to the forces of selection imposed by human users. The structure of a language is under intense selection because in its reproduction from generation to generation, it must pass through a narrow bottleneck: children's minds." (Deacon, 1997: 110)

Introduction

As language users humans possess a culturally transmitted system of unparalleled complexity in the natural world. Linguistics has revealed over the past 40 years the degree to which the syntactic structure of language in particular is strikingly complex. Furthermore, as Pinker and Bloom point out in their agenda-setting paper *Natural Language and Natural Selection* "grammar is a complex mechanism tailored to the transmission of propositional structures through a serial interface" (Pinker and Bloom, 1990: 707). These sorts of observations, along with influential arguments from linguistics and psychology about the innateness of language (see Chomsky, 1986; Pinker, 1994), have led many authors to the conclusion that an explanation for the origin of syntax must invoke neo-Darwinian natural selection.

> Evolutionary theory offers clear criteria for when a trait should be attributed to natural selection: complex design for some function, and the absence of alternative processes capable of explaining such complexity. Human language meets these criteria. (Pinker and Bloom, 1990: 707)

Since Pinker and Bloom made these arguments there have been many attempts to put forward a coherent evolutionary story that would allow us to derive known features of syntax from communicative selection pressures (e.g., Nowak, Plotkin and Jansen, 2000; Newmeyer, 1991; and discussion in Kirby, 1999a). One problem with this approach to evolutionary linguistics is that it often fails to take into account that biological natural selection is only one of the complex adaptive systems at work.

Language emerges at the intersection of three complex adaptive systems:

Learning. During *ontogeny* children adapt their knowledge of language in response to the environment in such a way that they optimize their ability to comprehend others and to produce comprehensible utterances.

Cultural evolution. On a historical (or *glossogenetic*) time scale, languages change. Words enter and leave the language, meanings shift, and phonological and syntactic rules adjust.

Biological evolution. The learning (and processing) mechanisms with which our species has been equipped for language, adapt in response to selection pressures from the environment, for survival and reproduction.

There are two problems with this multiplicity of dynamical systems involved in linguistic evolution. Firstly, we understand very little about how learning, culture, and evolution interact (though, see Belew, 1990; Kirby and Hurford, 1997; Boyd and Richerson, 1985), partly because language is arguably the only sophisticated example of such a phenomenon. There clearly *are* interactions: for example, biological evolution provides the platform on which learning takes place, what can be learnt influences the languages that can persist through cultural evolution, and the structure of the language of a community will influence the selection pressures on the evolving language users (see Figure 6.1).

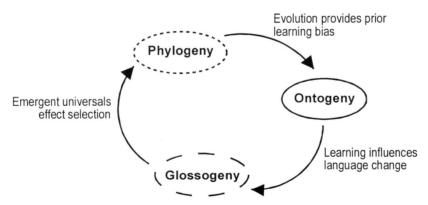

Figure 6.1 The three adaptive systems that give rise to language. Some of the interactions between these systems are shown.

Secondly, it is not clear what methodology we should use to study this problem. Mathematical techniques for looking at the interaction of dynamical systems and linguistic behavior are in their infancy (though, Nowak, Komarova and Niyogi, 2001, take a potentially valuable approach). We feel that computational modeling is currently the most appropriate methodology, but although simulations of language learning have a long history, and there are many methods from the A-life field that can be used for modeling evolution, models of the cultural transmission of learned behavior are relatively sparse (see Steels, 1997 for a review). This is unfortunate, since we will argue in this chapter that it is through this particular mechanism that the most basic features of human language syntactic structure can be explained.

To remedy this situation, we introduce here the Iterated Learning Model (ILM), a general approach to exploring the transmission over a glossogenetic time scale of observationally learned behavior. We will illustrate the ILM with a few examples of simulations that lead to two conclusions:

- There is a non-trivial mapping between the set of learnable languages (i.e., the languages allowed by our innate language faculty), and the set of stable languages (i.e., the languages we can actually expect to see in the world).

- Under certain circumstances, cultural evolution leads inevitably to recursively compositional (i.e., syntactic) languages.

The Iterated Learning Model

The central idea behind the ILM is to model directly the way language exists and persists via two forms of representation (Chomsky, 1986):

I-language. This is language as it exists internally, as patterns of neural connectivity, or more abstractly, as a grammar.

E-language. This is the external form of language as actual sets of utterances.

For a language, or a pattern within a language, to persist from one generation of language users to the next it must be mapped from I-language to E-language (through use) and from E-language back to I-language again (through learning). This is the bottleneck on transmission that Deacon mentions in the quotation that starts this chapter.

In this chapter (and in previous work) we are interested in a particular property of language that must persist in this way: the structure of the mapping from meanings to signals (and vice versa). In order to model this, the ILM needs four basic components:

1. A meaning space.

2. A signal space.

3. One or more language-learning agents.

4. One or more language-using adult agents.

Each iteration of the ILM involves an adult agent being given a set of randomly chosen meanings to produce signals for. The resulting meaning-signal pairs form training data (E-language) for one or more learning agents. After a learning period (the *critical period* to use terminology from language acquisition), the learning agents form their individual I-languages and thus become adults. New learners are added to replace the learners, and adults are removed in order to maintain population size[1]. This cycle is typically repeated several thousand times or until a stable point attractor in the dynamical system is reached. Importantly, a normal ILM simulation will be initialized with no linguistic system in place whatsoever. To put it another way, the initial adults have no I-language and at the start the community of agents have no E-language.

A simple ILM

To exemplify the iterated learning model, here we describe a simple simulation using neural networks as learners. Although the setup is very simple, there are still interesting properties that spontaneously emerge in the languages in the system, which will point the way to the simulations which we will describe in later sections.

In line with much of the work that we have done with the ILM, we will trade-off complex population dynamics against speed for the simulation. Accordingly, the population at any one time consists of a single adult and a single learner. The agents are feed-forward networks, with an $8 \times 8 \times 8$ structure. The networks map from an 8-bit binary number representing a signal to an 8-bit binary number representing a meaning, in other words, given appropriate input, the agents can learn to *parse* utterances.

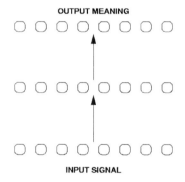

Figure 6.2 A network that maps from 8-bit binary signals to 8-bit binary meanings. Arrows indicate full connectivity.

Because the networks are unidirectional, we need some *production* mechanism, mapping back from meanings to signals. One way to do this is to use a version of the obverter learning strategy discussed in Oliphant and Batali (1997), Smith (in press; 2001a; 2001b). The idea behind *obverter* is for a communicative agent to produce signals that maximize the chance of the hearer understanding the correct meaning. Given that a speaker will not have access to the signal-to-meaning

---

[1] Note that this is a highly simplified population dynamic. See Briscoe (2000) for an alternative approach to population replacement. More sophisticated approaches to population dynamics are likely to be a prerequisite of a model of creolization, for example.

mapping of the hearer, obverter makes the simplifying assumption that the speaker's own mapping approximates that of the hearer. In practical terms this means that, in order to decide which signal to produce for a meaning, we need to search for the signal that would result in the desired meaning if parsed by the speaker.

We want to find a signal s given a meaning m.

$$(1) \qquad s_{\text{desired}} = \arg\max_{s} P(s|m)$$

$$(2) \qquad = \arg\max_{s} \frac{P(m|s)P(s)}{P(m)}$$

$$(3) \qquad = \arg\max_{s} P(m|s)$$

$$(4) \qquad = \arg\max_{s} C(m|s)$$

Where $C(m|s)$ is the confidence that the network has the mapping $s \to m$. In other words, find the signal that maximizes the network's confidence in the given meaning[2].

In order to calculate $C(m|s)$, we treat the real-numbered network outputs $o[1 \dots 8]$ as a measure of confidence in the meaning bits $m[1 \dots 8]$.

$$(5) \qquad C(m[1\dots8]|o[1\dots8]) = \prod_{i=1}^{8} C(m[i]|o[i])$$

$$(6) \qquad C(m[i]|o[i]) = \begin{cases} o[i] & \text{if } m[i] = 1, \\ (1 - o[i]) & \text{if } m[i] = 0. \end{cases}$$

In summary, the ILM in this case proceeds as follows:

1. An initial population is setup consisting of two randomly initialized networks, a speaker and a hearer.

2. A certain number of random meanings are chosen from the set of binary numbers 00000000 to 11111111, with replacement.

3. The speaker produces signals for each of these meanings by applying the obverter procedure.

4. This set of signal-meaning pairs is used to train the hearer network using the back-propagation of error learning algorithm[3].

---

[2] This is potentially a very computationally costly operation. However, the efficiency of obverter need not be a big issue if careful *memoization* is used in the network computation. As long as the training phase and the production phase of the agent are strictly separated, then network weights will be fixed throughout production. This means that, as results are computed, they can be stored in a lookup table (or cache). Notice that this kind of optimization is possible only because the ILM is generational. It would not work for the types of population model that Batali (2001, 1998) employs, for example.

5. The speaker is removed, the hearer is designated a speaker, and a new hearer is added (with randomly initialized weights).

6. The cycle repeats.

What happens in such a model? It turns out that there are three types of behavior. Which behavior emerges depends entirely on the number of utterances the hearer learns from. With a very small training set, the language evolves as shown in Figure 6.3. This graph shows the expressivity of the language (i.e., how much of the meaning space is covered by the 256 signals), and its stability (i.e., how different the hearer's language is from the speaker's after training). The graph shows the results with 20 randomly chosen meanings. The emergent language is inexpressive and unstable.

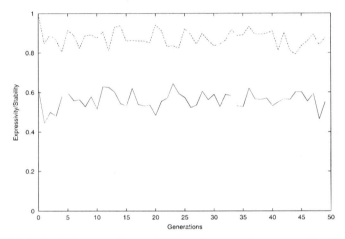

Figure 6.3 A simulation run where only 20 random meanings are produced by each generation's speaker. The dotted line corresponds to the size of the difference between the learners' and the adults' language after training. The solid line represents the proportion of the meaning space covered by the language. In this case the language is inexpressive and unstable.

For a very large training set the behavior is rather different. Figure 6.4 shows a longer run with 2000 randomly chosen meanings each generation. Eventually, a completely stable and completely expressive language is found. For a medium-sized training set, apparently similar behavior emerges, albeit faster. Figure 6.5 shows results for 50 randomly chosen meanings.

There is more to these results than initially meets the eye, however. The runs with a medium-sized training set have a feature that does not emerge in runs with either the very large or very small numbers of meanings in training. Whereas the language in the run with the large training set is completely expressive, the mapping from meanings to signals is essentially random. That is, there is no

---

[3] The learning algorithm used has a learning rate of 0.1 and no momentum term. Each learner is presented with 100 randomized epochs of the data set.

structure in the pairings[4].

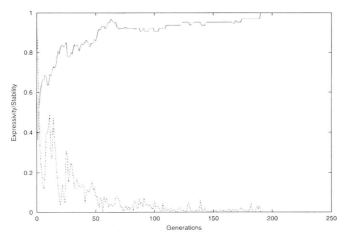

Figure 6.4 A simulation run where 2000 random meanings are produced by each generation's speaker. Here an expressive and stable system is eventually reached.

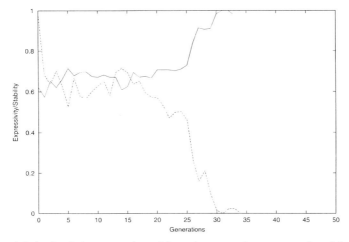

Figure 6.5 A simulation run where 50 random meanings are produced by each generation's speaker. In this run, an expressive stable system is reached relatively quickly.

---

[4] This sort of system has been described as a protolanguage system by Wray (1998). She uses this term rather differently from Bickerton (1990). Whereas Bickertonian protolanguage has some structure in the mapping from signals to meanings, in that it has words that are concatenated in a non-structured way to form proto-sentences, Wray argues that a better model of protolanguage of which there are current "living fossils" is one in which the meaning-signal mapping is completely holistic. The sentences in a protolanguage according to Wray cannot be broken down in any way which will give a clue to what the meaning of the sentence is.

Surprisingly, this random pairing never arises for runs with a medium-sized training set. In the run with 50 random meanings, the emergent meaning-signal mappings are highly structured. In fact, for the run shown in Figure 6.5, they can be captured by a very simple set of translation rules:

$$m_1 \leftrightarrow \neg s_3$$
$$m_2 \leftrightarrow \neg s_5$$
$$m_3 \leftrightarrow \neg s_6$$
$$m_4 \leftrightarrow \neg s_1$$
$$m_5 \leftrightarrow s_4$$
$$m_6 \leftrightarrow s_8$$
$$m_7 \leftrightarrow \neg s_2$$
$$m_8 \leftrightarrow s_7$$

where m_n is the nth bit of the meaning, s_n is the nth bit of the signal, and $\neg s_n$ is the logical negative of the nth signal bit. For example, the meaning 00110100 is expressed as 01101001.

It turns out that this kind of result *always* emerges as long as the number of training examples is not too small and not too large (the particular numbers seem to depend on the structure of the networks, and the size of the meaning and signal spaces – see Brighton & Kirby (2001a; 2001b) for an approach to quantifying the critical values for the ILM).

Recursive Compositionality out of Iterated Learning

Our second example of the ILM is covered in more detail elsewhere (Kirby, 2001a; 1999b; 1999c), so we will only describe it briefly here[5].

As mentioned earlier, perhaps the fundamental feature of human language that sets it apart from other animals' communication systems is the unique way in which the structure of a signal can be decomposed into separate meaning-bearing parts. Linguists refer to this as *compositionality*.

Compositionality. A compositional signaling system is one in which the meaning of a signal is some function of the meaning of the parts of that signal and the way in which they are put together.

Furthermore, the mapping from sentences to meanings in language is not only compositional, but recursive. Parts of sentences which can be ascribed a meaning can themselves be decomposed by the same compositional function. Ultimately, this leads to what has been called the "digital infinity" of language: as language users, we can make potentially infinite use of finite means by constructing meaning-bearing syntactic structures that themselves contain structures of the same type.

---

[5] The model described is an extension of an earlier model that was put forward in Kirby (2000).

A symbolic approach

It should be clear that the ILM described in the previous section produced a language that was compositional by the definition above with the correct number of training examples. However, it is impossible for the linguistic system also to be recursive using the same modeling methodology (for a start, the signal and meaning spaces are finite). The simulations described in this section are an attempt to get round this limitation.

Rather than using bit-vectors to represent meanings and signals, here we use a simple variant of predicate logic for meanings, and strings of characters as signals. For example, two possible meanings and their corresponding signals might be:

$$\text{loves(mary,john)} \leftrightarrow \texttt{marylovesjohn}$$
$$\text{knows(gavin,loves(mary,john))} \leftrightarrow \texttt{gavinknowsmarylovesjohn}$$

In the simulations reported here, there are five possible predicates each of which takes two arguments which may vary over five possible "people". There are also five predicates of propositional attitude which take one normal argument, and one recursive argument as in the example above.

In order to represent the mapping from meanings to strings we use a simple version of definite clause grammar (DCG). It is important here to note that this does not preclude non-compositional languages *a priori*. A dictionary-like holistic protolanguage can equally well be expressed in DCG notation as a recursively compositional language. So, the examples above could be generated by the non-compositional grammar. In this DCG representation, the material to the right of the slash indicates the meaning assigned to the syntactic category to the left of the slash.

$$S\!/\text{loves(mary,john)} \rightarrow \texttt{marylovesjohn}$$
$$S\!/\text{knows(gavin,loves(mary,john))} \rightarrow \texttt{gavinknowsmarylovesjohn}$$

Obviously, in this case, the same strings could be generated by a compositional grammar.

In order to implement the ILM, we need an inductive learning algorithm for these grammars. An ideal method might involve a search for optimal grammars using some metric such as Minimum Description Length. Indeed, we are pursuing work along these lines (Brighton & Kirby, 2001a; 2001b). However, one problem with the ILM lies in the necessity for very efficient learning methods (since any learning problem is necessarily scaled up many thousand-fold when generations of learners are required). Accordingly, the work we have undertaken with DCGs relies upon a heuristic-driven incremental grammar induction method described in detail in Kirby (2001b). Very briefly, this learning method initially incorporates rules into the grammar upon encountering each utterance, and then searches for possible generalizations over pairs of rules in the grammar that fit within a set of heuristic criteria which can replace those pairs of rules with a single one. In this way, the induction method ensures that the learner can always parse the data heard, but may also generalize to unseen examples if the generalization is justified by the data. (See, Tonkes and Wiles, in prep, for some criticism of this approach.)

In addition to a learning mechanism, the ILM in this case also needs some way of innovating new signals. Given that the initial agents will have no I-language, they will have no way of generating strings for any of the meanings they are called upon to produce. (This is not true of the networks in the previous example because a network will always produce an output for any input. In other words, there is no equivalent of a failure to generate.) The technique for generating novel strings is also described elsewhere (Kirby, 2001b) – it is essentially random, but aims to be somewhat "smart" in that it avoids generating random strings wherever there is clear compositional structure in the grammar that the agent currently possesses.

From protolanguage to recursive syntax

In each iteration of the ILM, the speaker is required to produce strings for 50 simple meanings (i.e., with no embedding), 50 degree-1 meanings (with one level of embedding), and 50 degree-2 meanings. If invention is employed by the speaker, both the speaker and the hearer add the resultant meaning-string pair to their linguistic knowledge. Otherwise, only the hearer carries out the induction processes mentioned above.

In the initial stages of the simulation, a protolanguage emerges. The language that is culturally evolving has some words for some meanings, but no structure. Here are some example sentences from an early language (with the meanings given as English glosses):

(7)   `ldg`
        "Mary admires John"

(8)   `xkq`
        "Mary loves John"

(9)   `gj`
        "Mary admires Gavin"

(10)   `axk`
        "John admires Gavin"

(11)   `gb`
        "John knows that Mary knows that John admires Gavin"

We can see no obvious structure here. There appears to be no compositional encoding of the meanings. In fact, given the symbolic nature of the system, we can inspect the agents' I-language directly:

$$S/\text{loves(john,mary)} \rightarrow \texttt{sdx}$$
$$S/\text{admires(mary,gavin)} \rightarrow \texttt{gj}$$
$$S/\text{admires(john,gavin)} \rightarrow \texttt{axk}$$
$$S/\text{admires(gavin,heather)} \rightarrow \texttt{nui}$$

S/loves(john,heather) → my
S/loves(mary,john) → xkq
S/admires(mary,john) → ldg
S/thinks(john,loves(mary,gavin)) → fi
S/thinks(heather,loves(heather,gavin)) → ad
S/thinks(john,admires(heather,gavin)) → xuy
S/knows(gavin,loves(gavin,mary)) → k
S/knows(gavin,loves(john,mary)) → ysw
S/thinks(mary,knows(gavin,loves(heather,john))) → pq
S/thinks(mary,knows(heather,loves(heather,john))) → rr
S/knows(john,knows(mary,admires(mary,john))) → lr
... (plus another 101 rules)

Early in the simulation, the process of producing input for the next generation learner involves a lot of random invention. As a result, for many generations communication systems appear to stay random and idiosyncratic. Surprisingly, however, at some point in every simulation run, the language suddenly changes[6]:

(12)   gj h   f tej   m
        John   Mary   admires
        "Mary admires John"

(13)   gj h   f tej   wp
        John   Mary   loves
        "Mary loves John"

(14)   gj qp   f tej   m
        Gavin Mary   admires
        "Mary admires Gavin"

(15)   gj qp   f h   m
        Gavin John   admires
        "John admires Gavin"

(16)   i h u   i tej   u   gj qp   f h   m
        John knows Mary   knows Gavin   John   admires
        "John knows that Mary knows that John admires Gavin"

In contrast to the previous example, there is obvious structure here. This is made even clearer by looking at the entire grammar of an agent in this simulation.

$S/p(x,y)$ → gj A/y f A/x B/p
$S/p(x,q)$ → i A/x D/p S/q
A/heather → dl

----

[6] The spaces are included here merely to aid comprehension. They are not available to the learner.

A/mary → tej
A/pete → n
A/gavin → qp
A/john → h
B/detests → b
B/loves → wp
B/hates → c
B/likes → e
B/admires → m
D/believes → g
D/knows → u
D/decides → ipr
D/says → p
D/thinks → m

Recursive compositionality has emerged in this simulation run after a few hundred generations, and is retained from that point onwards as a completely stable attractor in the dynamical system. Figure 6.6 gives a quantitative overview of the dynamics of the system.

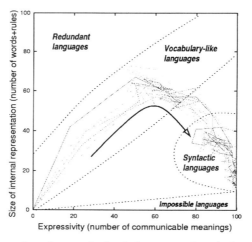

Figure 6.6 The results of several simulation runs plotted showing size and expressivity of the grammars. In order to plot this graph, meanings with embedding are ignored. In each run, the expressivity of the language increases to a maximum, but the size of the grammar initially grows during the protolanguage stage before reducing to a minimum, reflecting the emergence of syntax.

Irregularity and Frequency

Results like those described in the previous sections are encouraging in that they demonstrate the utility of the ILM as a mechanism for exploring how cultural transmission might explain some of the fundamental features of language that are the target of evolutionary linguistics. Ultimately we would like to widen the remit of this sort of glossogenetic evolutionary explanation, in order to see how much can be explained without appealing directly to natural selection-type functional arguments. In particular, there are a couple of problems with the type of language that emerges in the simulations reported so far. Although compositional structure is clearly very common in language, it is not universal.

If we look at the morphological systems of many languages, for example, we find that there are often subparts of a paradigm that appear to flout the compositional structure obvious in the rest of the language. A classic example of this type of behavior can be found by looking at the past tense of English verbs. Whilst the majority of verbs inflect quite regularly (by adding -*ed*), some have irregular past tenses (for example, the past tense of verbs like *go*, *have*, or *take*). This partial non-compositionality cannot be explained by the ILMs so far.

The second problem with the results shown previously is that the end results are always completely stable. However, one of the striking features of languages is that they are always changing. This has been termed "the logical problem of language change" (Niyogi and Berwick, 1995). Kirby (2001b) tries to address these issues within the context of the iterated learning model.

Laziness

In this version of the model, we include two extra features, both of which are motivated by our understanding of real language use. Firstly, learning is not the only mechanism at work in the transmission of language from generation to generation. When producing utterances, speakers may have multiple choices to make about how to express a particular meaning. As a simplification we assume that speakers are motivated at least in some part by a principle of *least effort*. To model this, agents in this simulation will always produce the shortest string for a given meaning[7]. In addition, we allow that the production of language may not always be perfect. In this simulation, therefore, we add a parameter that expresses a per-character probability of a character in the output string being dropped from the utterance. For the results reported here, this probability is 0.001.

Frequency

A second modification that we include is motivated by looking at which English verbs are irregular in the past tense. As Pinker (1999) notes, the top ten verbs of English ranked by frequency (*be*, *have*, *do*, *say*, *make*, *go*, *take*, *come*, *see*, *get*) are all irregular (Francis and Kucera, 1982). In the previous simulations, however, all the meanings have identical probability of being picked for production by the speakers.

---

[7] It is interesting to note that in an analysis of a simulation by Batali (1998), Goroll (1999) suggests that a least-effort principle is implicit in the production mechanism that Batali uses and may explain some of the features of his results.

To include a frequency bias, we define a non-uniform probability distribution over the meaning space that reflects the chance that a meaning will be picked. To make this easier to do, and to make the results more visualizable, we have simplified the meaning space for this simulation. Rather than a complex of predicates and arguments, each meaning can be thought of as a combination of just two components. These could correspond to two properties of an object or situation, for example shape and color, or action and tense. Here, we will refer to the meanings as having an *a* feature and a *b* feature, with the meanings ranging from (a_0,b_0) to (a_4,b_4). The probability of a particular meaning (a_i,b_j) is proportional to $(i+1)^{-1}(j+1)^{-1}$. This frequency distribution is inspired by the recognition that word usage is inversely proportional to its frequency rank (Zipf, 1936).

Frequency correlates with irregularity

The ILM was run with the same induction procedure as in the previous example, but with fewer training examples – 50 in this case. This reflects the fact that the meaning space is much smaller, with only 25 possible combinations as opposed to the potentially infinite number of predicates that could be formed recursively in the last section. It should be noted, however, that although the meaning space is small, each learner is still not guaranteed to see every meaning during learning. In fact, given the frequency distribution, it is vanishingly unlikely that all meanings will be present in a particular speaker's output.

To summarize the results: after several generations during which the language remains inexpressive and idiosyncratic (i.e., a protolanguage), structure emerges in the simulation. Here is an example of the complete language of the 256th speaker (Table 6.1).

Table 6.1 Example of emerged language

	a_0	a_1	a_2	a_3	a_4
b_0	**g**	**s**	kf	jf	uhlf
b_1	**y**	jgi	ki	ji	uhli
b_2	yq	jgq	kq	jq	uhlq
b_3	ybq	jgbq	kbq	jbq	uhlbq
b_4	yuqeg	jguqeg	kuqeg	juqeg	uhluqeg

Not only does this language have irregular forms (shown in bold in the table), but these correspond to the most frequent meanings (in the top left here). Furthermore, this language does not represent a stable attractor. There is a process of language change that continues throughout the simulation. Sometimes substrings corresponding to particular components of meaning will get shorter, other times completely new forms will emerge. However, the basic partially irregular compositionality is always in place. The irregular forms seem more stable than the regulars. For example, the irregular cluster here arose in generation 127 and lasted until generation 464, at which point y is reregularized to yi. It is also interesting that the length of the strings seems to be inversely correlated with frequency as has been observed in real languages (Zipf, 1936).

The only factor which can help an idiosyncratic meaning-form pair (i.e., one which does not conform to any generalization) to persist through the history of a language is increased frequency of use. An experiment described in Hurford (2000) corroborates the conclusion that frequency correlates with irregularity, but reveals an interesting difference from the example with irregular morphology just given. In this simulation, an iterated learning model was implemented, with a population of agents starting with no language at all, and over time a language emerged in the community in which there were completely general compositional rules for expressing a range of meanings represented as formulae in predicate logic. A variation on the basic simulation was then implemented in which one particular meaning was used with vastly greater frequency than any of the other available meanings. This inflated frequency held throughout the simulated history of the community. The result was that, as before, a language emerged in the population characterized by general compositional rules, but in addition, all speakers also had one special idiosyncratic stored fact pertaining to the highly frequent meaning.

The irregular treatment of the high-frequency item came about through the persistence of this one idiosyncratic meaning-form pairing from the earliest stages of the simulation. In the early stages, before any regular grammatical rules had emerged, all meanings were expressed idiosyncratically. Over time, the task of representing the other form-meaning pairings was taken over by general grammatical rules. But the single high-frequency meaning always (because of its frequency) happened to be encountered early in the learning experience of every agent, before the point where the agent had started to form any generalizations over observed examples. Thus the archaic idiosyncratic form-meaning pairing was perpetuated. No other meaning had the 'privilege' of assured early exposure to each new learner, and none of them ended up with an idiosyncratic irregular form. If this simulation were allowed to run on indefinitely, it is predicted that at some stage the eventual occurrence of an extremely improbable throw of the dice permitted by the random presentation of meanings would give a generation in which even this high-frequency meaning was unrepresented in examples given to the next generation, and its expression would then succumbs to regularization.

This model of the origins of irregularity differs from the model dealing with morphological irregularity given earlier, and in fact seems less realistic. It predicts that languages inexorably lose irregularities, albeit perhaps very slowly in the case of high-frequency items. Besides a Zipfian frequency distribution, the morphological simulation incorporated the realistic factor of random erosion of signals by noise during transmission. In this way, new forms not conforming to previously established regularities are constantly liable to enter the language. This mechanism is plausible as modeling the entrance of at least some of the irregular verbs of Modern English, such as *had* (from the more regular OE *haefde*), made (cf the corresponding regular German *machte*, and *says* and *said* with their irregularly shortened vowels (i.e., not rhyming with the phonologically regular *plays* and *played*).

Why Social Transmission Favors Linguistic Generalization

A theory of language transmission through iterated acquisition needs to take into account the capacities brought to the acquisition task by the child and the nature of the data to which the child is exposed. Clearly, humans are capable of acquiring and storing economical recursive grammars that generate infinite sets of sentences. Equally clearly, humans are also capable of acquiring and storing vast, but of course finite, inventories of arbitrary facts.

However many examples of meaning-form pairs a learner is exposed to, the number of examples is always finite. We refer to the finiteness of the examples presented to each generation of learners as a 'bottleneck'. In fact, various kinds of bottleneck in the transmission of linguistic information between generations are conceivable, and these are discussed in more detail in Hurford (2001a). The kind of bottleneck most crucial to the iterated learning model is a 'semantic bottleneck'. This refers to the fact that only a small proportion of all the available meanings are ever expressed during the learning experience of any learner; the learner is not exposed to an exhaustive review of all the meaning-form pairings of the language. In the interesting case of an infinite language, this is necessarily the case.

It is worth considering for a moment what would happen in an evolving population whose members were not capable of acquiring grammars expressing generalizations over data, but could only resort to storage of arbitrary individual meaning-form mappings. Such agents could only acquire, and subsequently pass on to the next generation, the meaning-form pairs they had actually observed. If the agents in a given generation happened never to be motivated to express a particular meaning for which they had acquired a signal, then that particular meaning-form pair would get lost in the historical process, and disappear from the language of the community. Only those meaning-form pairs which were guaranteed to be in the input to every new generation of learners would survive. If no meaning-form pairs were guaranteed such totally reliable exposure with each generation, no pairs would persist indefinitely through the history of the language. If some meanings were expressed more frequently than others each generation, these would tend to have historically more stable forms (not necessarily regular, as we have seen). Wherever a meaning-form mapping gets dropped in the historical process, the next time an agent is prompted to express that particular meaning, a new random form will be invented. The more such invention happens, the less stable is the language.

This simple scenario illustrates the idea of a language 'surviving' in a community of agents, and undergoing change which is predictable in terms of the learning capacity of the agents, the number of examples that each learner is exposed to, and the frequency distribution of the meanings. In this simple example, the agents only have recourse to a learning strategy which correlates examples observed during learning on a one-to-one basis with items of stored knowledge. Basically, such agents are capable only of internalizing a lexicon.

Now consider a situation in which agents have two kinds of strategy at their disposal. They can either store individual facts, as before, or they can *generalize* over examples, thus producing a many-to-one correlation between observed examples and items of stored knowledge. It does not matter at all how this generalization is done. Implicit in any notion of generalization in this context is the idea of *variables* ranging over (parts of) meanings or (parts of) forms. Also implicit, and essential here, is the idea that such generalization can be a form of overgeneralization, extending to some meaning-form pairs which the agent has not observed. This immediately builds into the agent a capacity to express meanings for which there were no exact precedents in its learning phase. Now there is the possibility of a particular form-meaning pair being absent from the language input to one generation of learners without it necessarily therefore being lost thereafter in the history of the language. A particular meaning-form pair can 'go underground', even for many generations, provided that all learners over the period acquire, from other examples, a generalization covering this particular meaning-form pair. But any meaning-form pair which is not covered by a generalization in the minds of learners/producers of the language, and which happens to be absent from the examples given to the next generation, will drop out of the language. Thus we see a difference in the survival potential of different types of meaning-form pairs in a language. Those that are covered by generalizations are much more likely to persist in the history of the language than those which are not covered by any generalization.

It follows that the strength or coverage of generalizations correlates with the survival potential of the meaning-form pairs they cover. Compare two generalizations made by a learning agent, one covering proportion n of the space of meanings and the other covering proportion m of the meaning space, where $0 < n < m < 1$. If this agent as an adult produces B distinct meaning-form pairs[8], in response to being prompted to express randomly chosen meanings, the probabilities that a learner in the next generation will observe at least one example covered by these generalizations are $1-(1-1/n)^B$ and $1-(1-1/m)^B$ respectively. Thus, in the simplified case where observation of just one example is sufficient to induce a generalization of any degree of coverage in a learner, a stronger generalization is more likely than a weaker generalization to be exemplified for the next generation, and thus transmitted onward in the history of the language. If we further take into account the fact that the number of examples actually required to induce a generalization of coverage n is likely to increase at most as some exponential root of n increases, the enhanced survival potential of stronger generalizations becomes even clearer. The favoring of linguistic generalizations by the very fact of social transmission (iterated learning) is explored in more detail in Hurford (2000); the mathematics of the required relationships between sample size and coverage of the generalization are discussed in more detail in Brighton & Kirby (2001a; 2001b).

The historical persistence of patterns of generalization in a language can be observed in its E-language, in a corpus of utterances produced by members of the population. In addition, native speaker investigators of a language can examine their own intuitions about made-up examples to test the generality of hypothesized rules. In real empirical linguistics, the actual representations of general rules in speakers' heads are inaccessible; they are hypothesized from E-language performance and grammatical intuitions. Generative linguistics has typically

---

[8] Assuming, for simplicity, no homonymy and no synonymy.

emphasized the role of child language acquirers as powerful generalizing machines. This infant drive for internal generalization has been taken to be the prime mover in causing the regularities observable in languages. And this conclusion is partially reinforced by the computational work reported here. Creatures with no drive at all to make mental generalizations from their observations would produce no historical E-languages with persistent regular patterns. But, as in evolution generally, one must distinguish between the evolutionary source of an attested phenomenon and the *reason for its persistence*. The evolutionary source of generalizations in languages is the child's innate capacity to generalize. The reason for the historical persistence of generalizations is the inherent advantage that general patterns have over idiosyncratic facts in being propagated across generations in the repeated spiral of acquisition and production (iterated learning). This inherent advantage of generalizations has nothing to do with the fitness of individual agents, nor with selection for ease of processing or usefulness in the community. Given the natural assumptions of the iterated learning model, it is a mathematical truism that languages characterized by very general patterns will emerge over time.

Linguists have tended to relegate any capacity for memorizing arbitrary knowledge items to facts which do not fall under generalizations, such as idioms and specific lexical items. Perhaps because the inherent advantage of generalizations in transmission across generations has not hitherto been recognized, it has been assumed that learners store economical, non-redundant grammars. If the organization of linguistic knowledge inside peoples' heads were messy, not capitalizing maximally on available generalizations, then, so the implicit argument goes, we should expect languages themselves to be messy. But languages are very regular (though admittedly not *completely* regular), and so the internal representation of grammars should be as regular and general as is compatible with the E-language data.

Computer simulations allow us an insight into the possible relationships between the linguistic knowledge stored in speakers' heads and the patterns of use in the community's E-language which exemplify generalizations. As in the examples of earlier sections, we can actually inspect the internal grammars of simulated agents. An experiment reported in Hurford (2000) shows that even in a population of agents which are significantly handicapped in their capacity to generalize over observed data, a regular, general E-language will emerge in the course of glossogeny. Agents in this experiment, like humans and all the artificial agents in the other simulations reported here, had both a capacity to represent lists of arbitrary unrelated facts and a capacity to make generalizations over facts perceived as related. But in this experiment, the agents' capacity for generalization was deliberately switched off for a random 75% of the training examples. In this model, agents either internalized general rules where possible (25% of the time), or simply rote-memorized the current example as a fact unrelated to any of the rest of their stored knowledge (75% of the time). The resultant internalized grammars of all the agents, were, as expected, very redundant, containing both general rules and lists of one-off records of form-meaning pairings. But in fact all of the one-off records were completely consistent with the general rules which the agents had also internalized. The general rules had become fixed in the language even though three-quarters of every instance of language-learning had been biased against the internalization of generalizations. An observer who could see only the actual utterances of the agents would have no reason to suspect that their internal representations redundantly duplicated facts by storing both general rules and many individual examples entirely consistent with them.

Extending the Scope and Context of ILM Models

We have presented the central features of the iterated learning model, and outlined the main results obtained so far. We believe the model illustrates a selective force inherent in the fact of the social transmission of languages, through their I-language and E-language phases across generations. As with any progressive research paradigm, new questions are raised and new avenues for investigation are immediately suggested. We review some of these related issues here.

The treatment of meanings

The implementations of the iterated learning model described here all make the simplifying assumption that children are presented with signals paired with whole meanings. Obviously, if the whole meaning of an utterance were actually observable in this way, there would be no need for the utterance in the first place. Language is used to convey meanings which are not wholly obvious from the context. However, in order for learning of a mapping between meanings and forms to be possible, at least some parts of intended meanings must be evident to the child. Several works have explored and implemented algorithms for learning meaning-form mappings from data in which the meanings are partially masked (Siskind, 1996; Hurford, 1999). It seems clear that the simplifying assumption of the child being given whole meanings does not make possible what would otherwise be impossible. Its principal effect is to simplify the implementations and considerably to shorten the experimental time in which results can be obtained. The masking of meanings in actual language acquisition can be safely idealized away from, as far as the implications of the ILM discussed here are concerned.

A related, and deeper, objection to the treatment of meanings in these ILM models is that they all assume meanings to be entirely inside the agents' heads, and moreover the 'same' meanings are identically represented in all agents. The meanings used here are not related in any way to aspects of an external environment in which the agents must survive, and about which they communicate. In ILM models it is taken for granted that agents learn meaning-form mappings as children and as adults produce utterances when prompted by particular meanings. The idealization away from considerations of fitness and survival was useful insofar as it revealed a mechanism by which languages adapt themselves to transmission across the generations of their host populations. But consideration of the wider context invites us to ask whether the treatment of meanings can be justified or explained.

Ongoing research at LEC (Smith, in press), following and extending the work of Steels (e.g., Steels, 1999), attempts to model the growth of conceptual representations inside agents' heads in response to a 'discrimination task' acted out in a (simulated) world of object external to the agents. In these simulations, objects in the world presented agents with gradient information about objects along several channels. Agents develop internal representations implemented as sets of trees, with one tree for each perceptual channel. The trees start as single nodes and get arborized or ramified more or less densely at various regions in the continuous ranges of the channels, to the point where every object in the environment can be distinguished by a combination of one or more node-addresses on the trees. This begins to approach the classical semantic distinction between sense and reference (Frege, 1892). The external objects correspond to the classical referents, while the developed internal trees correspond to agents' senses. Interestingly, under certain conditions, especially where there is no innate bias toward developing one channel/tree in preference to any other, populations emerge in which all agents are capable of making the same distinctions between objects, but may have distinct internal representations. Further work adds communication between agents to this system, and investigates whether the task of communicating about classes of objects in the environment results in more uniform internal representations across the population.

If it should turn out that agents do not necessarily share common internal semantic representations, this does not necessarily invalidate the basic results of ILM simulations, which so far have assumed common meaning representations. Undoubtedly, real humans share a common external environment, perceived through essentially similar sense organs, and there is surely some substantial overlap between representations in the minds of different individuals, even if there are also differences. The current ILM simulations can be taken as modeling the evolution of meaning-form mappings in the areas of overlap, which presumably are substantial. For areas in which different people conceptualize the world differently, we could conclude, with Wittgenstein, that "whereof one cannot speak, thereupon one must be silent".

The evolutionary story need not end on this Wittgensteinian note, envisaging a stable fixed boundary between the effable and the ineffable. Presumably, a population which manages to reduce the domain of "whereof one cannot speak" will reap certain benefits from being able to communicate about a wider set of experiences. Indeed, it is becoming clear that ILM models can shed light on how agents' conceptual representations adapt to the problem of fitting into a system of meaning-form mappings which can be transmitted across generations. The work surveyed above emphasizes that, for large or infinite meaning spaces, and with the natural severe bottleneck at the point of learning, only languages characterizable by general, compositional rules are stably transmitted across generations. Work by Brighton and Kirby (2001a; 2001b) relates the likelihood of achieving such stable compositional systems to a range of possible organizations of the space of meanings. They conclude

> If the perceptual space of the agent is not broken into multiple features or multiple values, then compositionality is not possible. However, even when

the conceptual space is cut up into a few features or values, compositionality is still unlikely. The principal reason for this is that a simple meaning space structure results in the rate of observation of feature values to be near the rate of observation of whole meanings. This situation would result in less of a stability advantage for compositionality. ... The more complex the meaning space, the more payoff in stability compositional language offers. However, too much complexity leads to a decrease in payoff. ... with a highly complex meaning space structure the meanings corresponding to the objects are scattered over a vast space, and as a result, regularity in the correspondence between signals and meanings is weakened. (Brighton & Kirby, 2001b)

Several of the simulations mentioned above have taken predicate logic representations 'off the shelf' as convenient ways of representing propositional meanings. Hurford, in recent work, has argued that neural correlates exist for a basic component of such logical formulae, PREDICATE(x), where x is an individual variable and PREDICATE is any one-place predicate constant. Such a formula represents the brain's integration of the two processes of sensory 'reference' to the location of an object, mapped in parietal cortex, and analysis of the object's properties by perceptual subsystems. The brain computes actions with a few 'deictic' variables pointing to objects, linking them with 'semantic' information about the objects, corresponding to logical predicates. Mental scene-descriptions pre-existed language phylogenetically, constituting a preadaptive platform for human language (Hurford, 2001b).

Which learning algorithms work?

The versions of the ILM reported here are the successful ones, where 'successful' means leading to the emergence of a stable communication system with at least some of the characteristics of a human language. We have concentrated here on models in which simple, but general and compositional, syntactic systems arise in the community. Earlier work naturally started with the emergence of simple vocabularies (e.g., Werner and Dyer, 1991; Ackley and Littman, 1994; MacLennan and Burghardt, 1994; Levin, 1995; Cangelosi and Parisi, 1998; Oliphant, 1996; Bullock, 1997; de Bourcier and Wheeler, 1997; Di Paolo, 1997; Werner and Todd, 1997; Noble, 1998). In both the lexical and the syntactic work, it transpires that a stable system only arises in the community if the learning algorithm has a certain property, which we have called 'obverter'. As we mention in the second section, an obverter learner is one whose acquired production behavior is guided by a consideration of how the learner itself would interpret a signal. The learner thinks something like this: "To express meaning M, I will use signal S, because I would understand S as meaning M." An obverter strategy for language learning anticipates (or constructs) the essential bi-directional nature of the linguistic sign, allowing use of the same body of knowledge both in speaking and in hearing.

Smith (2001b) has systematically investigated a class of simple neural nets which take representations of signals as input and give representations of meanings as output, and whose connection weights can be adjusted in response to training on input-output pairs. Smith classified the different net-types into a hierarchy of

'Constructors', 'Maintainers', 'Learners' and 'Nonlearners'. Constructors are nets embodying a learning algorithm which, if embedded in an ILM starting from no language at all, will result after many generations in an emergent efficient vocabulary code shared by the whole population. Maintainers, if used in an ILM, can manage to maintain an initially given code against a certain amount of noise in transmission across generations, but cannot contribute to the glossogenetic evolution of a code emerging from an initial 'zero' situation. Learners are able to acquire, and subsequently faithfully transmit a code to a succeeding generation, but only in the complete absence of noise. Non-learners cannot even faithfully acquire a code to which they are exposed. Only the constructor nets have the obverter property.

Because acquisition of syntax is much more complex than acquisition of vocabulary, it is hardly possible to conduct such a systematic comparison in the space of possible syntax acquisition algorithms. But it seems very likely that the only learning algorithms which will enable an ILM to evolve a stable syntactic system will also have the obverter property.

Coevolution

ILM simulations model a form of cultural evolution of language, which we call 'glossogeny'. As our introduction mentioned, the glossogenetic process is just one of several interacting evolutionary systems. ILM simulations work with specific learning algorithms. It is of interest to investigate how the two evolutionary processes, the phylogenetic and the glossogenetic, interact with each other. Complex simulations can be set up in which an ILM is embedded into a larger genetic algorithm which breeds language learners selected for good adult communication abilities.

Almost all simulations of the phylogeny of language associate fitness with the capacity to communicate. Combine this idea with the essential arbitrariness of the meaning-to-signal mappings that constitute languages, and a highly significant peculiarity of language evolution becomes apparent. A would-be communicating agent is born into a community with a pre-existing code of arbitrary meaning-form mappings. If fitness is correlated with the power to communicate, the agent's first priority is to become a fully participating member of the communicating population into which it was born. The language environment to which a human baby must adapt is not universal across the species, but is local and was constructed by previous generations of the local group solving the same survival problem. Now, if the language of the community is actually not an efficient code (say by having too much homonymy, or by being noncompositional), it is nevertheless more important for the newcomer to conform to that code than to attempt to improve it in any way. This implies that glossogenetic evolution can put a significant drag on phylogenetic adaptation. There are parallels with cultural evolution generally, where a society can perpetuate dysfunctional traits, such as approval of extreme forms of self-harm.

Kirby and Hurford (1997) describes a simulation in which an ILM is embedded into a genetic algorithm selecting over variants of an idealized learning algorithm, modeling the 'Principles and Parameters' version of generative linguistic language acquisition theory. A language learner can be innately endowed with more or less plasticity. In this theory, grammatical Principles correspond to genetically fixed aspects of language; Parameters represent aspects of language where the learner can adapt its acquired grammar to the ambient language of the community. The presence of mutations and crossover in the genetic algorithm allow innate language learning Principles to be replaced by Parameters, and vice versa. In this simulation, certain aspects of grammar were artificially and arbitrarily assigned a functional advantage, so that language learners could be influenced by a preference for languages with certain characteristics in preference to others. As expected, the emergent languages of the populations in the ILM embedded in this model tended over time to veer towards these preferred characteristics. After some time, the languages remaining in the simulated world were significantly skewed towards incorporating the designated positive functional characteristics. This represents a gradual skewing of all the language environments in the world into which a child can be born. Languages evolve glossogenetically in directions in which they are steered by functional pressures. Bearing in mind that it is always better for a child to learn the language of its community than to branch out on its own, this glossogenetic response to functional pressure is slow. But once such functional pressure has had some appreciable effect on the distribution of language universals, all human children will be born into linguistic environments biased in a certain direction. At this point it can be expected that phylogenetic evolution will begin to bite, and that there may be some evolution of the innate language learning device toward a preference for universals which originated as outcomes of the glossogenetic process (see also, Yamauchi, 1999).

Summary

In this chapter we have argued that there are three complex adaptive systems at work in the evolution of language: ontogeny, glossogeny and phylogeny. Our understanding of the interaction of these aspects of language evolution can be enhanced through a computational modeling approach. If, as we argue, much of the structure of language is emergent, then a modeling methodology is appropriate, since it is notoriously difficult to come up with reliable intuitions about emergent behaviors. To help approach the problem of modeling how language acquisition influences glossogenetic evolution, we have put forward the iterated learning model.

Using the ILM we have argued that some of the most fundamental features of human language can be best explained in terms of the pressures on language transmission. As Deacon suggests in the quotation that begins this chapter, the primary pressure on the evolution of languages (as opposed to language users) is the need to be learnt. If a language, or some part of a linguistic system, is not learnable then it will not persist. In a very real sense, the data a learner is exposed to is a bottleneck in the transmission of the knowledge of language. In one of our simulations, the meaning-signal mapping has a potentially infinite extent, but must nevertheless be recovered every generation from a finite sample of randomly chosen utterances. This provides a severe selection pressure (though a *linguistic* rather than *natural* one) on the behavior that is being replicated from generation to generation.

Fundamentally, the ILM demonstrates that we cannot draw a direct parallel between the innate properties of the language user and the structure of language. The early languages in the simulations are learnt and used by the agents – in other words they are not ruled out innately. However, they are not stable; only compositional languages have that property. From this we can conclude two things:

1. Even if we could examine the innate language faculty in minute detail, we could not directly read-off the structure of human language from it.

2. Conversely, given some universal property of language that we wish to explain, we should not necessarily place the whole burden of that explanation on an innate biological property.

Finally, if we consider what recursive compositionality gives us as a species it is tempting to conclude that communicative function must have a central role in its explanation. With it we can express an infinite range of ideas. We are not trapped in the prison of prior experience, forever constrained to convey only those things that have previously been conveyed to us. We do not deny here the obvious communicative advantage we have as a species that syntax buys us. However, counter to intuition, our explanation for syntactic structure does not make any reference to communication. It is the advantage that compositionality gives to language itself that drives its evolution, the benefits this has to us are merely a fortunate side-effect.

References

Ackley D, Littman M (1994) Altruism in the evolution of communication. In: Brooks R, Maes P (eds) *Artificial Life 4: Proceedings of the Fourth International Workshop on the Synthesis and Simulation of Living Systems.* Addison-Wesley, Redwood City CA, pp 40-48

Batali J (1998) Computational simulations of the emergence of grammar. In: Hurford J, Knight C, Studdert-Kennedy M (eds) *Approaches to the evolution of human language: Social and cognitive basis.* Cambridge University Press, Cambridge UK, 405-426

Batali J (2001). The negotiation and acquisition of recursive grammars as a result of competition among exemplars. In Briscoe EJ (ed) *Linguistic evolution through language acquisition: Formal and computational models.* Cambridge University Press, Cambridge UK

Belew R (1990) Evolution, learning, and culture: Computational metaphors for adaptive algorithms. *Complex Systems*, 4: 11-49

Bickerton D (1990) *Language and species.* University of Chicago Press

Boyd R, Richerson PJ (1985) *Culture and the evolutionary process.* University of Chicago Press

Brighton H, Kirby S (2001a) Meaning space structure determines the stability of culturally evolved compositional language. Submitted to the *Cognitive Science Society Conference 2001*

Brighton H, Kirby S (2001b) The survival of the smallest: stability conditions for the cultural evolution of compositional language. Submitted to *CogSci2001, the 23rd Annual Conference of the Cognitive Science Society*

Briscoe EJ (2000) Evolutionary perspectives on diachronic syntax. In: Pintzuk S, Tsoulas G, Warner A (eds) *Diachronic syntax: Models and mechanisms*. Oxford University Press, Oxford.

Bullock S (1997) An exploration of signalling behaviour by both analytic and simulation means for both discrete and continuous models. In: Husband P, Harvey I (eds) *Proceedings of the Fourth European Conference on Artificial Life*. MIT Press, Cambridge MA, pp 454-463

Cangelosi A, Parisi D (1998) The emergence of a 'language' in an evolving population of neural networks. *Connection Science*, 10: 83-97

Chomsky N (1986) *Knowledge of language*. Praeger

de Bourcier P, Wheeler M (1997) The truth is out there: The evolution of reliability in aggressive communication systems. In: Husband P, Harvey I (eds) *Proceedings of the Fourth European Conference on Artificial Life*. MIT Press, Cambridge MA, pp 444-453

Deacon TW (1997) *The symbolic species: The coevolution of language and human brain*. Penguin, London

Di Paolo E (1997) An investigation into the evolution of communication. *Adaptive Behaviour*, 6: 285-324

Francis N, Kucera H (1982) *Frequency analysis of English usage: Lexicon and grammar*. Houghton Mifflin, Boston

Frege G (1892) Uber sinn und bedeutung. *Zeitschrift fur Philosophie und Philosophische Kritik*, 100: 25-50

Goroll N (1999) *(The deep blue) Nile: Neuronal influences on language evolution*. Master's thesis, University of Edinburgh

Hurford JR (1999) Language learning from fragmentary input. In: Dautenhahn K, Nehaniv C (eds) *Proceedings of the AISB'99 Symposium on Imitation in Animals and Artifacts*. Society for the Study of Artificial Intelligence and the Simulation of Behaviour, pp 121-129

Hurford JR (2000). Social transmission favours linguistic generalization. In Knight C, Studdert-Kennedy M, Hurford JR (eds) *The evolutionary emergence of language: Social function and the origins of linguistic form*. Cambridge University Press, Cambridge, pp 324-352

Hurford JR (2001a) Expression/induction models of language evolution: dimensions and issues. In: Briscoe EJ (ed) *Linguistic evolution through language acquisition: Formal and computational models*. Cambridge University Press, Cambridge UK

Hurford JR (2001b) The neural basis of predicate argument structure. Submitted to *Behavioral and Brain Sciences*

Kirby S (1999a) *Function, selection and innateness: The emergence of language universals*. Oxford University Press, Oxford

Kirby S (1999b) Learning, bottlenecks and infinity: A working model of the evolution of syntactic communication. In: Dautenhahn K, Nehaniv C (eds) *Proceedings of the AISB'99 Symposium on Imitation in Animals and Artifacts*. Society for the Study of Artificial Intelligence and the Simulation of Behaviour, pp 121-129

Kirby S (1999c) Syntax out of learning: The cultural evolution of structured communication in a population of induction algorithms. In Floreano D, Nicoud J-D, Mondada F (eds) *Advances in Artificial Life (Number 1674 in Lecture Notes in Computer Science)*. Springer, Berlin

Kirby S (2000) Syntax without natural selection: How compositionality emerges from vocabulary in a population of learners. In Knight C, Studdert-Kennedy M, Hurford JR (eds) *The evolutionary emergence of language: Social function and the origins of linguistic form.* Cambridge University Press, Cambridge, pp 303-323

Kirby S (2001a) Learning, bottlenecks and the evolution of recursive syntax. In: Briscoe EJ (ed) *Linguistic evolution through language acquisition: Formal and computational models.* Cambridge University Press, Cambridge UK

Kirby S (2001b) Spontaneous evolution of linguistic structure: An iterated learning model of the emergence of regularity and irregularity. *IEEE Transactions on Evolutionary Computation and Cognitive Science*, 5(2): 102-110 (Special issue on Evolutionary Computation and Cognitive Science)

Kirby S, Hurford JR (1997) Learning, culture and evolution in the origin of linguistic constraints. In: Husband P, Harvey I (eds) *Proceedings of the Fourth European Conference on Artificial Life.* MIT Press, Cambridge MA, pp 493-502

Levin M (1995) The evolution of understanding: A genetic algorithm model of the evolution of communication. *Biosystems*, 36: 167-178

MacLennan B, Burghardt G (1994). Synthetic ethology and the evolution of cooperative communication. *Adaptive Behaviour*, 2: 161-187

Newmeyer F (1991) Functional explanation in linguistics and the origins of language, *Language and Communication*, 11: 3-28

Niyogi P, Berwick RC (1995) The logical problem of language change, Technical Report AIM-1516, MIT AI Lab

Noble J (1998) Evolved signals: Expensive hype vs conspiratorial whispers. In: Adami C, Belew R, Kitano H, Taylor C (eds) *Artificial Life 6: Proceedings of the Sixth International Conference on Artificial Life.* MIT Press, Cambridge MA

Nowak MA, Komarova NL, Niyogi P (2001) Evolution of universal grammar. *Science*, 291: 114-118

Nowak MA, Plotkin JB, Jansen VAA (2000) The evolution of syntactic communication. *Nature*, 404: 495-498

Oliphant M (1996) The dilemma of saussurean communication. *BioSystems*, 37: 31-38

Oliphant M, Batali J (1997) Learning and the emergence of coordinated communication. *Center for Research on Language Newsletter*, 11(1)

Pinker S (1994) *The language instinct.* Penguin

Pinker S (1999) *Words and rules.* Weidenfeld & Nicolson

Pinker S, Bloom P (1990) Natural language and natural selection. *Behavioral and Brain Sciences*, 13: 707-784

Siskind J (1996) A computational study of cross-situational techniques for learning word-to-meaning mappings. In: Brent MR (ed) *Computational approaches to language acquisition.* MIT Press, Cambridge MA pp 39-91

Smith K (2001a) Establishing communication systems without explicit meaning transmission. Submitted to the *European Conference on Artificial Life 2001*

Smith K (2001b) The evolution of learning mechanisms supporting symbolic communication. Submitted to *CogSci2001, the 23rd Annual Conference of the Cognitive Science Society*

Smith K (in press) Learners are losers: Natural selection and learning in the evolution of communication. *Adaptive Behaviour*

Steels L (1997) The synthetic modeling of language origins. *Evolution of Communication*, 1: 1-34

Steels L (1999) *The talking heads experiment (Volume I. Words and meanings)*. Antwerpen: Laboratorium, Special pre-edition

Tonkes B, Wiles J (in prep) Methodological issues in simulating the emergence of language. Submitted to the volume arising out of the *Third Conference on the Evolution of Language*, Paris 2000

Werner GM, Dyer MG (1991) Evolution of communication in artificial organisms. In: Langton C, Taylor C, Farmer J, Rasmussen S (eds) *Artificial Life 2*, Addison-Wesley, Redwood City CA, pp 659-687

Werner GM, Todd P (1997) Too many love songs: Sexual selection and the evolution of communication. In: Husband P, Harvey I (eds) *Proceedings of the Fourth European Conference on Artificial Life*. MIT Press, Cambridge MA, pp 434-443

Wray A (1998) Protolanguage as a holistic system for social interaction. *Language and Communication*, 18: 47-67

Yamauchi H (1999) *Evolution of the LAD and the Baldwin effect*. Master's thesis, University of Edinburgh

Zipf GK (1936) *The psycho-biology of language*. Routledge, London

Chapter 7

Population Dynamics of Grammar Acquisition

Natalia L. Komarova and Martin A. Nowak

Introduction

The most fascinating aspect of human language is grammar. Grammar is a computational system that mediates a mapping between linguistic form and meaning. Grammar is the machinery that gives rise to the unlimited expressibility of human language.

Children develop grammatical competence spontaneously without formal training. All they need is interaction with people and exposure to normal language use. The child hears a certain number of grammatical sentences and then constructs an internal representation of the rules that generate grammatical sentences. Chomsky realized that the evidence available to the child does not uniquely determine the underlying grammatical rules (Chomsky, 1965). This phenomenon is called the 'poverty of stimulus' (Wexler and Culicover, 1980). The 'paradox of language acquisition' (Jackendoff, 1997) is that children nevertheless reliably achieve correct grammatical competence. How is this possible?

As Chomsky pointed out, the child needs a pre-formed linguistic theory that can specify candidate grammars that might be compatible with the available linguistic data (Chomsky, 1965). He introduced the term Universal Grammar (UG) to denote this preformed 'linguistic theory', the initial pre-specification of the form of possible human grammars (Chomsky, 1972).

Hence, for language acquisition the child needs a mechanism for processing the input sentences and a *search space* of *candidate grammars* from which to choose the appropriate grammar. Chomsky's original concept is that UG is a rule system that generates the search space. More recent views use UG to encompass both the search space and the mechanism for evaluating input sentences. Therefore, UG has become almost synonymous with 'mechanism of language acquisition'.

The notion of an innate, genetically encoded, UG is controversial (Tomasello, 1995; Bates, 1984; Langacker, 1987; 1992). Much of the discourse, however, focuses on which specific linguistic features are innate (for example, phrase structure rules of X-bar theory, or lexical categories such as nouns and verbs) and to what extent UG is a specific syntactic module or simply uses general-purpose cognitive abilities. We do not participate in this controversy. Instead we choose a sufficiently general formulation of the process of language acquisition. Ultimately everybody agrees that human beings require some innate components for language acquisition. These innate components are what we call UG.

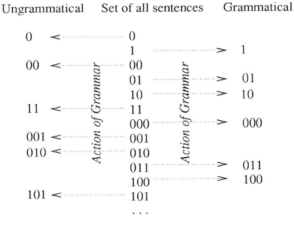

Figure 7.1 Mathematically, a grammar can be seen as a rule system that divides a countably infinite number of sentences into two subsets, grammatical and ungrammatical.

Various approaches to the mathematical description of language acquisition have been formulated by Hornstein and Lightfoot (1981), Osherson, Stob and Weinstein (1986), Manzini and Wexler (1987), Lightfoot (1991), Gibson and Wexler (1994), Niyogi (1998). The sentences of all languages can be enumerated. We can say that a grammar, G, is a rule system that specifies which sentences are allowed and which sentences are not allowed (see Figure 7.1). Universal grammar, in turn, contains a rule system that generates a set (or a search space) of grammars, $\{G_1, G_2, .., G_n\}$. These grammars can be constructed by the language learner as potential candidates for the grammar that needs to be learned. The learner cannot end up with a grammar that is not part of this search space. In this sense, UG contains the possibility to learn all human languages (and many more).

Figure 7.2 illustrates this process of language acquisition. The learner has a mechanism to evaluate input sentences and to choose one of the candidate grammars that are contained in his search space.

More generally, it is also possible to imagine that UG generates infinitely many candidate grammars, $\{G_1, G_2, ...\}$. In this case, the learning task can be solved if UG also contains a prior probability distribution on the set of all grammars. This prior distribution biases the learner toward grammars that are expected to be more likely than others. A special case of a prior distribution is one where a finite number of grammars are expected with equal probability and all other grammars are expected with zero probability, which is equivalent to a finite search space.

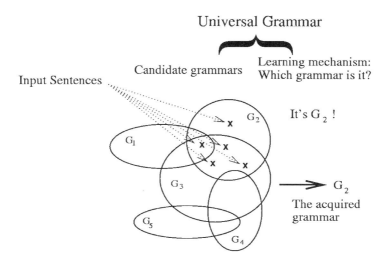

Figure 7.2 Universal grammar specifies the search space of candidate grammars and the learning procedure for evaluating input sentences. The basic idea is that the child has an innate expectation of grammar (for example a finite number of candidate grammars) and then chooses a particular candidate grammar that is compatible with the input.

A fundamental question of linguistics and cognitive science is what are the restrictions that are imposed by UG on human language. In other words, how much is innate and how much is learned in human language. In learning theory (Vapnik, 1995; Valiant, 1984) this question is studied in the context of an ideal speaker-hearer pair. The speaker uses a certain 'target grammar'. The hearer has to learn this grammar. The question is: what is the maximum size of the search space such that a specific learning mechanism will converge (after a number of input sentences, with a certain probability) to the target grammar?

In terms of language evolution, the crucial question is what makes a *population* of speakers converge to a coherent grammatical system. In other words, what are the conditions that UG has to fulfill for a population of individuals to evolve coherent communication? In the following, we will discuss how to address this question (see also Nowak, Komarova and Niyogi, 2001; Komarova, Niyogi, and Nowak, 2001). The material presented here is part of a larger effort to provide a mathematical formulation of the evolution of language (Hashimoto and Ikegami, 1996; Steels, 1997; Cangelosi and Parisi, 1998; Hurford, Studdert-Kennedy and Knight, 1998; Kirby, 1999; Cangelosi, 1999; Noble, 1999).

The rest of this chapter is organized as follows. In the next section we present the dynamical system, which describes the evolution of language learning by a population of individuals. In section "Coherence Threshold" we outline the main results concerning the dynamics of grammatical coherence. We derive the *coherence threshold* for the language system and analyze how it is related to the complexity of universal grammar. In particular, we specify how many sampling events children need to receive during the language acquisition period in order for the entire population to maintain a coherent communication system. In the Section of "Natural selection among variants of UG" we examine the evolutionary competition among different universal grammars. Conclusions are found in the final section.

The Language Dynamics Equations

In this section, we give a rigorous definition of grammar and put it in the context of population dynamics. The more technical material is moved to section "What is grammar?" and can be skipped at the first reading.

Population dynamics of learning

First of all, let us state that 'poverty of stimulus' has an elegant mathematical formulation known as Gold's theorem (Gold, 1967). Suppose there is a rule that generates a subset of all integers. A person is provided with a sample of integers that are generated by the rule. After some time the person is asked to produce other integers that are compatible with the rule. Gold's theorem states that this task cannot be solved. Any finite number of sample integers is not enough to determine uniquely the underlying rule. The person can only solve the task if she had a preformed expectation determining which rules are possible (or likely) and which are not. The sample integers correspond to the sentences presented to the child, the rule corresponds to the grammar used by the parents (or other speakers). The preformed expectation is universal grammar. Hence, in this sense 'poverty of stimulus' and the necessity of an innate universal grammar are not controversial issues, but mathematical facts.

Imagine a group of individuals that all have the same UG, given by a finite search space of candidate grammars, $G_1,..,G_n$, and a learning mechanism for evaluating input sentences. Let us specify the similarity between grammars by introducing the numbers s_{ij} which denote the probability that a speaker who uses G_i will say a sentence that is compatible with G_j; the rigorous way to compute the similarity matrix is given in the next section.

We assume there is a reward for mutual understanding. The payoff for someone who uses G_i and communicates with someone who uses G_j is given by

$$F(G_i, G_j) = (s_{ij} + s_{ji})/2$$

This is simply the average taken over the two situations when G_i talks to G_j and when G_j talks to G_i.

Denote by x_i the relative abundance of individuals who use grammar G_i. Assume that everybody talks to everybody else with equal probability. Therefore, the average payoff for all those individuals who use grammar G_i is given by

$$f_i = \sum_{j=1}^{n} x_j F(G_i, G_j).$$

We assume that the payoff derived from communication contributes to biological fitness; individuals leave offspring proportional to their payoff. These offspring inherit the UG of their parents and learn (possibly with mistakes) their grammar. In the setting of a language game with no learning mistakes, the function $F(G_i, G_j)$ defines the equilibrium states of the system. It is shown by Komarova and Niyogi (2001) that non-ambiguous languages (perhaps with some isolated synonym-like components) are the only evolutionarily stable states of the system.

To proceed, we need to include the process of language acquisition. Children receive language input (sample sentences) from their parents and develop their own grammar. In the most general approach, we do not specify a particular learning mechanism but introduce the stochastic matrix, $\{q_{ij}\}$, whose elements, q_{ij}, denote the probability that a child born to an individual using G_i will develop G_j. (For simplicity, we assume here that each child receives input from one parent. Models that allow input from several individuals are also possible.) The probability that a child will develop G_i if the parent uses G_i is given by q_{ii}. The quantities q_{ii} measure the accuracy of grammar acquisition. If $q_{ii}=1$ for all i (i.e., $\{q_{ij}\}$ is an identity matrix), then grammar acquisition is perfect for all candidate grammars. Some particular learning mechanisms that define the matrix $\{q_{ij}\}$ will be considered in the following section.

The population dynamics of grammar evolution are then given by the following system of ordinary differential equations, which we call the *language dynamics equations*:

$$\frac{dx_j}{dt} = \sum_{i=1}^{n} f_i q_{ij} x_i - \phi x_j, \qquad j = 1,...,n. \tag{1}$$

The term $-\phi x_j$ ensures that the total population size remains constant: the sum over the relative abundances, $\sum x_i$, is 1 at all times. The variable $\phi = \sum_{i=1}^{n} f_i x_i$ denotes the average fitness or *grammatical coherence* of the population. The grammatical coherence is given by the probability that a randomly chosen sentence of one person is understood by another person. It is a measure for successful communication in a population. If $\phi=1$, all sentences are understood and communication is perfect. In general, ϕ is a number between 0 and 1.

The language dynamics equation is reminiscent of the *quasispecies equation* of molecular evolution (Eigen and Schuster, 1979), but has frequency-dependent fitness values: the quantities f_i depend on the relative abundances $x_i,..,x_n$. In the limit of perfectly accurate language acquisition, $q_{ii}=1$, we recover the *replicator equation* of evolutionary game theory (Hofbauer and Sigmund, 1998). Thus, our model provides a connection between two of the most fundamental equations of evolutionary biology.

What is grammar?

In this section we give a rigorous definition of grammar and specify ways to calculate the similarity matrix, $\{s_{ij}\}$, and the payoff function, $F(G_i, G_j)$.

Consider a finite alphabet, that is a finite set of symbols, for example, $\{a,b,c\}$. Consider all possible strings (sentences) that can be formed with these symbols, $\{a, b, c, aa, ab, ac, ba,...\}$. There are a countably infinite number of such strings. Let us call the set of symbols Σ_1 and the set of all strings Σ_1^*. A grammar G_i is a rule system that divides the set of all strings into two subsets. One subset (the set G_i) consists of 'grammatical' strings, the other subset, $\Sigma_1^* \setminus G_i$, of 'ungrammatical' strings. This is illustrated in Figure 7.1.

This framework needs to be extended if we want to define a search space of candidate grammars. We want to calculate the probability, s_{ij}, that a speaker using grammar G_i says a sentence that is compatible with another grammar, G_j. In this context, a grammar G_i induces a measure (or probability distribution), μ_i, on the set of strings, Σ_1^*. The measure specifies the probabilities with which a speaker uses particular sentences. Clearly, $\mu_i(\Sigma_1^* \setminus G_i)=0$ and $\mu_i(G_i)=1$. The set G_i is the support of the function μ_i. For the similarity function, s_{ij}, we then have $s_{ij}=\mu_i(G_i \cap G_j)$.

So far, we have a purely syntactic formulation of grammar, but it is sometimes argued that a grammar should not only specify which sentences are correct and which are not, but also indicate what they mean. Hence, more generally, a grammar should mediate a mapping between form and meaning. Mathematically, this can be described in the following way.

Similar to the syntactic alphabet, Σ_1, we introduce a semantic alphabet, Σ_2, and assume that we can enumerate all possible meanings by the set of all semantic strings, Σ_2^*. Therefore, Σ_1^* is the set of all possible linguistic expressions and Σ_2^* is the set of all possible meanings. A grammar, G_i, generates a subset of $\Sigma_1^* \times \Sigma_2^*$, which is an infinite set of sentence-meaning pairs. In this case, G_i is specified by a measure μ_i on $\Sigma_1^* \times \Sigma_2^*$. As before, we can define $s_{ij}=\mu_i(G_i \cap G_j)$ to be simply the proportion of sentence meaning pairs that G_i and G_j have in common. Hence, s_{ij} is the probability that a user of G_i produces an utterance that a user of G_j can understand.

Finally we introduce a generalization that allows us to define the fitness of grammars. Each user of G_i is characterized by an encoding matrix P and a decoding matrix Q. Here, $p_{kl}=\mu(s_k, m_l)/\Sigma_i\mu(s_i, m_l)=\mu(s_k|m_l)$ which is simply the probability of using the expression s_k to convey the meaning m_l. Similarly, $q_{kl}=\mu(s_k, m_l)/\Sigma_j\mu(s_k, m_j)=\mu(m_l|s_k)$ is the probability of interpreting the expression s_k to mean m_l. The need to communicate meanings is related to events in the shared world of the linguistic community. Therefore, one can define a measure σ on the set of possible meanings (Σ_2^*) that speakers and hearers might wish to communicate with each other. Given this, we can define $s_{ij}=tr(P^{(i)}\Lambda(Q^{(j)})^T)$ (where Λ is a diagonal matrix such that $\Lambda_{ii}=\sigma(m_i)$). This is simply the probability that an event occurs and is successfully communicated from a user of G_i to a user of G_j.

Note that $F(G_i, G_i)$ is the probability that users of G_i will have a successful communication with each other. Communication might break down in one of two ways (i) *poverty*: an event happens whose meaning cannot be encoded by G_i, and (ii) *ambiguity*: an event happens whose meaning has an ambiguous encoding in G_i leading to a possibility of misunderstanding. $F(G_i, G_i)$ is a number between 0 and 1 and denotes the fitness of G_i. Maximum fitness, $F(G_i, G_i) = 1$, is achieved by grammars that can express every possible meaning (zero poverty) and have no ambiguities.

Coherence Threshold

The behavior of Equation (1) can be roughly described as follows. In general, the system admits multiple (stable and unstable) equilibria. For low accuracy of grammar acquisition (low values of q_{ii}, when the matrix $\{q_{ij}\}$ is far from identity), all grammars, G_i, occur with roughly equal abundance. There is no predominating grammar in the population, and the grammatical coherence is low. As the accuracy of grammar acquisition increases (i.e., the matrix $\{q_{ij}\}$ gets closer to identity), equilibrium solutions may arise where a particular grammar is more abundant than all other grammars. We will refer to such solutions as *one-grammar equilibria*, or one-grammar solutions. A coherent communication system emerges. This means that if the accuracy of learning is sufficiently high, the population will converge to a stable equilibrium with one dominant grammar. In the presence of multiple one-grammar solutions, the choice of the stable equilibrium depends on the initial conditions.

The accuracy of language acquisition (the closeness of the matrix $\{q_{ij}\}$ to identity) defines the level of coherence in the population. The accuracy of language acquisition, in its turn, depends on the following factors:

- the search space, $G_1, \ldots, G_n$,
- the learning mechanism,
- the number of learning examples, N, available to the child during the language acquisition stage.

Obviously, the less restricted the search space of candidate grammars is, the harder it is to learn a particular grammar. Depending on the specific values of s_{ij}, some grammars may be much harder to learn than others. For example, if a speaker using G_i has high probabilities formulating sentences that are compatible with many other grammars (s_{ij} close to 1 for many different j) then G_i will be hard to learn. In the limit $s_{ij} = 1$, G_i is considered unlearnable, because no sentence can refute the hypothesis that the speaker uses G_j. Also, an inefficient learning mechanism or one that evaluates only a small number of input sentences will lead to a low accuracy and hence prevent the emergence of grammatical coherence. We can therefore ask the crucial question:

Which properties must UG have such that a predominating grammar will evolve in a population of speakers?

In other words, which UG can induce grammatical coherence in a population? We will start by presenting two examples of the search space and then proceed by looking at specific learning algorithms.

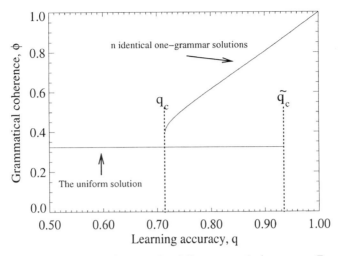

Figure 7.3 The bifurcation diagram for fully symmetrical systems, Example 1. Here, $s=0.3$ and $n=30$. At $q=q_c$, n identical one-grammar solutions appear. At $q = \tilde{q}_c$, the uniform (low coherence) solution becomes unstable.

Example 1: Fully symmetrical systems
Let us impose the following symmetry condition on the similarity matrix: $s_{ij}=s$ for all $i \neq j$, where s is some constant, $0<s<1$. We will further assume that the $\{q_{ij}\}$ matrix in this case is also symmetrical: we have $q_{ii}=q$ for all i, and $q_{ij}=(1-q)/(n-1)$. (Later on we will see that this assumption holds for some reasonable learning mechanisms.) When q is small (low learning accuracy), there is only one stable fixed point in the system, which we call the *uniform equilibrium*: all the grammars are represented in the population and have an equal abundance, and the coherence is low, see Figure 7.3. As q increases, a bifurcation occurs where n identical one-grammar equilibria appear, each of them corresponds to a particular dominant grammar. It is possible to calculate the error threshold, i.e., the value of q, q_c, such that for $q>q_c$ the one-grammar solutions exist and are stable. It is given by Komarova *et al.* (2001):

$$q_c = \frac{2\sqrt{s}}{1+\sqrt{s}} + O(1/n)$$

What is important, is that this value tends to a constant for large values of n, the size of UG. This means that no matter how large the search space is, there is a fixed threshold value for the accuracy of learning. We will refer to this interesting property as *universality* of universal grammar.

Example 2: Random similarity matrices

In this model, let us assume that the coefficients s_{ij} are independent random numbers drawn from some distribution between zero and one. In this case, we have the following results (Komarova, 2001). If the matrix s_{ij} is diagonally dominant in the sense that $s_{ii} > 1/2(s_{ij} + s_{ji})$ for all $j \neq i$, and if the matrix $\{q_{ij}\}$ is close enough to identity, then there exist exactly n stable one-grammar solutions in the system. In other words, if the accuracy of learning is high enough, then the population may find itself speaking any one of n grammars of UG (which one it is depends on the initial condition). The condition of diagonal dominance can be easily interpreted: we require that each of the n grammars understands itself better than it understands any other grammar, which is a very natural assumption.

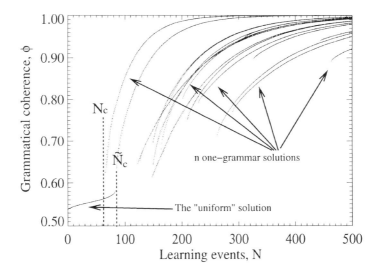

Figure 7.4 The bifurcation diagram for a system with a random similarity matrix, Example 2. Here $n=20$. At $N = N_c$, the first one-grammar solution appears. The learning accuracy matrix, $\{q_{ij}\}$, is calculated according to the memoryless learner algorithm, and depends on N, the number of learning events. At $N = \tilde{N}_c$ the low coherence solution equivalent to the uniform solution of the fully symmetrical system, disappears.

On the other hand, if for some k and m the diagonal dominance is violated, so that $s_{kk} < 1/2(s_{km} + s_{mk})$ and $s_{mm} < 1/2(s_{km} + s_{mk})$, then no matter how high the learning accuracy is, grammars G_k and G_k cannot become dominant. Instead, a mixture of the two grammars is possible as a stable *two-grammar equilibrium*. Note that such situations are in principle possible: highly ambiguous grammars might have a higher communicative efficiency when communicating with each other than they do with themselves (see Komarova and Niyogi, 2001).

An example of a bifurcation diagram for a random system is presented in Figure 7.4. There, the coefficients s_{ij} with $i \neq j$ are random uniformly distributed numbers and $s_{ii}=1$. The matrix $\{q_{ij}\}$ is parameterized with N, the number of learning events, according to the memoryless learner algorithm, see below. As N grows, the matrix $\{q_{ij}\}$ tends to identity. We observe that the error threshold phenomenon is also present in the random system; for complete analytical results refer to Komarova (2001). One particular feature is that if we assume that the matrix $\{q_{ij}\}$ is symmetrical (as defined in Example 1), then the error threshold does not depend on the size of the system. This means that the universality property holds in this case just like it does for fully symmetrical systems. There are preliminary results that the universality property also holds for a more general class of matrices $\{q_{ij}\}$.

Now let us look at some learning mechanisms which, given the search space, define the dependence of $\{q_{ij}\}$ on N, the number of learning events. We present results for two learning mechanisms that represent reasonable boundaries for the actual, unknown learning mechanism employed by humans.

The *memoryless learning* algorithm, a favorite with learning theorists, makes little demand on the cognitive abilities of the learner. It describes the interaction between a teacher and a learner. (The 'teacher' can be one or several individuals or the whole population.) The learner starts with a randomly chosen hypothesis (say G_i) and stays with this hypothesis as long as the teacher's sentences are compatible with this hypothesis. If a sentence arrives that is not compatible, the learner will at random pick another candidate grammar from his search space. The process stops after N sentences. The algorithm is called 'memoryless', because the learner does not remember any of the previous sentences nor which hypotheses have already been rejected. The algorithm works, primarily because once it has the correct hypothesis it will not change anymore (this is incidentally the definition of so called 'consistent learners').

The other extreme is a *batch learner* (resembling Jorge Louis Borges' man with infinite memory). The batch learner memorizes all the N sentences and at the end chooses the candidate grammar that is most compatible with the input.

Fully symmetrical systems

For this example, the memoryless learner algorithm leads to a symmetrical $\{q_{ij}\}$ matrix, as was assumed before. We have $q=1-\left(1-\frac{1-s}{n-1}\right)^N \frac{n-1}{n}$, For the memoryless learner, we can show that one-grammar equilibria exist if the number of input sentences, N, exceeds a constant times the number of candidate grammars,

$$N > C_1 n, \qquad (2)$$

where $C_1 = \log \frac{1+\sqrt{s}}{1-\sqrt{s}} /(1-s)$. This defines the value of N corresponding to the onset of high grammatical coherence.

For a batch learner (under some natural assumptions on the configuration of grammars in the search space) we have $q = \frac{1-(1-s^N)^n}{s^N n}$, and in order to maintain coherence, the number of input sentences has to exceed a constant times the logarithm of the number of candidate grammars,

$$N > C_2 \log n, \tag{3}$$

where $C_2 = \log^{-1}(1/s)$. Inequalities (2) and (3) define a *coherence threshold*, which limits the size of the search space relative to the amount of input available to the child. A UG that does not fulfill the coherence threshold does not lead to a stable, predominating grammar in a population.

Random similarity matrices

Let us first consider the following problem. A learner receives N sentences from the teacher and uses the memoryless learning mechanism to determine which grammar is the "correct" one. How many sample sentences does it typically take for the learner to guess the answer with the probability $1 - \Delta$? The following results have been obtained by Komarova and Rivin (2001a; 2001b). Let us assume that $s_{ii} = 1$ for all i. Then if the distribution of similarity coefficients, s_{ij}, is uniform for $i \neq j$, then it takes

$$N > c_1 n \log n \tag{4}$$

sample sentences for the learner to converge to the truth, where $c_1 \sim \log \Delta$. If the similarity coefficients have non-uniform distribution which favors small values, and values close to 1 almost never occur, then we have $N > c_1 n$, i.e., learning is faster. If, on the other hand, a large fraction of grammars are very similar to each other, then it takes longer to learn: $N > c_1 n^{1+\delta}$ for some positive δ (the details are given in Komarova and Rivin (2001a; 2001b)).

For a batch learner, in the case of a uniform distribution of s_{ij}, Rivin (2001) has shown that it takes

$$N > c_2 n \tag{5}$$

steps for an individual learner to be on target with a finite probability. The question then arises: how can these results for an individual learner be applied to the problem of population learning? There are no simple formulas in this case which would relate the accuracy of learning to the parameters of the system, as there were in the fully symmetrical case. Some perturbation methods must be used in order to obtain the stability conditions for one-grammar solutions. This work is still in progress. A preliminary result can be described as follows: if a learning algorithm satisfies the universality property, then estimates obtained for an individual learner will hold for the entire population. Numerical simulations suggest that the memoryless learner algorithm and the batch algorithm satisfy the universality property, and therefore results (4) and (5) are true for a population of learners. For the precise statement, refer to Komarova (2001).

To summarize so far, our mathematical modeling of the population dynamics of language acquisition suggests the following:

- Depending on the search space, it may be easier or harder to maintain coherence in the population. It is harder if a lot of the candidate grammars overlap and are difficult to distinguish. It is easier if they are well separated.

- For any reasonable search space, the batch learner will perform better than the memoryless learner. The learning mechanism used by humans will do better than the memoryless learner and worse than the batch learner. Hence it will have a coherence threshold somewhere between the bounds given above.

Natural Selection among Variants of Universal Grammar

So far we have assumed that all individuals have the same UG. Studying the biological evolution of UG, we need variation in UG and a system that describes natural selection among variants of UG.

Denote by x_i the fraction of individuals who use G_i of universal grammar U_1; denote by y_i the fraction of individuals who use G_i of universal grammar U_2. U_1 and U_2 contain, respectively, n_1 and n_2 candidate grammars. Some of the candidate grammars can be part of both universal grammars. The universal grammars, U_1 and U_2, can also differ in the number of sample sentences, N_1 and N_2, which are being considered. Therefore, we have to take into account the rate of producing offspring with grammatical communication; this rate is given by the declining function $r(N_i)$. The dynamics are described by

$$x_i = r(N_1)\sum_{j=1}^{n1} x_j f_j^{(1)} q_{ji}^{(1)} - \phi x_i, \quad i = 1,...,n_1,$$

$$y_i = r(N_2)\sum_{j=1}^{n2} x_j f_j^{(2)} q_{ji}^{(2)} - \phi y_i, \quad i = 1,...,n_2.$$

For $m=1, 2$, we have

$$f_i^{(m)} = \sum_{j=1}^{n1} x_j F(G_i^{(m)}, G_j^{(1)}) + \sum_{j=1}^{n2} x_j F(G_i^{(m)}, G_j^{(2)})$$

Grammatical coherence is given by

$$\phi = \sum_{i=1}^{n1} f_i^{(1)} x_i r(N_1) + \sum_{i=1}^{n2} f_i^{(2)} x_i r(N_2).$$

The superscripts 1,2 refer to U_1 and U_2 respectively.

Now, it is possible to compare the stability of, say, universal grammar U_1 against the invasion of U_2, simply by performing a linear stability analysis of the solution where $x_1,...,x_{n1}$ is the one-grammar equilibrium for the users of U_1, and $y_1=...=y_{n1}$ for the users of U_2. This framework can also be extended to more than two universal grammars, thus giving conditions for a universal grammar, U_*, being stable against invasion of *any* other UG (in a given class). Here we present some applications of this method.

First, let us consider universal grammars with the same search space and the same learning procedure, the only difference being the number of input sentences,

N (Komarova and Nowak, 2001). This quantity is proportional to the length of the learning period. We find that natural selection leads to intermediate values of N. For small N, the accuracy of learning the correct grammar is too low. For large N, the learning process takes too long (and thus the rate of producing children that have acquired the correct grammar is too low). This observation can explain why there is a limited language acquisition period in humans.

Second, consider universal grammars, U_1 and U_2, which differ in the size of their search space, n, but have the same learning mechanism and the same value of N. In general, there is selection pressure to reduce n. Only if n is below the coherence threshold, can the universal grammar induce grammatical communication. In addition, the smaller n, the larger is the accuracy of grammar acquisition. There can, however, also be selection for larger n: suppose universal grammar U_1 is larger than U_2 (that is $n_1 > n_2$). If all individuals use a grammar, G_1, that is in both U_1 and U_2, then U_2 is selected. Now imagine that someone invents a new advantageous grammatical concept, which leads to a modified grammar G_2 that is in U_1, but not in U_2. In this case, the larger universal grammar is favored. Hence there is selection both for reducing the size of the search space and for remaining open minded to be able to learn new concepts. For maximum flexibility, we expect search spaces to be as large as possible but still below the coherence threshold.

An interesting extension of the above model is obtained by assuming that UG is only very roughly defined by our genes. Randomness during the developmental process could give rise to variation in neuronal patterns in the brain and consequently to variation in UG. Hence it might be a reasonable assumption that individuals have slightly different UGs. Each individual could have a personal 'universal' grammar. An interesting question is how similar these UGs have to be such that a population achieves grammatical coherence. In this case, there is again selection for maintaining a large search space of candidate grammars, since the target grammar should be contained in each of the UGs.

Conclusions and Discussion

We have formulated a mathematical theory for the population dynamics of grammar acquisition. The key result here is a 'coherence threshold' that relates the maximum complexity of the search space to the amount of linguistic input available to the child and the performance of the learning procedure. The coherence threshold represents an evolutionary stability condition for the language acquisition device: only a universal grammar that operates above the coherence threshold can induce and maintain coherent communication in a population.

There are many ways in which the framework presented here can be extended. For instance, the language dynamics equations describe deterministic dynamics for a large population size. Smaller population sizes can play a role if we consider stochastic language dynamics. Computer simulations suggest that the equilibrium solutions of the deterministic system correspond to meta-stable states. Individual grammars will dominate for some time and then be replaced by other grammars. Such transitions are more likely to occur between similar grammars.

In a small population, the requirements imposed on UG are also slightly stronger. Grammatical coherence in a population will require a larger number of input sentences or smaller search spaces. A detailed mathematical study of the stochastic dynamics of our system is still outstanding.

Individual candidate grammars, G_i, can also differ in their performance. Some grammars can be less ambiguous or describe more concepts than others. In such a context, the language dynamics equation can describe a cultural evolutionary optimization of grammar within the space of grammars generated by UG (Niyogi and Berwick, 1996; 1997). It also provides a general framework for studying the dynamics of grammar change in the context of historical linguistics, where a direct comparison is possible between theoretically obtained results and the large corpus of available linguistic data.

Acknowledgments

Support from the Packard Foundation, the Leon Levy and Shelby White Initiatives Fund, the Florence Gould Foundation, the Ambrose Monell Foundation, the Alfred P. Sloan Foundation and NSF is gratefully acknowledged.

References

Bates E (1984) Bioprograms and the innateness hypothesis. *Behavioral and Brain Sciences*, 7: 188-190

Cangelosi A, Parisi D (1998) The emergence of a 'language' in an evolving population of neural networks. *Connection Science*, 10: 83-97

Cangelosi A (1999) Modeling the evolution of communication: From stimulus associations to grounded symbolic associations. In: Floreano D, Nicoud JD, Mondada F (eds) *Proceedings of ECAL99 the Fifth European Conference on Artificial Life (Lecture Notes in Artificial Intelligence)*. Springer-Verlag, Berlin

Chomsky N (1965) *Aspects of the theory of syntax*. MIT Press, Cambridge MA

Chomsky N (1972) *Language and mind*. Harcourt Brace Jovanovich, New York

Eigen M, Schuster P (1979) *The Hypercycle: A principle of natural self-organisation*. Springer, Berlin

Gibson E, Wexler K (1994) Triggers. *Linguistic Inquiry*, 25: 407-454

Gold EM (1967) Language identification in the limit. *Information and Control*, 10: 447-474

Hashimoto T, Ikegami T (1996) Emergence of net-grammar in communicating agents. *BioSystems*, 38: 1-14

Hofbauer J, Sigmund K (1998) *Evolutionary games and replicator dynamics*. Cambridge University Press

Hornstein NR, Lightfoot DW (1981) *Explanation in linguistics*. Longman, London

Hurford J, Studdert-Kennedy M, Knight C (eds) (1998) *Approaches to the evolution of language*. Cambridge University Press, Cambridge UK

Jackendoff R (1997) *The architecture of the language faculty*. MIT Press, Cambridge MA

Kirby S (1999) Syntax out of learning: The cultural evolution of structured communication in a population of induction algorithms In: Floreano D, Nicoud JD, Mondada F (eds)

Proceedings of ECAL99 the Fifth European Conference on Artificial Life (Lecture Notes in Artificial Intelligence). Springer-Verlag, Berlin, pp 694-703

Komarova NL (2001) *Population dynamics of language learning*. Manuscript in preparation

Komarova NL, Niyogi P, Nowak MA (2001) Evolutionary dynamics of grammar acquisition. *Journal of Theoretical Biology*, 209(1): 43-59

Komarova NL, Niyogi P (2001) Optimizing the mutual intelligibility of linguistic agents in a shared world. To be submitted to *Artificial Intelligence*

Komarova NL, Nowak MA (2001) Natural selection of the critical period for grammar acquisition, *Proceedings of the Royal Society B*, 268(1472): 1189-1196

Komarova NL, Rivin I (2001a) Harmonic mean, random polynomials and stochastic matrices. Submitted to *Advances in Applied Mathematics*. (arXiv.org preprint mathPR/0105236)

Komarova NL, Rivin I (2001b) Mathematics of learning, Submitted to *Electr Ann of the AMS* (arXiv.org preprint mathPR/0105235)

Langacker, R (1987) *Foundations of cognitive grammar (Volume 1)*. Stanford University Press, Stanford CA

Langacker R (1992) *Foundations of cognitive grammar (Volume 2)*. Stanford University Press, Stanford CA

Lightfoot D (1991) *How to set parameters: Arguments from language change*. MIT Press, Cambridge MA

Manzini R, Wexler K (1987) Parameters, Binding Theory, and learnability. *Linguistic Inquiry*, 18: 413-444

Niyogi P, Berwick RC (1996) A language learning model for finite parameter spaces. *Cognition*, 61: 161-193

Niyogi P, Berwick RC (1997) Evolutionary consequences of language learning. *Linguistics and Philosophy*, 20: 697-719

Niyogi P (1998) *The informational complexity of learning*. Kluwer Academic Publishers, Boston

Noble J (1999) Cooperation, conflict and the evolution of communication. *Adaptive Behavior*, 7(3/4): 349-370

Nowak MA, Komarova NL, Niyogi P (2001) Evolution of universal grammar. *Science*, 291: 114-118

Osherson D, Stob M, Weinstein S (1986) *Systems that learn*. MIT Press, Cambridge MA

Rivin I (2001) Unpublished

Steels L (1997) The synthetic modelling of language origins, *Evolution of Communication* 1: 1-34

Tomasello M (1995) Language is not an instinct. *Cognitive Development*, 10: 131-156

Valiant LG (1984) A theory of the learnable. *Communications of the ACM*, 27: 436-445

Vapnik V (1995) *The nature of statistical learning theory*. Springer, New York

Wexler K, Culicover P (1980) *Formal principles of language acquisition*. MIT Press, Cambridge MA

Chapter 8

The Role of Sequential Learning in Language Evolution: Computational and Experimental Studies

Morten H. Christiansen, Rick A.C. Dale,
Michelle R. Ellefson and Christopher M. Conway

Introduction

After having been plagued for centuries by unfounded speculations, the study of language evolution is now emerging as an area of legitimate scientific inquiry. Early conjectures about the origin and evolution of language suffered from a severe lack of empirical evidence to help rein in proposed theories. This led to outlandish claims such as the idea that Chinese was the original ur-language of humankind, surviving the biblical flood because of Noah and his family (Webb, 1669, cited in Aitchison, 1998). Or, the suggestion that humans have learned how to sing and speak from the birds in the same way as they would have learned how to weave from spiders (Burnett, 1773, cited in Aitchison, 1998). Given this state of the art, it was perhaps not surprising that the influential *Société Linguistique de Paris* in 1866 imposed a ban on papers discussing issues related to language origin and evolution, and effectively excluded such theorizing from the scientific discourse.

It took more than a century before this hiatus was overcome. Fueled by theoretical constraints derived from recent advances in the brain and cognitive sciences, the last decade of the twentieth century saw a resurgence of scientific interest in the origin and evolution of language. What has now become clear is that the study of language evolution must *necessarily* be an interdisciplinary endeavor. Only by amassing evidence from many different disciplines can theorizing about the evolution of language be sufficiently constrained to remove it from the realm of pure speculation and allow it to become an area of legitimate scientific inquiry. Nonetheless, direct experimentation is needed in order to go beyond existing data. As the current volume is a testament to, computational modeling has become the paradigm of choice for such experimentation. Computational models provide an important tool to investigate how various types of hypothesized constraints may affect the evolution of language. One of the advantages of this approach is that specific constraints and/or interactions between constraints can be studied under controlled circumstances.

In this chapter, we point to *artificial language learning* (ALL) as an additional, complementary paradigm for exploring and testing hypotheses about language evolution. ALL involves training human subjects on artificial languages with particular structural constraints, and then testing their knowledge of the language. Because ALL permits researchers to investigate the language learning abilities of infants and children in a highly controlled environment, the paradigm is becoming increasingly popular as a method for studying language acquisition (for a review, see Gomez and Gerken, 2000). We suggest that ALL can similarly be applied to the investigation of issues pertaining to the origin and evolution of language in much the same way as computational modeling is currently being used.

In the remainder of this chapter, we show how a combination of computational modeling and ALL can be used to elicit evidence relevant for the explanation of language evolution. First, we outline our theoretical perspective on language evolution, suggesting that the evolution of language is more appropriately viewed as the selection of linguistic structures rather than the adaptation of biological structure. Specifically, we argue that limitations on sequential learning have played a crucial role in shaping the evolution of linguistic structure. In support for this perspective we report on convergent evidence from aphasia studies, human and ape ALL experiments, non-human primate sequential learning studies, and computational modeling. We then present two case studies involving our own computational modeling and ALL research. The results demonstrate how constraints on basic word order and complex question formation can be seen to derive from underlying cognitive limitations on sequential learning. Finally, we discuss the current limitations and future challenges for our approach.

Language as an Organism

Languages exist only because humans can learn, produce, and process them. Without humans there would be no language (in the narrow sense of *human* language). It therefore makes sense to construe languages as organisms that have had to adapt themselves through natural selection to fit a particular ecological niche: the human brain (Christiansen, 1994; Christiansen and Chater, in preparation). In order for languages to "survive", they must adapt to the properties of the human learning and processing mechanisms. This is not to say that having a language does not confer selective advantage onto humans. It seems clear that humans with superior language abilities are likely to have a selective advantage over other humans (and other organisms) with lesser communicative powers. This is an uncontroversial point, forming the basic premise of many of the adaptationist theories of language evolution. However, what is often not appreciated is that the selection forces working on language to fit humans are significantly stronger than the selection pressure on humans to be able to use language. In the case of the former, a language can *only* survive if it is learnable and processable by humans. On the other hand, adaptation towards language use is merely *one out of many* selective pressures working on humans (such as, for example, being able to avoid predators and find food). Whereas humans can survive without language, the opposite is not the case. Thus, language is more likely to have adapted itself to its human hosts than the other way round. Languages that are hard for humans to learn simply die out, or more likely, do not come into existence at all.

The biological perspective on language as an adaptive system has a prominent historical pedigree. Indeed, nineteenth-century linguistics was dominated by an organistic view of language (e.g., for a review see McMahon, 1994). For example, Franz Bopp, one of the founders of comparative linguistics, regarded language as an organism that could be dissected and classified (Davies, 1987). More generally, languages were viewed as having life cycles that included birth, progressive growth, procreation, and eventually decay and death. However, the notion of evolution underlying this organistic view of language was largely pre-Darwinian. This is perhaps reflected most clearly in the writings of another influential linguist, August Schleicher. Although he explicitly emphasized the relationship between linguistics and Darwinian theory (Schleicher, 1863; cited in Percival, 1987), Darwin's principles of mutation, variation, and natural selection did not enter into the theorizing about language evolution (Nerlich, 1989). Instead, the evolution of language was seen in pre-Darwinian terms as the progressive growth toward attainment of perfection, followed by decay.

More recently the biological view of language evolution was resurrected by Stevick (1963) within a modern Darwinian framework, later followed by Nerlich (1989). Christiansen (1994) proposed to view language as a kind of beneficial parasite – a non-obligate symbiant – that confers some selective advantage onto its human hosts without whom it cannot survive. Building on this work, Deacon (1997) further developed this metaphor by construing language as a virus. The asymmetry in the relationship between language and its human host is underscored by the fact that the rate of linguistic change is far greater than the rate of biological change. Whereas it takes about 10 000 years for a language to change into a completely different "species" of language (e.g., from protolanguage to present day language, Kiparsky, 1976), it took our remote ancestors approximately 100-200 000 years to evolve from the archaic form of *Homo sapiens* into the anatomically modern form, *Homo sapiens sapiens* (e.g., Corballis, 1992). Consequently, it seems more plausible that the languages of the world have been closely tailored through linguistic adaptation to fit human learning, rather than the other way around. The fact that children are so successful at language learning is therefore best explained as a product of natural selection of linguistic structures, and not as the adaptation of biological structures, such as an innately specified linguistic endowment in the form of universal grammar (UG)[1].

---

[1] Many functional and cognitive linguists also suggest that the putative innate UG constraints arise from general cognitive constraints (e.g., Givón, 1998; Hawkins, 1994;

Cognitive constraints on language evolution and acquisition

From this perspective, it is clear that there exist innate constraints guiding language learning. Indeed, a recent population dynamics model by Nowak, Komarova, and Niyogi (2001) provides a mathematical setting for exploring language acquisition under constraints (such as UG), and evolutionary competition among them. This mathematical model is based on what the authors call a "coherence threshold". In order for a population to communicate successfully, all its members must acquire the same language. The coherence threshold is a property that UG or other potential constraints must meet for them to induce "coherent grammatical communication" in the linguistic community. When the authors used this mathematical framework to compare competing systems of constraints (different UGs), they found that complexity confers a fitness advantage upon them[2]. This is offered as an explanation for the emergence of complex, rule-based languages. Although UG is the purported object of study in Nowak *et al.*, there is little to preclude extending these findings to our own perspective. The innate constraints need not be language-specific in nature for the model's assumptions to be satisfied. The important question is therefore not about the existence of innate constraints on language – we take this to be given – but rather what the nature is of such constraints.

Given our perspective on language evolution, we suggest that many of these innate constraints derive from limitations on sequential learning. By "sequential learning" we here focus on the learning of hierarchically organized structure from temporally-ordered input, in which combinations of primitive elements can themselves become primitives for further higher-level combinations. For example, consider the case of following a recipe involving mixing separately one set of ingredients in one bowl and other ingredients in another bowl before mixing the contents of the two bowls together (possibly with additional ingredients). The preparation of certain plant foods by mountain gorillas (*Gorilla g. beringei*) in Rwanda, Zaire and Uganda provides another example of complex sequential learning (Byrne and Russon, 1998). Because their favorite foods are protected by physical defenses such as spines or stings, the gorillas learn hierarchical manual sequences with repeated subpatterns in order to collect the plant material and make it edible. Although sequential learning appears to be ubiquitous across animal species (e.g., Reber, 1993), humans may be the only species with complex

---

Langacker, 1987). Our approach distinguishes itself from these linguistic perspectives in that it emphasizes the role of sequential learning in the explanation of linguistic constraints. Another difference is our general emphasis on the acquisition of language, rather than the processing of language (cf. Hawkins, 1994).

[2] Nowak *et al.* (2001) also noted that when they varied the number of sentences available to the learners, they found that intermediate values maximized fitness. They claim this provides an explanation for the critical language acquisition period. Though the model is touted as an evolutionary framework for illuminating a supposedly biological property of our species (UG), this explanation for the critical period relies on an unbiological basis. Hypotheses of critical periods involve maturational issues of the learning mechanism, not the number of sentences offered by the environment.

sequential learning abilities flexible enough to accommodate a communication system containing several layers of temporal hierarchical structure (at the level of phonology, morphology and syntax). Next we present converging evidence from studies of aphasia, ALL, studies of non-human primates, and computational modeling – all of which points to the importance of sequential learning in the evolution language.

Language and Sequential Learning

Several lines of evidence currently support the importance of sequential learning in language evolution. This evidence spans a number of different research areas, ranging from sequential learning abilities of aphasic patients to computational modeling of language evolution. When these sources are considered within the framework argued for here, they converge in support of a strong association between sequential learning and language evolution, acquisition, and processing.

Evidence from aphasia studies

The first line of evidence comes from the study of aphasia. If language and sequential learning are subserved by the same underlying mechanisms, as we have suggested here, then one would expect that breakdown of language in certain types of aphasia to be associated with impaired sequential learning and processing. A large number of Broca's aphasics suffer from agrammatism. Their speech lacks the hierarchical organization we associate with syntactic structure, and instead appears to be a collection of single words or simple word combinations. Importantly, Grossman (1980) found that Broca's aphasics, besides agrammatism, also had an additional deficit in sequentially reconstructing hierarchical tree structure models from memory. He took this as suggesting that Broca's area subserves not only syntactic speech production, but also functions as a locus for supramodal processing of hierarchically structured behavior. Another study has suggested a similar association between language and sequential processing. Kimura (1988) reported that sign aphasics often also suffer from apraxia; that is, they have additional problems with the production of novel sequential hand and arm movements not specific to sign language.

More recently, Christiansen, Kelly, Shillcock, and Greenfield (in preparation) provided a more direct test of the suggested link between breakdown of language and breakdown of sequential learning. They conducted an ALL study using agrammatic patients and normal controls matched for age, socio-economic status, and spatial reasoning abilities. Artificial language learning experiments typically involve training and testing subjects on strings generated from a small grammar. The vocabulary of these grammars can consist of letters, nonsense words, or non-linguistic symbols (e.g., shapes). Because of the underlying sequential structure of the stimuli, the experiments can serve as a window onto the relationship between the learning and processing of linguistic and sequential structure. The subjects in the Christiansen *et al.* study were trained on an artificial language using a match-mismatch pairing task in which they had to decide whether two consecutively presented symbol strings were the same or different. After training, subjects were presented with novel strings, half of which were derived from the grammar and half not. Subjects were told that the training strings were generated by a complex set of rules, and asked to classify the new strings according to whether they followed these rules or not. The results showed that although both groups did very well on the pairing task, the normal controls were significantly better at classifying the new test strings in comparison with the agrammatic aphasics. Indeed, the aphasic patients were no better than chance at classifying the test items. Thus, the study indicates that agrammatic aphasic patients have problems with sequential learning in addition to their more obvious language deficits. Together, these experimentally observed sequential learning and processing deficits associated with agrammatic aphasia point to a close connection between the learning and processing of language and complex sequential structure.

Evidence from artificial language learning experiments

Our approach hypothesizes that many of the cognitive constraints that have shaped the evolution of language are still at play in our current cognitive and linguistic abilities. If this hypothesis is correct, then it should be possible to uncover the source of some of the universal linguistic constraints in human performance on sequential learning tasks. We therefore review a series of ALL studies with normal populations as a second line of evidence for the close relationship between language and sequential learning.

The acquisition and processing of language appears to be facilitated by the presence of multiple sources of probabilistic information in the input (e.g., concord morphology and prosodic information; see contributions in Morgan and Demuth, 1996). Morgan, Meyer, and Newport (1987) demonstrated that ALL is also facilitated by the existence of multiple information sources. They exposed adults to artificial languages with or without additional cue information, such as prosodic or morphological marking of phrases. Subjects provided with the additional cue information acquired more of the linguistic structure of the artificial language. More recently, Saffran (2001) studied the learning of an artificial language with or without the kind of predictive constraints found in natural language (e.g., the presence of the determiner, *the*, is a very strong predictor of an upcoming noun). She found that both adults and children acquired more of the underlying structure of the language when it incorporated the "natural" predictive constraints. Saffran (2000) has also demonstrated that the same predictive constraint is at play when subjects are exposed to an artificial language consisting of non-linguistic sounds (e.g., drum rolls, etc.), providing further support for the non-linguistic nature of the underlying constraints. In unison with our perspective, the authors of these ALL studies suggest that human languages might contain certain sequential patterns, not because of linguistic constraints, but rather because of the general learning constraints of the human brain.

The ALL studies with normal and aphasic populations together point to a strong association between language and the learning and processing of sequential structure. The close connection in terms of underlying brain mechanisms is further underscored by recent neuroimaging studies of ALL. Steinhauer, Friederici, and Pfeifer (2001) had subjects play a kind of board game in which two players were required to communicate via an artificial language. After substantial training, event-related potential (ERP) brainwave patterns were then recorded as the subjects were tested on grammatical and ungrammatical sentences from the language. The results showed the same frontal negativity pattern (P600) for syntactic violations in the artificial language as has been found for similar violations in natural language (e.g., Osterhout and Holcomb, 1992). Another study by Patel, Gibson, Ratner, Besson, and Holcomb (1998) further corroborates this pattern of results but with non-linguistic sequential stimuli: musical sequences with target chords either within the key of a major musical phrase or out of key. When they directly compared the ERP patterns elicited for syntactic incongruities in language with the ERP patterns elicited for incongruent out-of-key target chords, they found that the two types of sequential incongruities resulted in the same, statistically indistinguishable P600 components. In a more recent study, Maess, Koelsch, Gunter, and Friederici (2001) used magnetoencephalography (MEG) to localize the neural substrates that may be involved in the processing of musical sequences. They found that Broca's area in the left hemisphere (and the corresponding frontal area in the right hemisphere) produced significant activation when subjects listened to musical sequences that included an off-key chord. The ALL studies reviewed here converge on the suggestion that the same underlying brain mechanisms are used for the learning and processing of both linguistic and non-linguistic sequential structure, and that similar constraints are imposed on both language and sequential learning.

Evidence from non-human primate studies

The perspective on language evolution presented here suggests that language to a large extent "piggy-backed" on pre-existing sequential learning and processing mechanisms, and that limitations on these mechanisms in turn gave rise to many of the linguistic constraints observed across the languages of the world. If this evolutionary scenario is on the right track, one would expect to see some evidence of complex sequential learning in our closest primate relatives – and this is exactly what is suggested by the third line of evidence that we survey here.

A review of recent studies investigating sequential learning in non-human primates (Conway and Christiansen, 2001) indicates that there is considerable overlap between the sequential learning abilities of humans and non-human primates. For instance, macaque monkeys (*Macaca mulatta* and *Macaca fascicularis*) not only are competent list-learners (Swartz, Chen and Terrace, 2000) but they appear to encode and represent sequential items by learning each item's ordinal position (Orlov, Yakovlev, Hochstein and Zohary, 2000) rather than by a simple association mechanism. In addition, cotton-top tamarins (*Saguinus oedipus*) are able to successfully segment artificial words from an auditory speech stream by relying on statistical information in a manner similar to human infants (Hauser, Newport and Aslin, 2001; Saffran, Aslin and Newport, 1996). Finally, as mentioned earlier, a group of African mountain gorillas apparently observationally learn sequences of complex and hierarchically organized manual actions to bypass the natural defenses of edible plants (Byrne and Russon, 1998). However, despite these impressive sequential learning abilities, non-human primates also display certain limitations in comparison to humans. In some tasks, non-humans need considerably longer training in order to adequately learn sequential information (cf., Lock and Colombo, 1996). More importantly, non-human subjects often display sequential learning and behavior that is less complex and less developed compared to human children and adults (e.g., Oshiba, 1997), especially with regards to the learning of hierarchical structure (e.g., Johnson-Pynn, Fragaszy, Hirsh, Brakke and Greenfield, 1999; Spinozzi and Langer, 1999). We suggest that such limitations may help explain why non-human primates have not developed complex, human-like language.

The limitations of the non-human primates on sequential learning and processing are also likely to play a role in the explanation of the limited success of the numerous ape language learning experiments. Indeed, we see these experiments as complex versions of the ALL tasks used with humans[3]. Much like some human ALL experiments, the non-human primates must learn to associate arbitrary visual symbols (lexigrams), manual signs, or spoken words with objects, actions, and events. Some of these studies have shown that apes can acquire complex artificial languages with years of extensive training. Although some of the "stars" of these experiments – such as the female gorilla Koko (Patterson, 1978) and the male bonobo Kanzi (Savage-Rumbaugh, Shanker and Taylor, 1998) – have demonstrated remarkable abilities for learning the artificial language they have been exposed to, they nevertheless also seem to experience problems with complex sequential structures. Non-human primates, in particular the apes, possess sequential learning abilities of a reasonable complexity and appear to be able to utilize these abilities in complex ALL tasks. Yet the language abilities of these apes remain relatively limited compared to those of young children. On our account, the better sequential learning and processing abilities observed in humans are likely to be the product of evolutionary changes occurring after the branching point between early hominids and the ancestors of extant apes. These evolutionary improvements in sequential learning have then subsequently provided the basis for the evolution of language.

Evidence from computational modeling

An important question for all evolutionary accounts of language pertains to the feasibility of the underlying assumptions. For example, our approach emphasizes

---

[3] Early ape language experiments attempted to teach non-human primates actual human language (e.g., Kellogg and Kellogg, 1933). The animals were spoken to and treated in a manner similar to human infants and young children. However, this approach was subsequently abandoned because of lack of success and replaced by the artificial language methodology used today.

the role of linguistic adaptation over biological adaptation in the evolution of language. As we have mentioned earlier, computational modeling provides a very fruitful means with which to test the assumptions of a given approach. As a final line of evidence in support of our perspective on language evolution we therefore review some recent modeling efforts that demonstrate its computational feasibility[4].

Several recent computational modeling studies have shown how the adaptation of linguistic structure can result in the emergence of complex languages with features very similar to what is observed in natural languages. Batali's (1998) "negotiation" model explored the appearance of systematic communication in a social group of agents in the form of simple recurrent networks (SRN; Elman, 1990). An SRN is essentially a standard feed-forward neural network equipped with an extra layer of so-called context units. At a particular time step *t*, an input pattern is propagated through the hidden unit layer to the output layer. At the next time step, *t*+1, the activation of the hidden unit layer at time *t* is copied back to the context layer and paired with the current input. This means that the current state of the hidden units can influence the processing of subsequent inputs, providing a limited ability to deal with sequentially presented input incorporating hierarchical structure. Although these network agents were not initially equipped with a system of communication, the generated sequences gradually exhibited systematicity. Batali also demonstrated that this communication system enabled the agents to convey novel meanings. Importantly, there was no "biological" adaptation (e.g., selection of better learners); instead, the communication system emerged from linguistic adaptation driven by the social interaction of agents. Kirby offered a similar account for the evolution of typological universals (Kirby, 1998), and systematic communication in agents without prior grammatical encoding (Kirby, 2000; 2001). Using abstract rule-based descriptions of individual language fragments, Kirby demonstrated that fairly complex properties of language could arise under an adaptive interpretation of linguistic selection.

Livingstone (2000) and Livingstone and Fyfe (1999) used a similar technique to show that linguistic diversity can arise from an imperfect cultural transmission of language among a spatially organized group of communicating agents. In their simulations, neural network agents, able only to communicate with others in close proximity, exhibited a dialect continuum: intelligibility was high in clusters of agents, but diminished significantly as the distance between two agents increased. In a similar simulation without such spatial distribution (where any agent is equally probable to communicate with all others), diversity rapidly converged onto a global language. This work demonstrates how linguistic diversity may arise through linguistic adaptation across a spatially distributed population of agents, perhaps giving rise to different languages over time. Some of these emergent languages are likely to be more easily accommodated by sequential learning and processing mechanisms than other languages. This sequential learnability difference is, *ceteris*

---

[4] To keep our discussion brief, we focus on the computational modeling of linguistic adaptation, side-stepping the issue of the origin of language. For simulations relevant to this perspective, see e.g., Arbib (this volume) and Parisi and Cangelosi (this volume).

paribus[5], likely to result in different frequency distributions across languages. Simulations by Van Everbroek (1999) substantiate this hypothesis. He used a variation of the SRN to investigate how sequential learning and processing limitations might be related to the distribution of the world's language types. He constructed example sentences from 42 artificial languages, varying in three dimensions: word order (e.g., subject-verb-object), nominal marking (accusative vs. ergative), and verbal marking. The networks easily processed language types with medium to high frequency, while low frequency language types resulted in poor performance. These simulations support a connection between the distribution of language types and constraints on sequential learning and processing, suggesting that frequent language types are those that have successfully adapted to these learning and processing limitations.

The computational modeling results lend support to the suggestion that the evolution of language may have been shaped by linguistic adaptation to pre-existing constraints on sequential learning and processing. When these results are viewed together with the evidence showing a breakdown of sequential learning in agrammatic aphasia, the ALL demonstrations of linguistic constraints as reflections of sequential learning limitations with similar neural substrates, and the existence of relatively complex sequential learning abilities in apes, they all appear to converge on the language evolution account we have put forward here. Next, we present two case studies that provide further evidence for the idea that constraints on sequential learning may underlie many universal linguistic constraints.

Explaining Basic Word Order Constraints

Across the languages of the world there are certain universal constraints on the way in which languages are structured and used. These so-called *linguistic universals* help explain why the known human languages only take up a small fraction of the vast space defined by the logically possible linguistic subpatterns. From the viewpoint of the UG approach to language, the universal constraints on the acquisition and processing of language are essentially arbitrary (e.g., Pinker and Bloom, 1990). That is, given the Chomskyan perspective on language, these constraints appear arbitrary because it is possible to imagine a multitude of alternative, and equally adaptive, constraints on linguistic form. For instance, Piattelli-Palmarini (1989) contends that there are no (linguistic) reasons not to form yes-no questions by reversing the word order of a sentence instead of the normal inversion of subject and auxiliary. On our account, however, these universal constraints are in most cases *not* arbitrary. Rather, they are determined predominately by the properties of the human learning and processing mechanisms that underlie our language capacity. This can explain why we do not reverse the

---

[5] Of course, other factors are likely to play a role in whether or not a given language may be learnable. For example, the presence of concord morphology may help overcome some sequential learning difficulties as demonstrated by an ALL experiment by Morgan *et al.* (1987). Nonetheless, sequential learning difficulties are hypothesized to be strong predictors of frequency in the absence of such ameliorating factors.

word order to form yes-no questions; it would put too heavy a load on memory to store a whole sentence in order to be able to reverse it.

Head-order consistency

There is a statistical tendency across human languages to conform to a form in which the head of a phrase consistently is placed in the same position – either first or last – with respect to the remaining clause material. English is considered to be a head-first language, meaning that the head is most frequently placed first in a phrase, as when the verb is placed before the object noun phrase (NP) in a transitive verb phrase (VP) such as *"eat curry"*. In contrast, speakers of Hindi would say the equivalent of *"curry eat"*, because Hindi is a head-last language. Likewise, head-first languages tend to have *pre*positions before the NP in prepositional phrases (PP) (such as *"with a fork"*), whereas head-last languages tend to have *post*positions following the NP in PPs (such as *"a fork with"*). Within the Chomskyan approach to language (e.g., Chomsky, 1986) such head direction consistency has been explained in terms of an innate module known as X-bar theory which specifies constraints on the phrase structure of languages. It has further been suggested that this module emerged as a product of natural selection (Pinker, 1994). As such, it comes as part of the UG with which every child is supposedly born. All that remains for a child to "learn" about this aspect of her native language is the direction (i.e., head-first or head-last) of the so-called head-parameter.

The evolutionary perspective that we have proposed above suggests an alternative explanation in which head-order consistency is a by-product of non-linguistic constraints on the learning of hierarchically organized temporal sequences. In particular, if recursively consistent combinations of grammatical regularities, such as those found in head-first and head-last languages, are easier to learn (and process) than recursively inconsistent combinations, then it seems plausible that recursively inconsistent languages would simply "die out" (or not come into existence), whereas the recursively consistent languages should proliferate. As a consequence languages incorporating a high degree of recursive inconsistency should be far less frequent among the languages of the world than their more consistent counterparts. In other words, languages may need to have a certain recursive consistency across their different grammatical regularities in order for the former to be learnable by learning devices with adapted sensitivity to sequential information. Languages that do not have this kind of consistency in their grammatical structure may not be learnable, and they will, furthermore, be difficult to process (cf. Hawkins, 1994).

From this perspective, Christiansen and Devlin (1997) provided an analysis of the interactions in a recursive rule set with consistent and inconsistent ordering of the heads[6]. A recursive rule set is a pair of rules for which the expansion of one rule involves the second rule, and vice versa; e.g.,

---

[6] The fact that we use rules and (later) syntactic trees to characterize the language to be acquired should not be taken as suggesting that we believe that the end-product of the

$$A \rightarrow a \, (B)$$
$$B \rightarrow b \, A$$

This analysis showed that head-order inconsistency in a recursive rule set, such as,

$$A \rightarrow a \, (B)$$
$$B \rightarrow A \, b$$

creates center-embedded constructions, whereas a consistent ordering of heads creates right-branching constructions for head-first orderings and left-branching constructions for head-last orderings. Center-embeddings are difficult to process because constituents cannot be completed immediately, forcing the language processor to keep lexical material in memory until it can be discharged. For the same reason, center-embedded structures are likely to be difficult to learn because of the distance between the material relevant for the discovery and/or reinforcement of a particular grammatical regularity. This means that recursively inconsistent rule sets are likely to be harder to learn than recursively consistent rule sets.

To explore the notion of recursive inconsistency further, Christiansen and Devlin created the grammar skeleton shown in Table 8.1. The curly brackets around the constituents on the right-hand side of rules 1–5 indicate that the order of these constituents can be either as is (i.e., head-first) or the reverse (i.e., head-last). From this grammar skeleton, it is therefore possible to produce ($2^5=$) 32 different grammars with varying degrees of head-order consistency. There are two possibilities for recursive inconsistency: (a) the PP recursive rules set (rules 1 and 2), and (b) the PossP (possessive phrase) recursive rule set (rules 4 and 5). Since a PP can occur inside both NPs and VPs, an inconsistency within this rule set was predicted to impair learning more than an inconsistency violation within the PossP recursive rule set. Grammars that involved inconsistent PP recursive rule sets were therefore assigned an inconsistency penalty of 2 and grammars with inconsistent PossP recursive rule sets a penalty of 1. The top panel of Figure 8.1 shows the predicted learning difficulty of each grammar, ranging between 0 to 3.

Table 8.1 The grammar skeleton used by Christiansen and Devlin (1997). Curly brackets indicate that the ordering of the constituents can be either as is (i.e., head-first) or in reverse (i.e., head-last), whereas parentheses indicate optional constituents.

S	→	NP VP	
NP	→	{N (PP)}	(1)
PP	→	{adp NP}	(2)
VP	→	{V (NP) (PP)}	(3)
NP	→	{N (PossP)}	(4)
PossP	→	{Poss NP}	(5)

acquisition process is a set of rules. We merely use rules and syntactic trees as convenient descriptive devices, approximating the particular grammatical regularities that we are considering.

Connectionist simulations

In order to test the hypothesis that non-linguistic constraints on sequential learning restrict the set of languages that are easily learnable, Christiansen and Devlin conducted a series of connectionist simulations in which SRNs were trained on sentences generated from each of the 32 grammars. The networks were trained to predict the next lexical category in a sentence, using sentences generated by the 32 grammars. Each unit in the input/output layers corresponded to one of seven lexical categories or an end of sentence marker: singular/plural noun (N), singular/plural verb (V), singular/plural possessive genitive affix (Poss), and adposition (adp). Although these input/output representations abstract away from many of the complexities facing language learners, they suffice to capture the fundamental aspects of grammar learning important to our hypothesis. Network performance was measured in terms of the networks' ability to predict the probability distribution of possible next items given prior sentential context. The bottom panel of Figure 8.1 shows SRN performance, averaged over 25 networks, for each of the 32 different grammars. A comparison between the top and bottom panels in Figure 8.1 reveals that the grammars that were predicted to be harder to learn because of high recursive inconsistency are the ones that the SRNs showed decreased performance on. A regression analysis confirmed this observation, showing a strong correlation between the degree of head-order consistency of a given grammar and the degree to which the network had learned to master the grammatical regularities underlying that grammar: The higher the inconsistency, the more erroneous the final network performance was. The sequential biases of the networks made the corpora generated by consistent grammars considerably easier to acquire than the corpora generated from inconsistent grammars.

This is an important result because it is not obvious that the SRNs should be sensitive to inconsistencies at the structural level. The SRN did not have any built-in linguistic biases; rather, it was designed for the learning of complex sequential structure (e.g., Cleeremans, 1993). Moreover, recall that the networks only were presented with lexical categories one at a time, and that structural information about grammatical regularities had to be induced from the way the lexical categories combine in the input. No explicit structural information was provided, yet the networks were sensitive to the recursive inconsistencies. In this connection, it is worth noting that Christiansen and Chater (1999) have shown that increasing the size of the hidden/context layers (beyond a certain minimum) does not affect SRN performance on center-embedded constructions (i.e., structures which are recursively inconsistent structures). This suggests that Christiansen and Devlin's results may not be dependent on the specific size of the SRNs they used, nor is it likely to depend on the size of the training corpus.

Typological analyses by Christiansen and Devlin using the FANAL database (Dryer, 1992) with typological information about some 625 languages further corroborated our account. Languages that incorporated fragments that the networks found hard to learn tended to be less frequent among the languages of the world compared to languages the networks learned more easily. This suggests that constraints on basic word order may derive from non-linguistic constraints on the

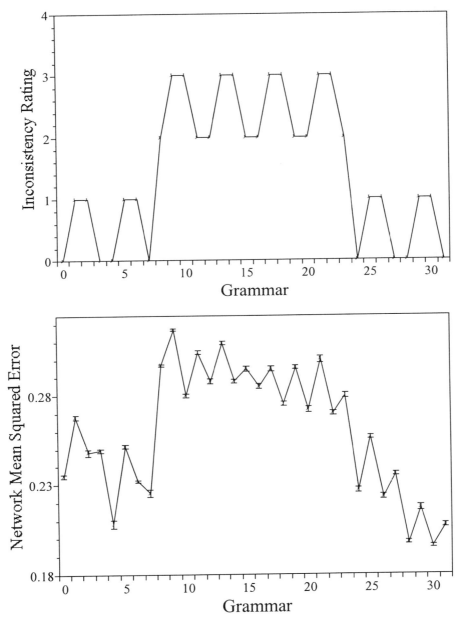

Figure 8.1 The predicted learning difficulty for the 32 grammars from Christiansen and Devlin (1997) (top panel) shown with the difficulty that the network experienced with each grammar (bottom panel). Error bars indicate standard error of the mean.

learning and processing of complex sequential structure, perhaps obviating the need for an innate X-bar module to explain such word order universals. Grammatical constructions incorporating a high degree of head-order inconsistency may be too hard to learn and would therefore tend to disappear.

Artificial language learning experiment

The final set of evidence supporting our explanation of basic word order universals comes from a recent ALL study by Christiansen (2000). In one experiment, Christiansen took two of the grammars that Christiansen and Devlin had used for their network simulation – a consistent and an inconsistent grammar (see Table 8.2) – and trained subjects on sentences (represented as consonant strings) derived from the two grammars. Training and test materials were controlled for length and differences in the distribution of bigram and trigram fragments. In the training phase of the experiment, subjects read and reproduced consonant strings on a computer. After training, subjects were informed that the strings were generated by a complex set of rules, and that they would be presented with additional strings; some of which were generated by the same rule set (i.e., grammatical), and some which were not (i.e., ungrammatical). They were then asked to decide which of the new strings were generated by the same rule set as before, and which were not. The results showed that the subjects trained on strings from the consistent grammar were significantly better at distinguishing grammatical from ungrammatical items than the subjects trained on the inconsistent grammar.

Together, Christiansen's ALL experiment and the three sets of evidence from Devlin and Christiansen converge in support of our claim that basic word order universals (head-ordering) can be explained in terms of non-linguistic constraints on sequential learning and processing. This research thus suggests that universal word order correlations may emerge from non-linguistic constraints on learning, rather than being a product of innate linguistic knowledge. In the next section, we show how constraints on complex question formation may be explained in a similar manner.

Table 8.2 The two grammars used for stimuli generation in Christiansen (2000). The vocabulary is: {X, Z, Q, V, S, M}

Consistent Grammar			Inconsistent Grammar		
S	→	NP VP	S	→	NP VP
NP	→	(PP) N	NP	→	(PP) N
PP	→	NP post	PP	→	pre NP
VP	→	(PP) (NP) V	VP	→	(PP) (NP) V
NP	→	(PossP) N	NP	→	(PossP) N
PossP	→	NP Poss	PossP	→	Poss NP

Subjacency through Linguistic Adaptation

According to Pinker and Bloom (1990), subjacency is one of the classic examples of an arbitrary linguistic universal that makes sense only from a linguistic perspective. Subjacency provides constraints on complex question formation. Informally, "Subjacency, in effect, keeps rules from relating elements that are 'too far apart from each other', where the distance apart is defined in term of the number of designated nodes that there are between them" (Newmeyer, 1991, p. 12). Consider the following sentences:

1. Sara heard (the) news that everybody likes cats.
   N   V   N comp   N   V   N

2. What (did) Sara hear that everybody likes?
   Wh   N   V comp   N   V

3. *What (did) Sara hear (the) news that everybody likes?
   Wh   N   V   N   comp   N   V

According to the subjacency principle, sentence 3 is ungrammatical because too many boundary nodes are placed between the noun phrase complement (NP-Comp) and its respective "gaps".

The subjacency principle, in effect, places certain restrictions on the ordering of words in complex questions. The movement of Wh-items (*what* in Figure 8.2) is limited with respect to the number of so-called bounding nodes that it may cross during its upward movement. In English, the bounding nodes are S and NP (circled in Figure 8.2). Put informally, as a Wh-item moves up the tree it can use comps as temporary "landing sites" from which to launch the next move. The subjacency principle states that during any move only a single bounding node may be crossed. Sentence 2 is therefore grammatical because only one bounding node is crossed for each of the two moves to the top comp node (Figure 8.2, top panel). Sentence 3 is ungrammatical, however, because the Wh-item has to cross two bounding nodes – NP and S – between the temporary comp landing site and the topmost comp, as illustrated in bottom panel of Figure 8.2.

Not only do subjacency violations occur in NP-complements, but they can also occur in Wh-phrase complements (Wh-Comp). Consider the following examples:

4. Sara asked why everyone likes cats.
   N   V   Wh   N   V   N

5. Who (did) Sara ask why everyone likes cats?
   Wh   N   V Wh   N   V   N

6. *What (did) Sara ask why everyone likes?
   Wh   N   V Wh   N   V

According to the subjacency principle, sentence 6 is ungrammatical because the interrogative pronoun has moved across too many bounding nodes (as was the case in 3).

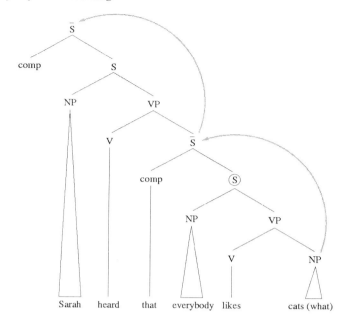

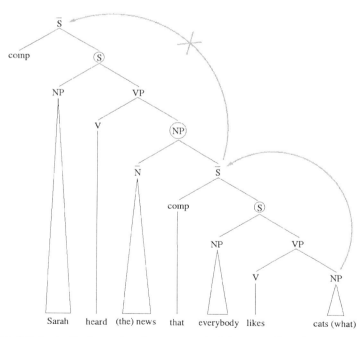

Figure 8.2 Syntactic trees showing grammatical (top panel) and ungrammatical (bottom panel) movement.

Table 8.3 The structure of the natural and unnatural languages in Ellefson and Christiansen (2000). The vocabulary is: {X, Z, Q, V, S, M}

Natural	Unnatural
N V N	N V N
Wh N V	Wh N V
N V N comp N V N	N V N comp N V N
N V Wh N V N	N V Wh N V N
Wh N V comp N V	*Wh N V N comp N V
Wh N V Wh N V N	*Wh N V Wh N V

Artificial language learning experiment

Ellefson and Christiansen (2000) explored an alternative explanation, suggesting that subjacency violations are avoided, not because of a biological adaptation incorporating the subjacency principle, but because language itself has undergone adaptations to root out such violations in response to non-linguistic constraints on sequential learning. They created two artificial languages to test this idea. As shown in Table 8.3, both languages consisted of six sentence types of which four were identical across the two languages. The two remaining sentence types involved complex question formation. In the natural language the two complex questions were formed in accordance with subjacency, whereas the two complex questions in the unnatural language violated the subjacency constraints. All training and test items were controlled for length and fragment information. As in the previous ALL experiment, subjects were not told about the linguistic nature of the stimuli until they received the instructions for the test phase.

The results showed that the subjects trained on the natural language had learned the language significantly better than the subjects trained on the unnatural language. Subjects in the natural condition performed marginally better than the subjects in the unnatural condition at classifying strings related to the two complex questions. Interestingly, the natural group was significantly better at classifying the remaining four sentence types in comparison with the unnatural group – despite the fact that both groups were trained on exactly the same items and saw exactly the same test items. The presence of the two unnatural question formation sentence types affected the learning of the other four test items. In other words, the presence of the subjacency violations in two of the sentence types in the unnatural language affected the learning of the language as a whole, not just the two complex question items. From the viewpoint of language evolution, languages such as this unnatural language would lose out in competition with other languages such as the natural language because the latter is easier to learn.

Connectionist simulations

In principle, one could object that the reason why Ellefson and Christiansen found differences between the natural and the unnatural groups is because the former in some way was able to tap into an innately specified subjacency principle when learning the language. Another possible objection is that the natural language follows the general pattern of English whereas the unnatural language does not, and that our human results could potentially reflect an "English effect" . To counter these possible objections, and to support the suggestion that the difference in learnability between the two languages is brought about by constraints arising from sequential learning, Ellefson and Christiansen conducted a set of connectionist simulations of the human data using SRNs – a sequential learning device that clearly does not have subjacency constraints built-in. They used one network for each subject, and found that the networks were significantly better at learning the natural language in comparison with the unnatural language. Thus, the simulation results closely mimicked the ALL results, corroborating the suggestion that constraints on the learning and processing of sequential structure may explain why subjacency violations tend to be avoided: These violations were weeded out because they made the sequential structure of language too difficult to learn. Thus, rather than having an innate UG principle to rule out subjacency violations, we suggest that they may have been eliminated altogether through linguistic adaptation.

General Discussion

In this chapter we have argued for a view of language evolution, acquisition, and processing that places these phenomena within the more general domain of sequential learning. We hypothesize that constraints on sequential learning help define a cognitive niche within which languages have had to adapt. A considerable amount of evidence that supports this view has been discussed. ALL studies of normal human subjects illuminate the importance of complexity and consistency in learning artificial languages. From our perspective, the experiments suggest that languages have evolved these and other properties to facilitate learning. Over time this process of linguistic adaptation has resulted in the structural constraints on language use that we observe today. The association between sequential learning and natural language is further evidenced by ALL experiments that demonstrate the accompanying breakdown of sequential skills in agrammatics (Christiansen *et al.*, in preparation). Related artificial language experiments have also demonstrated that non-human primates can achieve relative proficiency in complex sequential tasks. This is not surprising in our view, since the fundamental role of sequential learning implies a long phylogenetic history: Primate studies in natural contexts have readily provided evidence of these complex sequential abilities.

In a similar vein, computational models allow researchers to explore and test hypotheses about factors contributing to language evolution in maximally controlled circumstances. Many of the models discussed in this paper have incorporated sequential learning mechanisms that shape language evolution. Put simply, these models can be viewed as investigations into the constraints on language imposed by sequential learning in a social environment. The results of these computational efforts dovetail with our view. We discussed Christiansen and Devlin's (1997) simulation, which showed how constraints on sequential learning can explain basic word order constraints. Also, Ellefson and Christiansen (2000), provided an explanation, based on sequential learning constraints, for why subjacency violations tend to be avoided across the languages of the world. Together, the results from all these computational models suggest that constraints arising from general cognitive processes, such as sequential learning and processing, are likely to play a larger role in language evolution than has traditionally been assumed. What we observe today as linguistic universals may be stable states that have emerged through an extended process of linguistic evolution.

As is customary in many scientific endeavors, the sources of evidence reported here abstracts away from many potentially important details. First, ALL experiments provide a highly idealized window into real language acquisition and processing. The artificial languages are highly simplified and often lack the social context in which language is normally acquired. Nonetheless, ALL studies have yielded important insights into the acquisition of language (for a review, see Gomez and Gerken, 2000). The computational modeling of language evolution shares many of the same limitations as ALL. For example, most of the modeled communication systems are equally simplified and embedded within a social context that is often reduced to a collection of abstract semantic features. However, we see the limitations on current use of ALL and computational modeling as unavoidable growing pains associated with a field very much in its infancy. Both lines of research methodologies are essential sources of information for furthering our understanding of language evolution. Their limitations underscore the necessity of an interdisciplinary approach. Only by amassing converging evidence from multiple lines of investigation can evolutionary hypotheses be supported, or discarded. Computational modeling and ALL experiments, rendered more sophisticated and naturalistic, hold considerable promise as two essential sources of evidence for studying language evolution.

References

Aitchison J (1998) On discontinuing the continuity-discontinuity debate. In: Hurford J, Studdert-Kennedy M, Knight C (eds) (1998) *Approaches to the evolution of language.* Cambridge University Press, Cambridge UK, pp 17-29

Batali J (1998) Computational simulations of the emergence of grammar. In: Hurford J, Studdert-Kennedy M, Knight C (eds) (1998) *Approaches to the evolution of language.* Cambridge University Press, Cambridge UK, pp 405-426

Burnett J (1773) *The origin and process of language (vols I-IV).* A Kincaid, Edinburgh

Byrne RW, Russon AE (1998) Learning by imitation: A hierarchical approach. *Behavioral and Brain Sciences*, 21: 667-721

Chomsky N (1986) *Knowledge of language.* Praeger, New York

Christiansen MH (1994) *Infinite languages finite minds: Connectionism learning and linguistic structure.* Unpublished doctoral dissertation, Centre for Cognitive Science, University of Edinburgh UK

Christiansen MH (2000) Using artificial language learning to study language evolution: Exploring the emergence of word order universals. In: Dessalles JL, Ghadakpour L (eds)

The Evolution of Language: 3rd International Conference. Ecole Nationale Superieure des Telecommunications, Paris, pp 45-48

Christiansen MH, Chater N (1999) Toward a connectionist model of recursion in human linguistic performance. *Cognitive Science*, 23: 157-205

Christiansen MH, Chater N (in preparation) *Language as an organism: Language evolution as the adaptation of linguistic structure.* Manuscript in preparation

Christiansen MH, Devlin JT (1997) Recursive inconsistencies are hard to learn: A connectionist perspective on universal word order correlations. In: *Proceedings of the 19th Annual Cognitive Science Society Conference.* Lawrence Erlbaum Associates, Mahwah NJ, pp 113-118

Christiansen MH, Kelly L, Shillcock RC, Greenfield K (in preparation) *Artificial grammar learning in agrammatism.* Manuscript in preparation

Cleeremans A (1993) *Mechanisms of implicit learning: Connectionist models of sequence processing.* MIT Press, Cambridge MA

Conway C, Christiansen MH (2001) *Sequential learning in non-human primates.* Manuscript submitted for publication

Corballis MC (1992) On the evolution of language and generativity. *Cognition*, 44: 197-226

Davies AM (1987) 'Organic' and 'organism' in Franz Bopp. In: Hoenigswald HM, Wiener LF (eds) *Biological metaphor and cladistic classification.* University of Pennsylvania Press, Philadelphia PA, pp 81-107

Deacon TW (1997) *The symbolic species: The co-evolution of language and the brain.* WW Norton, New York

Dryer MS (1992) The Greenbergian word order correlations. *Language*, 68: 81-138

Ellefson MR, Christiansen M H (2000) Subjacency constraints without universal grammar: Evidence from artificial language learning and connectionist modeling. In: *The Proceedings of the 22nd Annual Conference of the Cognitive Science Society*, Erlbaum, Mahwah NJ, pp 645-650

Elman JL (1990) Finding structure in time. *Cognitive Science*, 14: 179-211

Givón T (1998) On the co-evolution of language, mind and brain. *Evolution of Communication*, 2: 45-116

Gomez RL, Gerken L (2000) Infant artificial language learning and language acquisition. *Trends in Cognitive Science*, 4(5): 178-186

Grossman M (1980) A central processor for hierarchically structured material: Evidence from Broca's aphasia. *Neuropsychologia*, 18: 299-308

Hauser MD, Newport EL, Aslin RN (2001) Segmentation of the speech stream in a non-human primate: Statistical learning in cotton-top tamarins. *Cognition*, 78: 53-64

Hawkins JA (1994) *A performance theory of order and constituency.* Cambridge University Press, Cambridge UK

Johnson-Pynn J, Fragaszy DM, Hirsh EM, Brakke KE, Greenfield PM (1999) Strategies used to combine seriated cups by chimpanzees (Pan troglodytes), bonobos (Pan paniscus) and capuchins (Cebus apella). *Journal of Comparative Psychology*, 113: 137-148

Kellogg WN, Kellogg LA (1933) *The ape and the child: A study of environmental influence on early behavior.* Hafner, New York

Kimura D (1988) Review of what the hands reveal about the brain. *Language and Speech*, 31: 375-378

Kiparsky P (1976) Historical linguistics and the origin of language. *Annals of the New York Academy of Sciences*, 280: 97-103

Kirby S (1998) Fitness and the selective adaptation of language. In: Hurford J, Studdert-Kennedy M, Knight C (eds) (1998) *Approaches to the evolution of language.* Cambridge

University Press, Cambridge UK, pp 359-383

Kirby S (2000) Language evolution without natural selection: From vocabulary to syntax in a population of learners. In: Knight C, Hurford JR, Studdert-Kennedy M (eds) *The evolutionary emergence of language: Social function and the origins of linguistic form.* Cambridge University Press, Cambridge UK, pp 303-323

Kirby S (2001) Spontaneous evolution of linguistic structure: an iterated learning model of the emergence of regularity and irregularity. *IEEE Transactions on Evolutionary Computation and Cognitive Science*, 5(2): 102-110

Langacker RW (1987) *Foundations of cognitive grammar: Theoretical perspectives Vol 1.* Stanford University Press, Stanford CA

Livingstone D (2000) A modified-neural theory for the evolution of linguistic diversity. Paper presented at *The Evolution of Language: 3rd International Conference*, Paris, France

Livingstone D, Fyfe C (1999) Modelling the evolution of linguistic diversity. In In: Floreano D, Nicoud JD, Mondada F (eds) *Proceedings of ECAL99 the Fifth European Conference on Artificial Life (Lecture Notes in Artificial Intelligence)* Springer-Verlag, Berlin, pp 704-708

Lock A, Colombo M (1996) Cognitive abilities in a comparative perspective. In: Lock A, Peters CR (eds) *Handbook of human symbolic evolution.* Clarendon Press, Oxford, pp 596-643

Maess B, Koelsch S, Gunter TC, Friederici AD (2001) Musical syntax is processed in Broca's area: An MEG study. *Nature Neuroscience*, 4: 540-545

McMahon AMS (1994) *Understanding language change.* Cambridge University Press, Cambridge UK

Morgan JL, Demuth K (eds) (1996) *Signal to syntax: Bootstrapping from speech to grammar in early acquisition.* Lawrence Erlbaum, Mahwah, NJ

Morgan JL, Meier RP, Newport EL (1987) Structural packaging in the input to language learning: Contributions of prosodic and morphological marking of phrases to the acquisition of language. *Cognitive Psychology*, 19: 498-550

Nerlich B (1989) The evolution of the concept of 'linguistic evolution' in the 19th and 20th century. *Lingua*, 77: 101-112

Newmeyer F (1991) Functional explanation in linguistics and the origins of language. *Language and Communication*, 11: 3–28

Nowak MA, Komarova NL, Niyogi P (2001) Evolution of universal grammar. *Science*, 291: 114-118

Oshiba N (1997) Memorization of serial items by Japanese monkeys a chimpanzee and humans. *Japanese Psychological Research*, 39: 236-252

Orlov T, Yakovlev V, Hochstein S, Zohary E (2000) Macaque monkeys categorize images by their ordinal number. *Nature*, 404: 77-80

Osterhout L, Holcomb PJ (1992) Event-related brain potentials elicited by syntactic anomaly. *Journal of Memory and Language*, 31:785-806

Patel AD, Gibson E, Ratner J, Besson M, Holcomb PJ (1998) Processing syntactic relations in language and music: An event-related potential study. *Journal of Cognitive Neuroscience*, 10: 717-733

Patterson FG (1978) The gestures of a gorilla: Language acquisition in another pongid. *Brain and Language*, 5: 72-97

Percival WK (1987) Biological analogy in the study of languages before the advent of comparative grammar. In Hoenigswald HM, Wiener LF (eds) *Biological metaphor and cladistic classification.* University of Pennsylvania Press, Philadelphia PA, pp 3-38

Piattelli-Palmarini M (1989) Evolution selection and cognition: From "learning" to parameter setting in biology and in the study of language. *Cognition*, 31: 1-44

Pinker S (1994) *The language instinct: How the mind creates language*. William Morrow, New York

Pinker S, Bloom P (1990) Natural language and natural selection. *Behavioral and Brain Sciences*, 13: 707-727

Reber AS (1993) *Implicit learning and tacit knowledge: An essay on the cognitive unconscious*. Oxford University Press, New York

Saffran JR (2000) Non-linguistic constraints on the acquisition of phrase structure. In: Gleitman LR, Joshi AK (eds) *Proceedings of the 22nd Annual Conference of the Cognitive Science Society*. Erlbaum, Mahwah NJ, pp 417-422

Saffran JR (2001) The use of predictive dependencies in language learning. *Journal of Memory and Language*, 44: 493-515

Saffran JR, Aslin RN, Newport EL (1996) Statistical learning by 8-month-old infants. *Science*, 274: 1926-1928

Savage-Rumbaugh ES, Shanker SG, Taylor TJ (1998) *Apes language and the human mind*. Oxford University Press, New York

Schleicher A (1863) *Die Darwinsche Theorie und die Sprachwissenschaft*. Böhlau, Weimer

Spinozzi G, Langer J (1999) Spontaneous classification in action by a human-enculturated and language-reared bonobo (Pan paniscus) and common chimpanzees (Pan troglodytes). *Journal of Comparative Psychology*, 113 : 286-296

Steinhauer K, Friederici AD, Pfeifer E (2001) ERP recordings while listening to syntax errors in an artificial language: Evidence from trained and untrained subjects. Poster presented at the *14th Annual CUNY Conference on Human Sentence Processing*, Philadelphia PA

Stevick RD (1963) The biological model and historical linguistics. *Language*, 39: 159-169

Swartz KB, Chen S, Terrace HS (2000) Serial learning by rhesus monkeys: II. Learning four-item lists by trial and error. *Journal of Experimental Psychology: Animal Behavior Processes*, 26: 274-285

Van Everbroeck E (1999) Language type frequency and learnability: A connectionist appraisal. In: *Proceedings of the 21st Annual Cognitive Science Society Conference*. Erlbaum, Mahwah NJ, pp 755-760

Webb J (1669) *An historical essay endeavouring the probability that the language of the empire of china is the primitive language*. London

Part IV

GROUNDING OF LANGUAGE

Chapter 9

Symbol Grounding and the Symbolic Theft Hypothesis

Angelo Cangelosi, Alberto Greco and Stevan Harnad

The Origin and Grounding of Symbols

Scholars studying the origins and evolution of language are also interested in the general issue of the evolution of cognition. Language is not an isolated capability of the individual, but has intrinsic relationships with many other behavioral, cognitive, and social abilities. By understanding the mechanisms underlying the evolution of linguistic abilities, it is possible to understand the evolution of cognitive abilities. Cognitivism, one of the current approaches in psychology and cognitive science, proposes that symbol systems capture mental phenomena, and attributes cognitive validity to them. Therefore, in the same way that language is considered the prototype of cognitive abilities, a symbol system has become the prototype for studying language and cognitive systems. Symbol systems are advantageous as they are easily studied through computer simulation (a computer program is a symbol system itself), and this is why language is often studied using computational models.

A symbol system is made up by a set of arbitrary "physical tokens" (i.e., symbols) that can be manipulated on the basis of explicit rules (i.e., syntax). Some of the main properties of such a symbol system are: (a) *compositeness*, that is symbols and rules can be recursively composed; and (b) *semantic interpretability*, specifying that the entire system and its parts can be systematically assigned a meaning (Pylyshyn, 1984; Harnad, 1990). Some significant issues arise when studying such symbol systems as a direct metaphor and model of language. These will also have direct implications for the study of the origins and evolution of language. The first issue is to establish exactly what a symbol is, by giving a clear and unambiguous definition of it. Subsequently, the process of how symbols take their meanings needs to be understood, for example by studying the symbol grounding problem. Finally, questions regarding the evolution of symbols and symbol manipulation abilities need to be addressed.

Definition of a symbol

The definition of a symbol is a yet open and highly debatable issue. Although it is possible to give a precise definition of a symbol in a computational symbol system, it is more difficult when we use this term in the context of language and communication systems. Historically, a semiotic distinction was made between the different constituents of a communication system: icons, indices, and symbols. This distinction, originally introduced by Peirce (1978), is based on the type of reference existing between objects and components of a communication system. Peirce's distinction between icons, indices, and symbols is based on the fact that (1) an "icon" is associated with an object because of its physical resemblance to it, (2) an "index" is associated with an object because of time/space contiguity, and finally (3) a "symbol" is associated with an object due to social convention or implicit agreement and it has an arbitrary shape, with no resemblance to its referent.

Recently, similar distinctions have been proposed. For instance, Deacon (1997) uses a hierarchy of referencing systems based on icons, indices, and symbols. He distinguishes three types of relationships between the means of communication and their referents in the external world and/or in the same communication system. Icons have associations with entities in the world because of stimulus generalization and conventional similarity. Indices are associated with world entities by spatio-temporal correlation or part-whole contiguity. These indexical references are commonly used in animal communication systems. Symbols are characterized by the fact that they have double referential relationships. One type of relationship is based upon the indexical link of a symbol with a referent in the world. The second type of association connects logical and combinatorial relationships with other symbols. For example, in English the verb "to give" is a symbol because it refers to an action, and as a verb it is also associated with nouns that can be used as subject, nouns that can be used as patients, etc. Deacon's definition of symbols is not restricted to language, although symbols express their best potentials in language. There are non-linguistic symbolic tasks, such as the ability to combine elements together using logical combination rules, and general mathematical tasks.

Harnad (1990) distinguishes between three types of mental representations: iconic, categorical, and symbolic. The first two are internal to the individual and non-symbolic. Iconic representations are analogical representations of the proximal sensory projections of distal objects and events. Categorical representations are learned (or innate) feature-detectors that pick out the invariant features of object and event categories from their sensory projections. Elementary symbols are the names of these objects and event categories, assigned on the basis of their non-symbolic categorical representations. Higher-order symbolic representations, grounded in these elementary symbols, consist of symbol strings (i.e., propositions) mainly describing category membership relationships.

These more recent definitions of symbols share the fact that the real symbolic feature of a communication system relies on the fact that each symbol is part of a wider and more complex system. This system is mainly regulated by compositional rules, such as syntax. In this chapter we will use this characterization of symbols, and in particular we will focus on grounded symbols.

The symbol grounding problem

The symbol systems' property of systematic semantic interpretability implies that any part of the system, and the whole system itself, can be assigned a meaning. Therefore a fundamental question must be asked: How is a symbol given a meaning? This is the problem of symbol grounding. The type of link that exists between symbols and objects is of central importance when using symbol systems as models of language and cognition. Cognitivists avoid this problem by ignoring it or trivializing it. They claim that the autonomous functional module of the symbol system will lately be connected to peripheral devices, in order to see the world of objects to which the symbols refer (Fodor, 1976). In practice, cognitivists often resolve this problem by creating their computational models with another level of yet-to-be-grounded "semantic symbols" that supposedly stand for objects, events, and state of affairs in the world. For example, in a cognitivist model of language it is sufficient to define the set of basic symbols/words (e.g., "John", "Mary", "loves") and some syntactic rules to connect them. Subsequently, each basic symbol will be assigned a meaning (e.g., the meaning of "John" is "the-boy-with-blue-eyes"). This approach is subject to the problem of infinite regression: where does the meaning of the meaningless yet-to-be-grounded "semantic symbols" ("the", "boy", "with", "blue", "eye") come from? It is not enough to have simply a parasitic link of symbols with the meanings in our heads.

This situation is similar to the paradox of the Chinese Room argument (Searle, 1982; Harnad, 1990). Suppose you don't speak Chinese and you are given the task of replying to some questions asked to you in Chinese. If you used a Chinese–Chinese dictionary alone, you could try to solve this task by looking at the symbols defining the Chinese query words and using them to select (i.e., to chain) a new set of Chinese symbols for your answer. In reality, this trip through the dictionary would amount to a merry-go-round, passing endlessly from one meaningless symbol or symbol-string (the *definientes*) to another (the *definienda*), never coming to a halt on what anything meant (Harnad, 1990). Even if we you were able to do this, you would still not have understood Chinese the same way you understand the meaning of English words. In order to use and fully understand Chinese, it is essential that you link (i.e., ground) at least some essential words to your native language[1]. Therefore, we cannot use this task as an experiment for studying

---

[1] There is a hard version of the Symbol Grounding problem in which the user of a Chinese–Chinese dictionary does not previously know any other language. This person would have to

Chinese linguistic abilities, nor as an experiment of general linguistic abilities. For the same reason, we cannot use a non-grounded symbol system as a model of linguistic and cognitive abilities.

In order to address the problem of symbol grounding, and to propose workable and plausible solutions, a model needs to include an intrinsic link between at least some basic symbols and some objects in the world. A system must use symbols that are directly grounded through cognitive representations, such as categories. This way symbol manipulation can be constrained and governed not by the arbitrary shapes of the symbol tokens, but by the non-arbitrary shapes of the underlying cognitive representations.

The evolutionary origin of symbols

In language origin research it is important to loot at the issues of the evolutionary acquisition of symbol manipulation abilities and their role in the evolution of language. Deacon (1997) has proposed an integrated neural and cognitive theory of the evolution of symbolic and linguistic abilities. His explanation of the origin of language is based on the evolution of his hierarchical referencing system. This theory relies on the symbol acquisition problem. Under normal circumstances[2], only humans have an ability to acquire symbols and language. Animal communication systems are only based on indexical references, i.e., simple object-signal associations. These associations are mostly innate (e.g., monkeys' calls) and can be explained by mere mechanisms of rote learning and conditional learning. Instead, the symbolic associations of human languages have double references, one between the symbol and the object, and the second between the symbol itself and other symbols[3]. These associations between symbols are reflected by the syntactic rules of human languages. When a complex set of logical and syntactical relationships exist between symbols, we can call these "words" and distinguish grammatical classes of words. A language-speaking individual knows that a word refers to an object and implicitly knows that the same word has grammatical relationships with other words. This combinatorial interrelationship between words can lead to an exponential growth of references. When a new word is learned, it can be combined with other pre-existing words to exponentially increase the overall number of meanings that can be expressed.

Deacon (1997) also gives a neural explanation for this distinctive difference between non-symbolic communication in animals and symbolic languages in humans. He uses neurodevelopmental and neuropsychological data to show that

---

solve the task of connecting Chinese symbols between themselves, and also the task of learning to associate meaning to symbols (Harnad, 1990)

[2] Deacon admits that under specific experimental circumstances, some species of animals, mainly apes, can acquire some type of symbol manipulation abilities. For example, in ape language experiments the acquisition of symbolic communication systems has been shown (Savage-Rumbaugh and Rumbaugh, 1978).

[3] Deacon's use of the term "reference" for the association between symbols has been highly criticized (Hurford, 1998). In semiotics, reference is mainly used to indicate the association between a symbol and the entity it *refers* to.

the enlarged prefrontal cortex in humans allows them extra processing abilities, such as symbol acquisition and symbol manipulation abilities.

In the next sections we will present a theoretical and computational framework for explaining the cognitive mechanisms for symbol grounding and symbol acquisition. In the second part of the chapter we will present a model for the evolution of language based on grounded symbols.

Cognitive Theories and Models for Symbol Grounding and Symbol Acquisition

This section describes a cognitive theory that explains the mechanisms for symbol grounding. It is based on a general psychological theory that sees our basic ability to build categories of the world as the groundwork for language and cognition (Harnad, 1987). This theory focuses on hierarchical mental representations and their role in grounding language. It starts from the principle that symbolic representations must be grounded bottom-up in non-symbolic iconic and categorical representations. This system of hierarchical representations has significant advantages. It restricts the problem of direct symbol grounding to a smaller set of elementary symbols. Any combination of these symbols, through syntactic rules, will inherit the semantic grounding from its low-level elementary symbols. Consider the case of learning a new concept from a pure linguistic definition of it. Let's suppose that you do not know what a zebra is, but are familiar with what horse and stripe patterns look like, because you have seen many real horses and striped patterns. You also know two symbols (names): "horse" for the category of horses and "stripe" for the category of striped patterns. Suppose that the following linguistic definition of zebra: `"zebra"="horse"+"stripe"` is introduced. You can immediately understand that the symbol "zebra" must correspond to a combination of (the categorical representation of) horses with (the categorical representation of) stripes. Moreover, when you see an individual zebra, you will be able to identify it as a member of the linguistically learned category zebra. This example shows how easy it is to learn new categories and new grounded names through the combination of the directly grounded names of basic categories. You would be able to fully understand any English sentence simply by knowing a relatively low number of English words[4] and by using an English–English dictionary to look up unknown words (i.e., grounding the meaning of new words in the known basic words used as definitions).

The cognitive mechanism at the core of this hierarchy of representations and bottom-up groundings is categorical perception. Our ability to build categories results in categorical representations that are a "warped" transformation of iconic representations. This feature filtering ability compresses within-category differences and expands between-category distances in similarity space so as to allow a reliable category boundary to separate members from non-members. Categorical perception consists of this compression/expansion effect (Harnad

---

[4] About 2000 words is the size of the vocabulary of an average English speaking person.

1987). It has been shown to occur in both human subjects (Goldstone 1994; Andrews, Livingstone and Harnad, 1998; Pevtzow and Harnad 1998) and neural networks (Harnad, Hanson and Lubin, 1991; 1995; Tijsseling and Harnad 1997) during the course of category learning.

Connectionism, a recent theoretical and methodological development in psychology and cognitive sciences, proposes the use of artificial neural networks as cognitive models. Neural network models are based on some general structural and functional properties of the brain and permit the modeling of behavioral and cognitive tasks, such as categorization and language (Rumelhart and McClelland, 1986). Various neural networks have proved particularly good at tasks that require the classification of input patterns into separate categories. More neural network models of language have also been developed (Christiansen and Chater, 1999; Elman, 1990). Therefore, connectionism is the natural candidate for learning the invariant features underlying categorical representations and connecting names to the proximal projections of the distal objects they stand for (Harnad, 1990). In this way connectionism can be seen as a complementary component in a hybrid non-symbolic/symbolic model of the mind. Such a hybrid model would not necessarily need an autonomous symbolic module. The symbolic functions could emerge as a consequence of the bottom-up grounding of categories' names in their sensory representations. In this way symbol manipulation would be governed not only by the arbitrary shapes of the symbol tokens, but also by the non-arbitrary shapes of the icons and category invariants in which they are grounded.

In the next two sections we will describe some connectionist models for the phenomena of categorical perception and symbol grounding. Subsequently, we will focus on models of the acquisition of language in which lexicons are directly grounded into sensory and categorical representations.

Models of categorical perception and symbol grounding

We have argued that in a plausible cognitive model of symbol origin, symbolic activity should be conceived as some higher-level process, which is not stand-alone but takes its raw material from non-symbolic representations, i.e., analogue sensori-motor (iconic) in the first instance and then categorical representations. This shift from non-symbolic to symbolic processes is one of the most fascinating aspects to be explained when considering language origins. In this section, we provide a detailed description of the mechanisms for the transformation of categorical perception (CP) into grounded low-level labels, and subsequently into higher-level symbols. Finally, we will describe how new symbols can be acquired from just the combination of already-grounded symbols, a phenomenon called grounding transfer. We shall also show how all such processes can be implemented into a single neural network model.

Neural networks can readily discriminate between sets of stimuli, extract similarities, and categorize. More importantly, networks exhibit the basic CP effect, whereby members of the same category "look" more similar (there is a compression of within-category distances) and members of different categories look more different (expansion of between-categories distances). One of the early models of CP is Harnad, Hanson and Lubin (1991, 1995). They trained three-layer feed-forward networks to sort lines into categories according to their length. Such lines were represented by 8 input units using two basic coding schemes, iconic (e.g., a length-4 line could be coded as "11110000") vs. positional (e.g., the same line coded as "00010000"). Single bit values could also be more or less discrete (coarse representations such as .1 for 0, or .9 for 1 were used, and in some cases boundaries were enhanced by using more distant values for opposite adjacent units). Given that CP is defined as a decrease in within-category inter-stimulus distances and an increase in between-category distances, a baseline for assessing such decreasing or increasing movements is required. The first step in this simulation is simply to allow networks to "discriminate" between different stimuli (to tell pairs of stimuli apart) using a pre-categorization task with auto-associative learning (i.e., networks were trained to produce exactly the same input pattern in the output units). The hidden unit activation vectors were examined and the baseline distances were calculated for each pair of input patterns. After this task the networks were finally trained to sort lines into three categories (short, middle, long) using the back-propagation algorithm.

Such networks not only exhibit successful categorization, which – as we said – is a relatively easy task for neural networks, but they also exhibit the same natural side-effect revealed by human categorization, i.e., CP. In other words, within-category compression and between-categories expansion can be observed both in humans and networks. Another point of interest from CP simulation is that a close scrutiny of hidden representations allows us to propose hypotheses about the factors upon which CP is based. Harnad, Hanson and Lubin (1995) found that the distances between hidden unit pattern representations are already maximized during auto-association (by effect of the baseline discrimination): this could be one source of the maximal interstimulus separation in CP. This separation, however, is not always so clear-cut as to allow linear separability[5], which is a clear-cut categorization, so in some cases there are "bad" or unclear representations, which happen to be close to the plane separating the categories. The back-propagation algorithm, which simulates category learning through supervised feedback, has the effect of "pushing" such unclear representations away from this plane. The result is an improved separation between categories and, at the same time, a smaller distance between representations for the same category; in other words, the CP effect.

Cases where linear separability between categories is attained more easily are not random, but this effect is mostly observed with iconic stimuli. Tijsseling and Harnad (1997), who replicated these results, suggested that CP is strongly related to factors, like the similarity between stimuli. This can lead to different possibilities for the linear separability of representations resulting from simple discrimination (the auto-association phase in the described simulations). When there is either extreme nonseparability or extreme separability, the CP effect is not

---

[5] Given a space where points represent stimulus dimension values, linear separability is the possibility of drawing a line (in two-dimensional space), a plane (in three-dimensional space), or a hyperplane (in n-dimensional space) to separate points belonging to different categories. In the simulation described, three hidden units were used to represent activation values in a three-dimensional space.

observed. In the former case, due to the fact that the task is too difficult, already at the discrimination level, and in the latter case because it is too easy and there is no need for category learning because categories exist.

CP is a very strong and ubiquitous effect. For example, it was observed in the human categorization of speech sound and of colors (Berlin & Kay, 1969). Nakisa and Plunkett (1998) recently observed it in a simulation of phonological learning. They showed that neural networks could categorize spectral sounds into the phonemes of English. The inputs were sounds, sampled from a single language (the network "native" language) chosen from among 14 natural languages. Nakisa and Plunkett found that networks form similar representations regardless of the particular native language, and such representations exhibit a CP partitioning in the spectral continuum. These networks were trained using genetic algorithms thus showing that CP is not an artifact of particular forms of categorical learning.

The functional role of CP in symbol grounding is clearer as an interaction between discrimination and identification. To discriminate is to distinguish some (undefined) pattern in sensors ("there is something"), and it is a relative judgment because something and something else are logically implied in any distinction. To identify is to assign a stable identity to what has been discriminated; this is revealed by a consistent system reaction when the "same" pattern is presented again. Identification is an absolute judgment, and – since it necessarily comes from a "sameness" judgment – it has a categorical nature. The CP process attains the identification result precisely by acting upon discriminability or separation between different categories. Subsequently, CP is a basic mechanism for providing more compact representations, compared with the raw sensory projections where feature-filtering has already done some of the work in the service of categorization. Identification does not presuppose naming. To react consistently to some category of stimuli does not require being able to say what such stimuli are, e.g., by using labels that act as names for them. However, compact CP representations are more suitable than the sensory ones in the subsequent process of learning labels for categories. These labels, or names for categories, can be further combined into propositions and become symbols.

So far, the process we have described is based on a direct sensorimotor interaction with the environment. Symbols derived from this can be called "grounded symbols". There is, however, a different way of acquiring new categories, namely by combining grounded symbols. The previous example of learning that a zebra is a horse with stripes should be recalled to illustrate this point. Cangelosi and Harnad (in press) called the first method of acquiring categories "sensorimotor toil" and the second method "symbolic theft", to stress the benefit of not being forced to learn from a direct sensory experience for every new category.

A recent model by Cangelosi, Greco and Harnad (2000) simulated this overall process of CP, subsequent acquisition of grounded names, and learning of new high-order symbols from grounded ones (grounding transfer). Three-layer feed-forward neural networks were used (see Figure 9.1), having two groups of input units: 49 units simulating a retina and 6 units simulating a linguistic input. The networks had five hidden units, and two groups of output units replicating the organization of input (retina and verbal output). The retinal input depicted geometric images (circles, ellipses, squares, rectangles) with different sizes and positions[6]. The activation of each of the verbal input units corresponded to the presentation of a particular category name. The training procedure had four learning stages. In the prototype sorting task, the networks were trained to categorize figures: from input shapes they had to produce the categorical prototype as output. The same networks were subsequently given the task of associating each shape with its name. This task is called "entry-level naming". An imitation learning cycle was also used for the linguistic input and output units. Names acquired this way, however simple, can be considered grounded because they were explicitly connected with sensory retinal inputs.

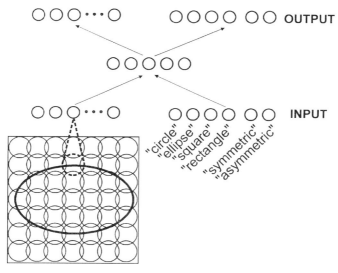

Figure 9.1 Network architecture in the Cangelosi, Greco and Harnad (2000) model. The encoding of output units exactly reflects the structure of input units.

The most interesting part of the simulation is the final stage, where the same networks learn the conjunction of such grounded names (for example, "square" or "rectangle") with new arbitrary names (e.g., "symmetric" or "asymmetric"). This higher-level learning was accomplished by simple imitation learning of the combination of names. It is like teaching "square [is] symmetric" or "rectangle [is] asymmetric". The retinal part of the network was not used during this task. After this procedure, networks were used for the grounding test. The shapes were presented as retinal inputs and output names were checked. In 80% of cases, the networks were able to produce the basic name of each shape and its high-order name (symmetric/asymmetric). As this learning comes from the association of grounded names with new names, the grounding is "transferred" to names that did not have such a property. This is why the process is called "grounding transfer".

---

[6] Pixels were pre-processed in order to compress information using the receptive fields technique (Jacobs and Kossylyn, 1994).

This model has been extended to use the combination of the grounded names of basic features in order to learn higher-order concepts. The same architecture and training procedure were kept. In this case, networks first learned to sort prototypes of shapes depicting turtles, spots, horses and stripes, and then associate such shapes with their names, thereby grounding them. They were then taught to associate the conjunction of grounded names ("horse"+"stripes" or "turtle"+"spots") with a new name ("zebra" or "sportoise"). Networks were able to name the pictures of zebras they had never seen before with the name "zebra". This was achieved solely on the basis of a linguistic combination of grounded names, which in fact can be considered the effect of a true prepositional definition.

Models of the acquisition of grounded symbols

The path we have followed, starting from stimulus discrimination and leading to categories and grounded names for them, actually describes the first stages of language acquisition. Language is not a common sensorial input. It is not like commonly perceived objects, but has something special, because it acts like a "comment" on the world. We shall now focus on some models of the acquisition of grounded lexicons. The most natural source of inspiration for this kind of simulation is language acquisition in children.

A plausible task to be modeled is the presentation of words along with their referents, something like what happens when a parent shows her child a ball while uttering "ball". One example of such a kind of simulation is the work of Plunkett, Sinha, Møller and Strandsby (1992). They designed a network that had to associate simple pictures with labels. The network architecture was similar to the one used by Cangelosi, Greco and Harnad (2000) and described earlier (Figure 9.1). There were two distinct sensory modalities (retinal and verbal) in the input and output layers, and two hidden layers. An auto-associative learning task was used. During testing, only either the verbal or retinal input was given and the net was requested to give the corresponding other output. As often happens with neural networks, they were not able to correctly perform these tasks at all training stages; performance was obviously poor at early stages and better with intensive training. The interesting result was that performance was not linearly related to the extent of training. It suddenly improved at some point, exhibiting something like a "vocabulary spurt" without any apparent reason. This happened both for comprehension and for production, but at different times. This exactly reflects what is observed in children, or in adults when learning a new language: comprehension precedes production. In other words, at some stage the net was able to "understand" what image a name referred to, but not yet to produce this name when given the corresponding image. But at a later stage a new, sudden improvement would be also observed in production.

A similar network architecture and learning scheme using auto-association were used in a model by Schafer and Mareschal (2001). Networks learned to associate names (coded as phonemes) with objects (arbitrary binary vectors). In this study, the network's capability to distinguishing different names when associated with the same objects was tested. This task models an observation coming from studies with infants, reporting that younger (8-month-old) infants are able to produce finer phonetic discriminations than older (14-month-old) infants. Schafer and Mareschal use this model to claim that there is a possibility of obtaining similar developmental discontinuities without necessarily hypothesizing a difference in processing strategies. The disruption in low-level processing (discrimination) does not necessarily arise from a higher cognitive load in higher-level (semantic) processing. Note that this model considers the role of discrimination in language acquisition. In this case, a hindrance to name (phonetic) discrimination is subsequent to a well-established association with some object pattern. The perspective can also be reversed, with names seen as emerging from a need for discrimination, as long as they are able to capture differences in perceived objects. This fundamental property reveals the critical role of language as a symbolic tool.

A model by Greco and Cangelosi (1999) investigates the role of linguistic labels in categorization. It focuses on the feature-extraction process which is affected when names already exist for perceived patterns. They trained four-layered networks to associate names with pictures. Names referred to different features of the input (name, color, function) and there were three input conditions (visual features, name, features+name). Analyses of hidden activation show that representations were different in the feature+name condition and these strongly depended on the name. For example, the visual input of a blue pen presented together with the label "blue" activated the *blue* units in the network, while the name "pen" activated the *pen* units. This shows the mediating role of language in categorization. The same model was successfully extended to display this knowledge explicitly by using a further module that re-described the hidden-layer representations using a competitive learning algorithm.

All these toy models show that simulations can fit observed behavioral results or generate reliable predictions about them, but they are obviously simplifications of the real lexicon acquisition task in children. Inasmuch as these models investigate how representations are constructed of name-objects associations, they are also symbol grounding models. However, words acquired by children are not always associated with their referents, but mostly associated with other words. Such models should also be extended to the expressive functions of language, as we know them from developmental psychology studies.

Evolving Grounded Languages

We have established the importance of direct grounding in models of categorization and in language acquisition. Models of the evolution of symbols and language should also include the grounding of symbols into the world. The computational study of the evolution of language has the additional objective of understanding how and when symbol acquisition abilities originated and how the ability to ground symbols in real world meaning emerged. By using models with emerging symbol grounding properties the researcher is released from the task of deciding which meanings to input to the system in the different evolutionary stages. For example, a non-grounded approach would have significant limitations in investigating the possible existence (or not) of sequential stages of syntactic complexity in the evolution of language. The researcher would have to define an a priori series of stages of semantic complexity upon which syntax would be biased to gradually develop. Instead, in a symbol grounding approach, other autonomous factors, such as the emergence of different stages of behavioral complexity during an organism's adaptation, would be free to affect (or not) the evolution of different stages of syntax complexity.

Current computational models of the evolution of language deal with the symbol grounding problem in different ways. Some models simply avoid the problem by ignoring it, or by assuming that this is not a real problem because it is easily solved in later stages. They think, as cognitivists do, that the researcher will connect the symbols in the simulated communication system with the meaning in the real world. This is the case of models that use a self-referential system where some symbols are used for communication and other symbols are used to represent semantics (e.g., when a list of words is used to denote the list of semantic categories). For example, in models that study the auto-organization of signal-meaning tables (e.g., Steels, 1996; Oliphant and Batali, 1997), the researchers provide the system with a fixed list of N symbols denoting "meanings" and M symbols denoting communication signals. In other models (e.g., Kirby 2000), the symbols used for communication vary whilst an invariant semantic layer is provided by means of a list of names of semantic categories (e.g., "John", "Mary", "love"). This represents an intermediate layer between the real referents (Mr. John, Ms Mary, the feeling of love) and the communicating symbols associated to them ("blap", "blop", "blup"). However, the missing link between the real feeling of love and the semantic category "love" is what makes symbol grounding interesting. We cannot ignore the implication of this (cognitive) process in the investigation of the evolution of language.

A different group of computational models of language evolution deals directly with the symbol grounding problem using simulated languages with grounded semantics. An example is the embodied approach to the evolution of communication between robots (Steels and Vogt, 1997; this volume). Robots interact in a real environment with physical entities (walls, obstacles, other robots) through sensorimotor devices (video cameras, radio receivers, wheels, arms). This experience constitutes the basis for extracting meanings to communicate. Recently, robotic models have been extended to the Internet and to communication with humans. In Steels and Kaplan's (1999) "Talking Heads" experiment, two robotic agents have the task of describing the location of colored geometrical shapes in a whiteboard. Through various Internet sites, human subjects can be remotely "embodied" in one of the robotic talking heads and can participate in language games. A similar methodology has been applied to the evolution of direct communication between robots and humans. The SONY entertainment robot AIBO is being trained to evolve a lexicon for communicating with humans (Kaplan, 2000).

Additionally, direct grounding of symbols can be obtained through simulation, such as artificial life simulations (Parisi, 1997). This type of model achieves symbol grounding by explicitly simulating the environment in which the communicating agents live and interact. Simulated agents can perform foraging tasks by learning to classify different sources of energy (e.g., mushroom types) and to communicate their attributes. The agents' behavior is controlled by neural networks, which we have shown to be ideal candidates for dealing with categorization and symbol grounding. This categorization of food provides the basic meaning upon which agents will ground their communication symbols. A detailed example of this approach is presented in the following section. A specific theory of the origin of language based on hearsay and symbolic theft will be tested using the symbol grounded metaphor of a "mushroom world" (Cangelosi and Harnad, in press).

The symbolic theft hypothesis of the origins of words and language

We have already discussed categorical perception and the ability to build categories of objects, events and states of affair in the world. These constitute the groundwork of cognition and language. There are two opposite ways of acquiring categories. First, we can use "sensorimotor toil", in which new categories are acquired through real-time, feedback-corrected, trial and error experience. Secondly, we can use "symbolic theft", in which new categories are acquired through language, based on hearsay from propositions (e.g., through boolean combinations of symbols describing them). In competition, symbolic theft always outperforms sensorimotor toil. It is more efficient than toil because only one propositional description of a new category is enough to learn it. In contrast, repeated experience is required to learn a category by sensorimotor toil. Due to this significant advantage, it has been hypothesized that symbolic theft is the basis of the adaptive advantage of language (Harnad, 1996). However, some basic categories must still be learned by toil to avoid an infinite regress in the symbol grounding problem. The picture of language origins and evolution that emerges from this hypothesis is that of a powerful hybrid symbolic/sensorimotor capacity. Initially, organisms evolved an ability to build some categories of the world through direct sensorimotor toil. They also learned to name such categories. Subsequently, some organisms must have experimented with the propositional combination of the names of these categories and discovered the advantage of this new way of learning categories, stealing their knowledge by hearsay. The benefits of the symbolic theft strategy must have given these organisms the adaptive advantage in natural language abilities. This is infinitely superior to its purely sensorimotor precursors, but still grounded in and dependent on them.

To test this hypothesis of language origin Cangelosi and Harnad (in press) developed a computational model which simulates a community of foraging organisms. They rely on learning categories of foods to survive. Category formation is achieved through toil or theft strategies. The model tests the prediction that acquiring categories through symbolic theft is more adaptive than acquiring them through sensorimotor toil. Moreover, the model should help us to understand the mechanisms central to symbol grounding. For example, it should show that new categories learnt by theft inherit their grounding from the low-level categories.

Computer simulation

The computational model uses the mushroom world scenario (Harnad 1987) to simulate the behavior of virtual organisms that forage among the mushrooms, learning what to do with them. For example, mushrooms with feature A (i.e., those with black spots on their tops) are to be eaten; mushrooms with feature B (i.e., a dark stalk) are to have their location marked, and mushrooms with both features A and B are to be eaten, marked and returned to. All mushrooms have three irrelevant features (C, D and E) that the foragers must learn to ignore. When organisms approach a mushroom, they emit a call associated with their functionality (EAT, MARK). Both the correct action pattern (eat, mark) and the correct call (EAT, MARK) are learned during the foragers' lifetime through supervised learning (sensorimotor toil). Under some conditions, the foragers also receive the call of another forager as input. This will be used to simulate theft learning of the return behavior.

The behavior of organisms is controlled by neural networks that process the sensory information about the closest mushroom and activates the output units corresponding to the movement, action and call patterns. For each action, the forager first produces a movement and an action/call output using the information about the physical features of the mushroom. The network's action and call outputs are compared with their expected output and this difference is then backpropagated to adjust connection weights. In this way the forager learns to categorize the mushrooms by performing the correct action and call. In the second spread of activation the forager also learns to imitate the call. It only receives the correct call for that kind of mushroom as input, which it must imitate on its call output units. This learning is likewise supervised by back-propagation.

The population of foragers is also subject to selection and reproduction through a genetic algorithm (Goldberg, 1989). The initial population consists of 100 neural networks with a random weight matrix. During the forager's lifetime, the fitness is computed by assigning points for each time a forager reaches a mushroom and performs the correct action on it (eat/mark/return). At the end of their life-cycles, the 20 foragers with the highest fitness in each generation are selected and allowed to reproduce by engendering five offspring each. The population of newborns is subject to random mutation of their initial connection weights.

Adaptive advantages of Theft versus Toil learning

Our hypothesis is that the theft strategy is more adaptive (i.e., results in greater fitness and more mushroom collection) than the toil strategy. To test this, we compare foragers' behavior for the two learning conditions. In the first simulation, two experimental groups were directly compared: Toil and Theft. In the first 200 generations, all organisms learn through sensorimotor Toil to eat mushrooms with feature A and to mark mushrooms with feature B. They also learn the names of the basic categories: EAT and MARK. The return behavior, and its name are not yet taught. From generation 200 to 210, organisms live for a longer life stage. In the second part of their lifetime, they are divided into the two groups of Toilers and Thieves. Toil foragers go on to learn to return to AB mushrooms in the same way they had learned to eat and mark them through honest toil. In contrast, Theft foragers learn to return on the basis of hearing the vocalization of the mushrooms' names. They rely completely on other foragers' calls to learn to return as they do not receive the feature input. To test the adaptive advantage of Theft versus Toil learning, we compare foragers' behavior for the two conditions by counting the number of AB mushrooms that are correctly returned to. Thieves successfully return to more AB mushrooms (55) than Toilers (44). This means that learning to return from the grounded names EAT and MARK is more adaptive than learning it through direct toil based on sampling the physical features of the mushrooms.

A more direct way to study the adaptive advantage of Theft over Toil is to see how they fare in competition against one another. We performed some competitive simulations. Again, foragers live for two life stages. In the first, all learn to eat and mark through Toil. In the second life stage, the foragers are randomly divided into 50 Thieves and 50 Toilers who must all learn to return. Direct competition only occurs at the end of the life cycle, in the selection of the fittest 20 foragers to reproduce. In the present ecology, the assumption is that mushrooms are abundant and that fitness efficiency only affects the selection of the top 20 foragers. Figure 9.2 shows the proportion of Thieves in the overall population of Theft vs. Toil. Thieves gradually come to outnumber Toilers, so that in less than 10 generations the whole population is made up of Thieves.

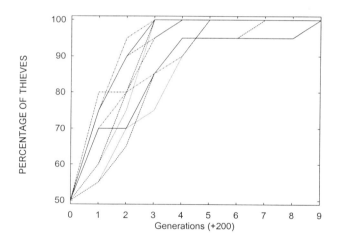

Figure 9.2 Proportion of Thieves in the ten competitive simulations.

The direct competition between Toilers and Thieves has been studied in more detail in other simulations. In one study we varied the availability of mushrooms to see the effects of scarcity/abundance of mushrooms for competition. In another simulation, kinship relationships determined the choice of the listener organism to which the names of mushrooms was vocalized. Results show that when the scarcity of the mushrooms is varied, Theft beats Toil provided there are plenty of mushrooms for everyone. However, when the mushrooms are scarce and vocalizing risks losing the mushroom to the Thief, Toil beats Theft and the foragers are mute. Further studies analyzing kinship showed that under conditions of scarcity vocalizing only to relatives beats vocalizing to everyone.

All these results support the original hypothesis that a Theft learning strategy, based on language, is much more adaptive that a Toil strategy. This adaptive advantage could be basis for the origin of language and its adaptive advantage.

Categorical perception effects

These computational models are also useful in the investigation of the changes that communication and linguistic abilities cause in the organism. We have already stressed the importance of internal categorical representation in the grounding of symbols. Previously we showed the compression of within-category distances and the expansion of between-category distances in categorization and naming tasks. Now we will show how these phenomena are also present in the model of the evolution of Theft learning and communication. We will study the changes in the foragers' hidden-unit representations for the mushrooms to determine internal changes during Toil and Theft. We compare categorical representations in four different experimental conditions: (1) Pre-learning, for random-weight networks before learning; (2) No-return, for foragers' networks that were only taught to eat and to call EAT, and to mark and to call MARK, (3) Toil-return, for networks that also learned to return and to call RETURN with feature input, and (4) Theft-return for learning to return from calls alone.

We recorded the Euclidean distances between and within categories using the coordinates of the five hidden unit activations. At the end of each simulation, the five fittest foragers in each condition were tested based on the measurement of within- and between-category distances. For each type of distance, there are four means for the distances between the internal representations of the Do-nothing (neither Mark nor Eat nor Return), Eat only, Mark only, Eat+Mark+Return. The average within-category distances in three experimental conditions are shown in Table 9.1. Statistical tests on these data suggest that within-category distances decrease significantly from Pre-learning to No-return to Toil. As expected, the greatest decrease is between the (random) Pre-learning and all the post-learning nets. When we compared the four types of categories, all means differed from each other except the Eat and Mark within-distances. That is, the within-category distance for Eat and Mark are the same, whereas the within distance of Do-nothing is the greatest and that of Return the smallest. These results are consistent with categorical perception effects. There is a compression of the category from the pre-learning condition to all other post-categorization cases.

Table 9.1 Table of means for the within-category distances. Values for the Theft condition are not reported because the distance is always 0 (all ten samples of mushrooms use the same call input)

CATEGORY	Pre-learning	No-return	Toil-return
Do-nothing	0.34	0.16	0.14
Eat	0.32	0.14	0.12
Mark	0.30	0.13	0.12
Eat+Mark(+Return)	0.29	0.11	0.09

The between-category distances amongst the four categories in the different experimental conditions are described more clearly in Figure 9.3. The quadrilateral represents the 2D projections of the between-category distances in the four conditions (= four sides of the quadrilateral shape). All distances, except Eat vs. Mark, are directly comparable and reflect the actual Euclidean distances between categories. The figure shows an expansion of the between-category distances from Pre-learning to Theft Learning. The thin dashed rectangle refers to the between-category distances before learning (random weights). The thick dashed line represents what they look like after Toil learning of Eat and Mark without Return. The thin continuous line refers to the Toil learning of Eat and Mark, with Return, and the thick continuous line is for Theft learning of Return. First, this chart shows the expected categorical perception effect of differentiation between (the centers of) categories from the Pre-learning condition to the other post-learning cases. What is new and more important is the effect of Theft learning relative to the other learning conditions/categories. The distances between Return and all the other categories (Return vs. Eat, Return vs. Mark, Return vs. Do-nothing) are the highest in the Theft condition. This suggests that Theft learning not only is more advantageous for survival (Thieves collect more return mushrooms that Toilers), but it also "optimizes" the internal categorical representation by the categorical effects of within-category compression and between-category expansion. That is, language learning is based on categorization, but in return it improves categorical learning.

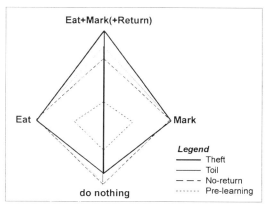

Figure 9.3 Two-dimensional projections of the between-category distances (quadrilateral sides) in the four learning conditions.

Summary

In this chapter we have focused on the simulation of the acquisition and evolution of grounded languages. We have given a definition of symbol based mainly on cognitive (Harnad, 1987) and neurally-related semiotic factors (Deacon, 1997).

The real symbolic feature of communication relies on the fact that each symbol is part of a wider and more complex system. This is mainly regulated by compositional rules, such as syntax. Subsequently, the problem of symbol grounding in cognitive models was illustrated. Psychologically plausible models of language and cognition should include an intrinsic link between at least some basic symbols and some objects in the world. These basic symbols must be directly grounded in cognitive representations, such as categories. This way symbol manipulation can be constrained by the non-arbitrary shapes of the underlying cognitive representations.

Various computational models of categorization, symbol grounding transfer, and language acquisition have been described. They are primarily based on the use of neural networks. These models can easily abstract from similarities between stimuli and achieve categories. Moreover, they can associate names with categories. They exhibit the basic categorical perception effect, whereby internal representations of members of the same category look more similar and members of different categories look more different. In the evolutionary model of the symbolic theft acquisition of categories and language, it has been shown that such cognitive factors for category learning and symbol grounding can be integrated to test hypotheses on the evolution of language.

We suggest that the inclusion of direct grounding in simulation models of the evolution of syntax can improve their potential to explain the emergence of linguistic and cognitive abilities. The simulation approach proposed here, and other methodologies such as the robotic modeling of the evolution of communication (see next chapter), are clear examples of how symbols can directly and autonomously ground their meaning.

References

Andrews J, Livingston K, Harnad S (1998) Categorical perception effects induced by category learning. *Journal of Experimental Psychology: Learning, Memory, and Cognition*, 24: 732-753

Berlin B, Kay P (1969) *Basic color terms: Their universality and evolution.* University of California Press, Berkley

Cangelosi A, Greco A, Harnad S (2000) From robotic toil to symbolic theft: Grounding transfer from entry-level to higher-level categories. *Connection Science*, 12: 143-162

Cangelosi A, Harnad S (in press) The adaptive advantage of symbolic theft over sensorimotor toil: Grounding language in perceptual categories. *Evolution of Communication*

Christiansen MH, Chater N (1999) Connectionist natural language processing: The state of the art. *Cognitive Science*, 23: 417-437

Deacon TW (1997) *The Symbolic Species: The coevolution of language and human brain* London: Penguin

Elman JL (1990) Finding structure in time. *Cognitive Science*, 14: 179-211

Fodor JA (1976) *The Language of thought*, Thomas Y Crowell, New York

Goldberg DE (1989) *Genetic algorithms in search, optimization, and machine learning.* Addison-Wesley, Reading MA

Goldstone R (1994) Influences of categorization of perceptual discrimination. *Journal of Experimental Psychology: General*, 123: 178-200

Greco A, Cangelosi A (1999) Language and the acquisition of implicit and explicit knowledge: A pilot study using neural networks. *Cognitive Systems*, 5: 148-165

Harnad S (ed) (1987) *Categorical perception: The groundwork of cognition*. Cambridge University Press, New York

Harnad S (1990) The symbol grounding problem. *Physica D*, 42: 335-346

Harnad S (1996) The origin of words: A psychophysical hypothesis. In: Velichkovsky BM, Rumbaugh DM (eds) *Communicating meaning: The evolution and development of language*. Lawrence Erlbaum Associates, Mahwah NJ

Harnad S, Hanson SJ, Lubin J (1991) Categorical perception and the evolution of supervised learning in neural nets. In: Powers DW, Reeker L. (eds) *Proceedings of the AAAI Spring Symposium on Machine Learning of Natural Language and Ontology*

Harnad S, Hanson SJ, Lubin, J (1995) Learned categorical perception in neural nets: Implications for symbol grounding. In: Honavar V, Uhr L (eds) *Symbol processors and connectionist network models in artificial intelligence and cognitive modelling: Steps toward principled integration*. Academic Press, pp 191-206

Hurford J (1998) Review of Terrence Deacon, 1997 The Symbolic Species: The co-evolution of language and the human brain. *The Times Literary Supplement*, October 23rd, 1998, 34

Jacobs RA, Kosslyn SM (1994) Encoding shape and spatial relations: The role of receptive field size in coordinating complementary representations. *Cognitive Science*, 18: 361-386

Kaplan F (2000) Talking AIBO: First experimentation of verbal interactions with an autonomous four-legged robot. In: Nijholt A, Heylen D, Jokinen K (eds) *Learning to behave: Interacting agents*. CELE-TWENTE Workshop on Language Technology, pp 57-63

Kirby S (2000) Syntax without natural selection: How compositionality emerges from vocabulary in a population of learners. In: Knight C, Studdert-Kennedy M, Hurford J (eds) *The evolutionary emergence of language: Social function and the origins of linguistic form*. Cambridge University Press, pp 303-323

Nakisa RC, Plunkett K (1998) Evolution of a rapidly learned representation for speech. *Language and Cognitive Processes*, 13: 105-127

Oliphant M, Batali J (1997) Learning and the emergence of coordinated communication *Centre for Research in Language Newsletter*, 11(1)

Parisi D (1997) An Artificial Life approach to language. *Mind and Language*, 59, 121-146

Peirce CS (1978) *Collected Papers (Vol II: Element of Logic)*. In: Hartshorne C, Weiss P (eds), Belknap Cambridge, MA

Pevtzow R, Harnad S (1997) Warping similarity space in category learning by human subjects: The role of task difficulty. In: Ramscar M, Hahn U, Cambouropolos E, Pain H (eds) *Proceedings of SimCat 1997: Interdisciplinary Workshop on Similarity and Categorization*. Department of Artificial Intelligence, Edinburgh University, pp 189-195

Plunkett K, Sinha C, Møller MF, Strandsby O (1992) Symbol grounding or the emergence of symbols? Vocabulary growth in children and a connectionist net. *Connection Science*, 4: 293-312

Pylyshyn ZW (1984) *Computation and cognition*. MIT Press, Bradford Books, Cambridge MA

Rumelhart DE, McClelland JL, the PDP Research Group (eds) (1986) *Parallel distributed processing: Explorations in the microstructure of cognition (Vol 1: Foundations)*. MIT Press, Cambridge MA

Savage-Rumbaugh S, Rumbaugh DM (1978) Symbolization, language, and Chimpanzees: A theoretical reevaluation on Initial language acquisition processes in four Young Pan troglodytes. *Brain and Language*, 6: 265-300

Schafer G, Mareschal D (2001) Modeling infant speech sound discrimination using simple associative networks. *Infancy*, 2(1)

Searle JR (1982) The Chinese room revisited. *Behavioral and Brain Sciences*, 5: 345-348

Steels L (1996) Self-organising vocabularies. In: Langton CG, Shimohara K (eds) *Proceedings of the ALIFE V*. MIT Press, Cambridge MA, pp 179-184

Steels L, Kaplan F (1999) Collective learning and semiotic dynamics. In: Floreano D, Nicoud JD, Mondada F (eds) *Proceedings of ECAL99 the Fifth European Conference on Artificial Life (Lecture Notes in Artificial Intelligence)*. Springer-Verlag, Berlin, pp 679-688

Steels L, Vogt P (1997) Grounding adaptive language games in robotic agents. In: Husband P, Harvey I (eds) *Proceedings of the Fourth European Conference on Artificial Life*. MIT Press, Cambridge MA, pp 474-482

Tijsseling A, Harnad S (1997) Warping similarity space in category learning by backprop nets. In: Ramscar M, Hahn U, Cambouropolos E, Pain H (eds) *Proceedings of SimCat 1997: Interdisciplinary Workshop on Similarity and Categorization*. Department of Artificial Intelligence, Edinburgh University, pp 263-269

Chapter 10

Grounding Symbols through Evolutionary Language Games

Luc Steels

Introduction

To explain the origins of language, we need to explain three puzzles: First how it has been possible for a group of agents, i.e., our early human ancestors, to develop a shared repertoire of sounds with the complexity of human languages. Our species was not the first to do so, because birds have also complex evolving sound repertoires. But it is one distinguishing feature with respect to other hominid species such as chimpanzees (Lieberman, 1991). Second, we must explain how sounds, gestures, or other physical signs can be given meaning, in other words how a semiotic system may arise. The songs of birds do not carry meaning. Even though gestures or sounds used by chimpanzees might, they do not appear to productively associate an open-ended set of meanings with an open-ended set of forms. Third, we must explain the emergence of grammar: how compositional form can be associated with compositional meaning. This chapter focuses only on the second part, the emergence of semiotic systems.

From a traditional point of view, semiotic systems establish a relation between a sign (such as a word in natural language), a concept, and an object in the external world. As we are interested not only in a description of semiotic relations from the viewpoint of an observer but in the question how a cognitive agent might acquire a semiotic system we must also worry about how concepts are represented, how perception and categorization processes may establish the relation between the world and its internal representation, and how words are produced and recognized as physical tokens. It follows that the problem of the emergence of cognitive systems goes beyond that of acquiring the relation between words and concepts. There is also the problem how concepts can be grounded through perception and action in the real world, and how the development of conceptual representations and the development of words to express them may bootstrap each other.

Defining the Problem

Given the terminological confusion in the cognitive sciences it is worthwhile to define more precisely the problems we try to approach.

1. In computer science, a *representation* is a physical state of a machine (computer memory for example) which acts as a "stand in" for something else. The physical state then becomes a means to store information and physical processes operating over the state can implement whatever representational transformations we wish to enact. Thus a representation of a number in a computer is a configuration of digital states. Calculation takes place by changing these states. Objects, concepts, actions, etc. can likewise be represented in an artificial agent by postulating internal states for each of these. Cognitive processes like decision-making, language parsing, object recognition, etc. can then be conceived as physical operations over these internal states.

A representation (both the physical medium chosen and the convention used to map information onto this medium) is arbitrary with respect to what one wants to represent. The only requirement is that the mapping is systematic and that processes operating over a representation are consistent with respect to the mapping that has been adopted. Thus we can not only use binary representations of numbers but also hexadecimal representations, and make use of marks on a laser disk as well as electromagnetic states of an electronic circuit as the state medium. Note that once we are at the level of physical states and physical processes, no additional homunculus is involved to "interpret" representations.

In neuroscience parlance, the equivalent of the computer science notion of a representation is the notion of a neural correlate. This is a biological state (for example the activation of a neuron or set of neurons) which stands for something else, like a control signal for an arm, the recognition of a concept, experience of the color red, etc. Neural processes operating over these physical states, usually thought to take the form of the selective propagation of signals through a network, are the neural correspondence of the physical operations carried out over computational states.

So even if computational implementations are very different from biological implementations, the notion of representation is similar in AI and (cognitive) neuroscience. Using representations and operations over representations to explain cognitive functions seem now so natural and obvious that it is difficult to follow philosophers who claim that cognition does not involve representations. Perhaps they simply have not understood what representations are or are using the notion of representation in another way.

2. We say that a representation is *grounded* when there is an autonomous process that transforms sensations (i.e., data flowing from sensors or motors into

internal states) into internal representations and transforms internal representations into motor activations. Through these grounding processes, the agent can coordinate his activities with the world and other agents. The representation need not be an exact, full, veridical representation of the world, and it can be analog or categorial, but it needs to be sufficiently detailed and faithful to support the agent's interaction with the world and others.

Grounding is trivially achieved for devices like a calculator. The user pushes buttons which directly activate internal representations. It is obviously much more complex for representations about the world. Sensors reflect physical properties of the environment which are not necessarily those that the agent needs to focus on. The information is hidden in the sensory-motor data and requires complex processing to get out. Often there is not enough information in the sensory data and so representations have to be hypothesized in a top-down fashion and mapped onto the sensory-motor data.

In a lot of (pre-1990) AI work the problem of grounding was (temporarily) abstracted out by supplying the representations directly to the computer and only focusing on the processing aspect. This was a useful strategy for a while but it has been rightfully criticized because not all the representations assumed by early AI programs can be grounded on a physically embodied robot (Brooks, 1999; Steels and Brooks, 1995). For example, it is far from obvious that abstract geometric representations about the world, as envisioned by David Marr (1982), can be extracted from real-world images given the available resources. This has led to a healthy move away to simpler representations and better exploitation of bodily interaction (Pfeifer and Scheier, 1999). We should nevertheless keep in mind that many experiments which involve complex representations – even from the very beginning of AI – have considered the problem of grounding these representations. A typical example is the SRI Shakey robot (Nilson, 1984) which was a model of many subsequent robotics efforts. So there is nothing in the notion of representation that makes them inherently not groundable. It is only that the grounding of representations is a very non-trivial and difficult technical problem involving a whole arsenal of statistical and pattern recognition techniques (Marco *et al.*, 2000) and that not every abstract representation can be grounded (Horswill, 1993). Most importantly it is not possible to program representation grounding by hand, partly because of the complexity involved and partly because the perception and categorization process must adapt themselves to the real-world situations that are effectively encountered.

3. Representations are *symbolized* when there are external tokens (speech sounds, gestures, scratches on a piece of paper, configurations on a display) that are associated with the representation and used for external communication with another agent. The relationship is entirely conventional. Sender and receiver must agree, but there are in principle endless possibilities. The process of relating a representation to its symbolization and vice versa must be carried out autonomously by each agent. We say that a symbol is grounded if its representation is grounded.

The relations considered so far are summarized in the semiotic square depicted in Figure 10.1. By sensation, I mean the perceptual or motor data streams that directly connect the agent to the world. By representation I mean a conceptual representation useful for decision making, language or other cognitive tasks. The semiotic square is reminiscent of the semiotic triangle familiar from the semiotic literature (Eco, 1968) which relates world, concept, and symbol. The relation between a symbol and the world is the *reference relation*. The relation between a symbol and a representation is the *meaning relation*. In the philosophy of language literature, the reference relation and the meaning relation are studied theoretically, independently of how this relation is established by a cognitive agent. This kind of research in formal semantics is of interest when one wants to investigate how a symbol system can in principle be related to the world, but is a very different topic from the one considered in this paper.

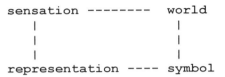

Figure 10.1 The semiotic square summarizes the relation between world, sensation, representation, and symbol.

The problem of symbolization is trivially solved by a calculator which transforms the internal representation of a number into an external representation on a display and which displays on the buttons the conventional representations of numbers so that users know which button to push. It is extraordinarily difficult in the case of natural language, because the conventions are not universal, they are open-ended, and involve a non-trivial multi-level mapping from representations to symbols.

Just for clarification, it is perhaps important to sharpen in this context the notion of symbolic processing, as it has been used in AI. Symbolic processing means that a set of symbols, possibly without any relation to the world, such as postulated in a logical calculus, is mapped onto internal representations and the internal representations are processed to conform to given symbol manipulation rules, for example the rules of natural deduction of the predicate calculus. The outcome of processing is then translated back into symbols. A logic programming system such as PROLOG or a functional programming language like LISP supports this kind of operation. They can handle millions of symbols and their compilers optimize for fast symbolic processing. This technology is of enormous value for building non-trivial cognitive agents but does not address in itself grounding nor learning.

4. *Learning grounded representations* means that the agent allocates states for certain representations but also – and more importantly – that the agent learns to use the representation appropriately, more specifically (1) the ability to relate the representation to the world through the sensori-motor apparatus, and (2) the use of the representation for some purpose, such as making a decision about the next action to take.

Learning a symbolization means to acquire the relation between representations and symbols as required for communication in a specific community. The agent must acquire the ability to activate the intended internal representations given a set of symbols or to select a set of symbols to externalize a particular representation.

Learning a semiotic system means to acquire both grounded representations and their symbolizations, i.e., the relation between world, sensation, representation, and symbol. This is the problem that the child faces when growing up in a language community and the problem that concerns us further in this paper. We take this problem to be equivalent to the *symbol grounding problem*.

Two Approaches

The first approach to the symbol grounding problem is to follow a divide-and-conquer strategy. It assumes that there is on the one hand a process that learns grounded representations. Once the representations are in place, it then postulates a second independent process that associates symbols with the already acquired representations. It has been suggested that this is the way symbol grounding happens in humans (Harnad, 1990) and various experiments have been done following this approach (de Jong, 1999; Cangelosi, Greco and Harnad, 2000).

However a second approach is possible, as first suggested in Steels (1997a; 1999), in which there is a strong *structural coupling* between the two. This means that learning representations and learning their symbolization go hand in hand and influence each other. A representation that has been learned can be the subject of symbolization but communication through symbols provides important feedback to representational learning.

In my opinion the structural coupling approach is the only viable way to explain the massive build up of representations and symbols that humans use and it can be profitably used in artificial systems. This approach seems paradoxical at first because instead of solving two difficult problems one by one, we try to solve both of them at the same time, which intuitively seems to be even more difficult. Here are the reasons why I nevertheless believe that a structural coupling approach is better.

There are many ways to learn grounded representations, but broadly speaking mechanisms fall into two classes: unsupervised or supervised learning. In the case of unsupervised learning, clustering techniques (possibly implemented as neural networks such as the Kohonen network) extract from a series of data invariances that are then associated with internal representations. However, we note two things: (1) not all representations of interest to a cognitive agent are reflected as invariants in sensori-motor data, and (2) often there is more than one possible way to cluster the data depending on the dimensions that are considered or the parameter settings of the clustering algorithm.

This generates the problem that if different agents each independently develop representations about the world, there is no guarantee that they arrive at mutually compatible representations. Today the experimenter carefully designs the features that are input to the learning system, carefully selects appropriate example sets, and then tweaks parameters until an appropriate clustering comes out. This is not quite the autonomous learning that we would hope for, but using sensory data in their raw form, i.e., bitmaps captured by a camera, motor states, the audio signal directly coming from a microphone, etc. does not leave any other choice.

In the case of supervised learning, the agent is given a series of cases as well as feedback whether the representations being developed are appropriate with respect to some task. Thus if the task is classification, the agent would be given examples and counterexamples, if the task is action in the world, the agent gets a feedback signal whether the action was successful (as in reinforcement learning algorithms). Because the task can incorporate some form of coordination with other agents, it is in principle possible to steer the acquisition of representations in such a way that they are compatible with those used by others, by incorporating in the feedback some element that is related to representation sharing. But the critical question here is: Where does the feedback come from? In real-world circumstances, feedback is never direct and obvious, specifically not concerning internal representations. Feedback comes only through the *use* of a representation. If the designer has to carefully determine feedback and prepare the example sets, then we are missing something fundamental. By letting the use of a representation generate feedback, the learning setup becomes self-supervised.

There are many users of representations. For example for planning actions, particularly at a microlevel (like for grasping an object), the agent needs adequate categorizations of reality dedicated to that task. So action execution can be a possible source of feedback. Language is another big user of representations because before anything can be said the world must be conceptualized in the way that has been lexicalized and grammaticalized in the language (and this can differ substantially from one language to another (Talmy, 2000)). But language is not only useful because it provides representational feedback, it also helps a community of agents to settle on similar representations.

The next thing we need is a good framework to study these issues concretely. I claim that evolutionary language games are such framework. We started to work on this around 1995 (Steels, 1996) and have since shown in an increasing number of papers and experiments that the framework is a rich foundation for studying both language formation and concept acquisition. The remainder of this paper reports these developments in some more detail.

Evolutionary Language Games

Evolutionary games are now widely used to study issues in evolution (Maynard Smith, 1982) or economics (Aumann and Hart, 1994). They form part of the larger framework of game theory. A game is an interaction between two agents according to certain rules and having a certain outcome. Games can be adversary (like the Prisoner's Dilemma (Axelrod, 1984)) or cooperative. A game is evolutionary if the players change their internal states in order to be more successful in the future (Lindgren and Nordahl, 1995).

In the past few years, we have done several experiments exploring the co-evolution of language and meaning. Although we have also studied multi-word expressions and the emergence of syntax within the same experimental context

(Steels, 1998) the present paper only discusses single word utterances so that we can focus completely on the issue of grounding meaning.

The main experiment that will be briefly described here is known as the Talking Heads experiment. The robotic setup consists of a set of 'Talking Heads' connected through the Internet. Each Talking Head features a Sony EVI-D31 camera with controllable pan/tilt motors for horizontal and vertical movement (Figure 10.2), a computer for cognitive processing (perception, categorization, lexicon lookup, etc.), a screen on which the internal states of the agent currently loaded in the body are shown, a TV-monitor showing the scene as seen through the camera, and devices for audio in- and output. Agents can load themselves in a physical Talking Head and teleport themselves to another Head by traveling through the Internet. By design, an agent can only interact with another one when it is physically instantiated in a body located in a shared physical environment. The experimental infrastructure also features a commentator which reports and comments on dialogs, displays measures of the ontologies and languages of the agents and game statistics, such as average communicative success, lexical coherence, average ontology and lexicon size, etc.

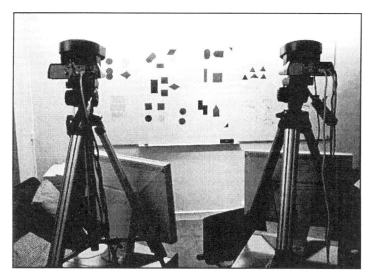

Figure 10.2 Two Talking Head cameras and associated monitors showing what each camera perceives.

For the experiments reported in this paper, the shared environment consists of a magnetic white board on which various shapes are pasted: colored triangles, circles, rectangles, etc. Although this may seem a strong restriction, we have learned that the environment should be simple enough to be able to follow and experimentally investigate the complex dynamics taken place in the agent population.

The guessing game

The interaction between the agents consists of a language game, called the guessing game. The guessing game is played between two visually grounded agents. One agent plays the role of *speaker* and the other one then plays the role of *hearer*. Agents take turns playing games so all of them develop the capacity to be speaker or hearer. Agents are capable of segmenting the image perceived through the camera into objects and of collecting various sensory data about each object, such as the color (decomposed in RGB channels), average gray-scale or position. The set of objects and their data constitute a *context*. The speaker chooses one object from this context, further called the *topic*. The other objects form the *background*. The speaker then gives a linguistic hint to the hearer.

The linguistic hint is an utterance that identifies the topic with respect to the objects in the background. For example, if the context contains [1] a red square, [2] a blue triangle, and [3] a green circle, then the speaker may say something like "the red one" to communicate that [1] is the topic. If the context contains also a red triangle, he has to be more precise and say something like "the red square". Of course, the Talking Heads do not say "the red square" but use their own language and concepts which are never going to be the same as those used in English. For example, they may say "malewina" to mean [UPPER EXTREME-LEFT LOW-REDNESS].

Based on the linguistic hint, the hearer tries to guess what topic the speaker has chosen, and he communicates his choice to the speaker by pointing to the object. A robot points by transmitting in which direction he is looking in his own agent-centered coordinates. The other robot is calibrated in the beginning of the experiment to be able to convert these coordinates into his own agent-centered coordinates. The game succeeds if the topic guessed by the hearer is equal to the topic chosen by the speaker. The game fails if the guess was wrong or if the speaker or the hearer failed at some earlier point in the game. In case of a failure, the speaker gives an extra-linguistic hint by pointing to the topic he had in mind, and both agents try to repair their internal structures to be more successful in future games.

The architecture of the agents has two components: a conceptualization module responsible for categorizing reality or for applying categories to find back the referent in the perceptual image, and a verbalization module responsible for verbalizing a conceptualization or for interpreting a form to reconstruct its meaning. Agents start with no prior designer-supplied ontology nor lexicon. A shared ontology and lexicon must emerge from scratch in a self-organized process. The agents therefore not only play the game but also expand or adapt their ontology or lexicon to be more successful in future games.

The conceptualization module

Meanings are categories that distinguish the topic from the other objects in the context. The categories are organized in discrimination trees (Figure 10.3) where each node contains a discriminator able to filter the set of objects into a subset that satisfies a category and another one that satisfies its opposition. For example, there might be a discriminator based on the horizontal position (HPOS) of the center of an object (scaled between 0.0 and 1.0) sorting the objects in the context in a bin for the category 'left' when HPOS < 0.5 (further labeled as [HPOS-0.0,0.5]) and one for 'right' when HPOS > 0.5 (labeled as [HPOS-0.5,1.0]). Further subcategories are created by restricting the region of each category. For example, the category 'very left' (or [HPOS-0.0,0.25]) applies when an object's HPOS value is in the region [0.0,0.25]. For the experiments in this paper, the agents have only channels for horizontal position (HPOS), vertical position (VPOS), color (RGB indicated as RED, GREEN, BLUE), and grayscale (GRAY). The system is open to exploit any channel with additional raw data, such as audio, or results from more complex image processing.

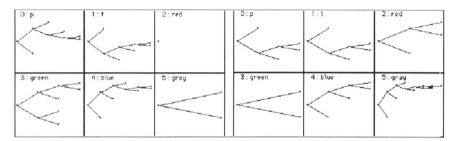

Figure 10.3 The discrimination trees of two agents.

A distinctive category set is found by filtering the objects in the context from the top in each discrimination tree until there is a bin which contains only the topic. This means that only the topic falls within the category associated with that bin, and so this category uniquely filters out the topic from all the other objects in the scene. Often more than one solution is possible, but all solutions are passed on to the lexicon module.

The discrimination trees of each agent are formed using a growth and pruning dynamics coupled to the environment, which creates an ecology of distinctions. Discrimination trees grow randomly by the addition of new categorizers splitting the region of existing categories. Categorizers compete in each guessing game. The use and success of a categorizer is monitored and categorizers that are irrelevant for the environments encountered by the agent are pruned. More details about the discrimination game can be found in (Steels, 1997a)

Verbalization module

The lexicon of each agent consists of a two-way association between forms (which are individual words) and meanings (which are single categories). Each association has a score. Words are random combinations of syllables, although any set of distinct word symbols could be used. When a speaker needs to verbalize a category, he looks up all possible words associated with that category, orders them and picks the one with the best score for transmission to the hearer. When a hearer needs to interpret a word, he looks up all possible meanings, tests which meanings are applicable in the present context, i.e., which ones yield a possible single referent, and uses the remaining meaning with the highest score as the winner. The topic guessed by the hearer is the referent of this meaning.

Based on feedback on the outcome of the guessing game, the speaker and the hearer update the scores. When the game has succeeded, they increase the score of the winning association and decrease the competitors, thus implementing lateral inhibition. When the game has failed, they each decrease the score of the association they used. Occasionally new associations are stored. A speaker creates a new word when he does not have a word yet for a meaning he wants to express. A hearer may encounter a new word he has never heard before and then store a new association between this word and the best guess of the possible meaning. This guess is based on first guessing the topic using the extra-linguistic hint provided by the speaker, and on performing categorization using his own discrimination trees as developed thus far. These lexicon bootstrapping mechanisms have been explained and validated extensively in earlier papers (Steels and Kaplan, 1998) and are basically the same as those reported by Oliphant (1996).

The conceptualization module proposes several solutions to the verbalization module which prefers those that have already been lexicalized. Agents monitor success of categories in the total game and use this to target growth and pruning. The language therefore strongly influences the ontologies agents retain. The two modules are structurally coupled and thus are coordinated without a central coordinator.

Examples

Here is the simplest possible case of a language game. The speaker, **a1**, has picked a triangular object at the bottom of the scene as the topic. There is only one other rectangular object in the scene, nearer to the top. Consequently, the category [VPOS-0.0,0.5]$_{a1}$, which is valid when the vertical position *VPOS* < *0.5*, is applicable because it is valid for the triangle but not for the rectangle. Assuming that **a1** has an association in his lexicon relating [VPOS-0.0,0.5]$_{a1}$ with the word "lu", then **a1** will retrieve this association and transmits the word "lu" to the hearer, which is agent **a2**.

Now suppose that **a2** has stored in his lexicon an association between "lu" and [RED-0.0,0.5]$_{a2}$. He therefore hypothesizes that [RED-0.0,0.5]$_{a2}$ must be the meaning of "lu". When he applies this category to the present scene, in other words when he filters out the objects whose value for the redness channel (RED) do not fall in the region [0.0,0.5], he obtains only one remaining object, the triangle. Hence **a2** concludes that this must be the topic and points to it. The speaker recognizes that the hearer has pointed to the right object and so the game succeeds.

The complete dialog is reported by the commentator as follows:

```
Game 125.
a1 is the speaker.
a2 is the hearer.
```

```
a1 segments the context into 2 objects
a1 categorizes the topic as [VPOS-0.0,0.5]
a1 says:  "lu"
a2 interprets "lu" as [RED-0.0,0.5]
a2 points to the topic
a1 says: "OK"
```

This game illustrates a situation where the speaker and the hearer picks out the same referent even though they use a different meaning. The speaker uses vertical position and the hearer the degree of redness in RGB space.

Here is a second example, The speaker is again **a1** and he uses the same category and the same word "lu". But the hearer, **a3**, interprets "lu" in terms of horizontal position [HPOS-0.0,0.5]$_{a3}$ (left of the scene). Because there is more than one object satisfying this category in the scene the agents look at, the hearer is confused. The speaker then points to the topic and the hearer acquires a new association between "lu" and [VPOS-0.0,0.5]$_{a3}$, which starts to compete with the one he already had. The commentator reports this kind of interaction as follows:

```
Game 137.
  a1 is the speaker. a3 is the hearer.
  a1 segments the context into 2 objects
  a1 categorizes the topic as [VPOS-0.0,0.5]
  a1 says: "lu"
  a3 interprets "lu" as [HPOS-0.0,0.5]
  There is more than one such object
  a3 says: "lu?"
  a1 points to the topic
  a3 categorizes the topic as [VPOS-0.0,0.5]
  a3 stores "lu" as [VPOS-0.0,0.5]
```

The Table 10.1 below shows part of a vocabulary of a single agent after 3 000 language games. The table shows also the score.

Table 10.1 Vocabulary of a single agent

Form	Meaning	Score	Form	Meaning	Score
wovota	[RED-0.0,0.125]	1.0	sogavo	[GREEN-0.5,1.0]	0.0
tu	[GRAY-0.25,0.5]	0.0	naxesi	[GREEN-0.5,1.0]	0.0
gorepe	[VPOS-0.0,0.5]	0.3	ko	[GREEN-0.5,1.0]	0.0
zuga	[VPOS-0.0,0.5]	0.1	ve	[GREEN-0.5,1.0]	0.0
lora	[VPOS-0.25,0.5]	0.1	migine	[GREEN-0.5,1.0]	0.0
wovota	[VPOS-0.25,0.5]	0.2	zota	[GREEN-0.5,1.0]	0.9
di	[VPOS-0.25,0.5]	0.0	zafe	[GREEN-0.5,1.0]	0.1
zafe	[VPOS-0.0,0.25]	0.2	zulebo	[HPOS-0.0,1.0]	0.0
wowore	[VPOS-0.0,0.25]	0.9	xi	[HPOS-0.0,1.0]	0.0
mifo	[HPOS-0.0,1.0]	1.0			

We see in this table that for some meanings (such as [RED-0.0,0.125]) a single form "wovota" has firmly established itself. For other meanings, like [GRAY-0.25,0.5], a word was known at some point but is now no longer in use. For other meanings, like [VPOS-0.0,0.5], two words are still competing: "gorepe" and "zuga". There are words, like "zafe", which have two possible meanings [VPOS-0.0,0.25] and [GREEN-0.5,1.0].

Natural languages are clearly not totally coherent even in the same language community, and languages developed autonomously by physically embodied agents will not be fully coherent either.

1. Different agents may prefer a different word for the same meaning. These words are said to be *synonyms* of each other. An example is "pavement" versus "sidewalk". The situation arises because an agent may construct a new word not knowing that one is already in existence. Synonymy is often an intermediate stage for new meanings whose lexicalization has not stabilized yet. Natural languages show a clear tendency for the elimination of synonyms. Accidental synonyms tend to specialize, incorporating different shades of meaning from the context or reflecting socio-linguistic and dialectal differences of speaker and hearer.

2. The same word may have different preferred meanings in the population. These words thus become *ambiguous*. This situation may arise completely accidentally, as in the case of "bank" which can mean river bank and financial institution. These words are then called *homonyms*. The situation may also arise whenever there is more than one possible meaning compatible with the same situation. An agent on hearing an unknown word may therefore incorrectly guess its meaning. Ambiguity also arises because most words are *polysemous*: The original source meaning has become extended by metaphor and metonymy to cover a family of meanings (Victorri and Fuchs, 1996). Real ambiguity tends to survive in natural languages only when the contexts of each meaning is sufficiently different, otherwise the hearer would be unable to derive the correct meaning.

3. The same meaning may denote different referents for different agents *in the same context*. This is the case when the application of a category is strongly situated, for example 'left' for the speaker may be 'right' for the hearer. Deictic terms like "this" and "that" are even clearer examples from natural language. In natural languages, this *multi-referentiality* is counter-acted by verbalizing more information about the context or by avoiding words with multi-referential meanings when they may cause confusion.

4. It is possible and very common with a richer categorial repertoire, that a particular referent in a particular context can be conceptualized in more than one way. For example, an object may be to the left of all the others, *and* much higher positioned than all the others. In the same situation different agents may therefore use different meanings. Agents only get feedback about whether they guessed the object the speaker had in mind, not whether they used the same meaning as the speaker. This *indeterminacy* of categories is a cause of ambiguity. A speaker may mean 'left' by "bovubo", but a hearer may have inferred that "bovubo" meant 'upper'.

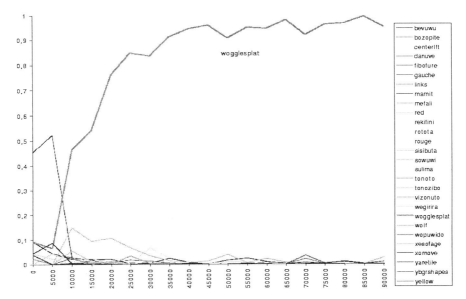

Figure 10.4 A meaning-form diagram which graphs for a specific meaning all the possible forms and their score. A winner-take-all situation is clearly observed. X-axis shows language games and y-axis the score of forms. There is a steadily growing population reaching 1500 agents towards the end.

So, although circumstances cause agents to introduce incoherence in the language system, there are at the same time opposing tendencies, attempting to restore coherence. Synonyms tend to disappear and ambiguity is avoided. In the remainder of this paper, we want to show that the dynamics of the guessing game, particularly when it is played by situated embodied robotic agents, leads unavoidably to incoherence, but that there are tendencies towards coherence as well. Both tendencies are emergent properties of the dynamics. There is no central controlling agency that weeds out synonyms or eliminates ambiguity, rather they get pushed out as a side-effect of the collective dynamics of the game. Let us examine some of the dynamics that came out of the experiments.

First it can be shown that agents indeed construct and acquire conventions in a group given the behaviors outlined above (see Steels, 1997b). Moreover if agents take turn being speaker and hearer and a speaker is allowed to invent a new symbol occasionally when he does not have an association yet to symbolize a particular representation, a set of conventions can establish itself from scratch in the population (Figure 10.4). This is due to a self-organizing positive feedback loop. Associations that are successful become even more so because their score goes up, so that they become used even more frequently and hence propagate in the rest of the population. The dynamics is similar to that of ant societies self-organizing a path or to increasing returns market phenomena as studied in non-equilibrium economics.

Second we see that agents indeed coordinate their categorial repertoires because of the tight integration between meaning and language. This is illustrated more sharply by an experiment with the guessing game (carried out by Tony Belpaeme, 2001) where a categorial repertoire for color was evolved first with and then without coupling to language. The bottom graph shows the similarities of the repertoires when the agents played discrimination games but without communicating about them. The second graph (top) shows the result when the discrimination game was coupled to a naming game. The results are shown in Figure 10.5. This demonstrates the basic point of the paper, namely that categorial coherence is reached when there is a co-evolution of language and meaning.

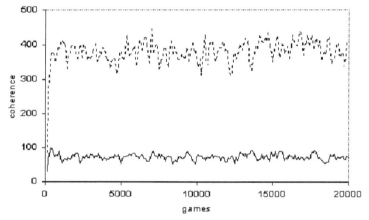

Figure 10.5 The top graph shows the categorial coherence in a group of agents when they play discrimination games coupled to naming games. The bottom graph shows categorial coherence with only discrimination games.

Conclusions

This paper advocates a tight structural coupling between processes for learning grounded representations and learning symbolizations of them. Both constrain each others' degrees of freedom and enable the learner to get feedback about the adequacy of a representation. Our experiments in simulation and on real robots have sufficiently demonstrated that applications can be constructed in a straightforward way using these principles. At the moment we are specifically targeting research on language games for humanoid robots.

This work raises many additional interesting issues. For example, there has been a longstanding debate between nativists who claim that language learning amounts to learning labels for existing categories and relativists such as Whorf who claim that each language implies a different categorization of reality. The structural coupling of concept formation and language acquisition advocated in this paper explains how a relativistic view is not only possible but unavoidable. If language enables and influences the learning of representations then it is easy to see how representations can become language specific. Of course representations are still strongly constrained by the world and tasks carried out in the world as well – they are not completely conventional or arbitrary. But they need not be innate to explain how they can become shared.

Acknowledgement

This paper was triggered by discussions with Stevan Harnad as part of the committee concerning the Ph.D. thesis by Paul Vogt and by discussions with Erik Myin on the Wittgensteinian view of language. It is based on a large amount of technical work conducted in collaboration with Tony Belpaeme, Edwin De Jong, Frederic Kaplan, Angus Mcintyre, Joris Van Looveren and Paul Vogt.

References

Aumann RJ, Hart S (eds) (1994) *Handbook of game theory with economic applications (Volume 2)*. North-Holland, Amsterdam

Axelrod R (1984) *The evolution of cooperation*. Basic Books, New York

Belpaeme T (2001) Simulating the formation of color categories. Submitted to *ECAL-2001 European Conference on Artificial Life*

Brooks R (1999) *Cambrian Intelligence: The early history of the new AI*. MIT Press, Cambridge MA

Cangelosi A, Greco A, Harnad S (2000) From robotic toil to symbolic theft: Grounding transfer from entry-level to higher-level categories. *Connection Science*, 12: 143-162

de Jong ED (1999) Analyzing the evolution of communication from a dynamical systems perspective. In: *Proceedings of the European Conference on Artificial Life ECAL'99*. Springer-Verlag, Berlin, pp 689-693

Eco U (1968) *La struttura assente*. Milano

Harnad S (1990) The symbol grounding problem. *Physica D*, 42: 335-346

Horswill I (1993) Polly: A vision-based artificial agent. In: *Proceedings of the Eleventh National Conference on Artificial Intelligence (AAAI-93)*. MIT Press, Washington DC

Lieberman P (1991) *Uniquely Human. The evolution of speech, thought and selfless behavior*. Harvard University Press, Cambridge Ma.

Lindgren K, Nordahl MG (1995) Cooperation in artificial ecosystems In: Langton CG (ed) *Artificial Life: An overview*. MIT Press, Cambridge MA, pp 15-37

Marco R, Sebastiani P, Cohen PR (2000) Bayesian analysis of sensory inputs of a mobile robot. In: *Proceedings of the Fifth International Workshop on Case Studies in Bayesian Statistics (Lecture Notes in Statistics)*. Springer, New York

Marr D (1982) *Vision*. Freeman, San Francisco

Maynard Smith J (1982) *Evolution and the theory of games*. Cambridge University Press, Cambridge UK

Nilson NJ (1984) Shakey the robot. SRI AI Center Technical Note, 323

Oliphant M (1996) The dilemma of Saussurean communication. *Biosystems*, 37: 31-38

Pfeifer R, Scheier C (1999) *Understanding intelligence*. MIT Press, Cambridge MA

Steels L (1996) Self-organising vocabularies. In Langton CG, Shimohara K (eds) *Proceedings of the ALIFE V*. MIT Press, Cambridge MA, pp 179-184

Steels L (1997a) Constructing and sharing perceptual distinctions. In: van Someren M, Widmer G (eds) *Proceedings of the European Conference on Machine Learning*. Springer-Verlag, Berlin

Steels L (1997b) The synthetic modeling of language origins. *Evolution of Communication*, 1: 1-35

Steels L (1998) The origins of syntax in visually grounded robotic agents. *Artificial Intelligence*, 103: 1-24

Steels L (1999) How language bootstraps cognition. In: Wachsmutt I, Jung B (eds) *KogWis99 Proceedings der 4 Fachtagung der Gesellschaft fuer Kognitionswissenschaft* Infix, Sankt Augustin, pp 1-3

Steels L, Brooks R (eds) (1995) *The Artificial life route to artificial intelligence: Building embodied, situated agents*. Lawrence Erlbaum Assoc, New Haven

Steels L, Kaplan F (1998) Situated grounded word semantics. In: *Proceedings of IJCAI-99*. Morgan Kauffman Publishing, Los Angeles, pp 862-867

Talmy L (2000) *Toward a cognitive semantics: Concept structuring systems (Language, speech, and communication)*. MIT Press, Cambridge MA

Victorri B, Fuchs C (1996) *La polysemie construction dynamique du sens*. Hermes, Paris

Part V

BEHAVIORAL AND NEURAL FACTORS

Chapter 11

Grounding the Mirror System Hypothesis for the Evolution of the Language-ready Brain

Michael A. Arbib

Introduction

What was the evolutionary path whereby humans came to have brains so well-suited for language? There has been much debate as to whether language evolved by the modification of systems present in non-human primates, or by some more radical, discontinuous evolutionary change. Noam Chomsky, perhaps the most influential linguist of the 20th century, has often suggested that there is nothing remotely resembling human language anywhere in the animal repertoire but that our genetic constitution includes a Universal Grammar – the basic ground plan for the grammar of all possible human languages. Certainly, the human brain and body evolved in such a way that we have hands, larynx, facial mobility and the brain mechanisms needed to produce and rapidly perceive and generate sequences of gestures that can be used in language. However, rather than accept Chomsky's hypothesis, Arbib (2001) offers the counter-hypothesis that the human brain and body are *language-ready* in the sense that the first *Homo sapiens* used a form of vocal communication which was but a pale approximation of the richness of current languages, and that these languages evolved *culturally* as a more or less cumulative set of "inventions".

Language-readiness includes many properties, but our approach starts with parity – the property that *What counts for the speaker must count in much the same way for the listener*. The processes of production and perception must somehow be linked. We understand when one individual is attacking another or when someone is peacefully eating an apple, that is, we perceive the intention. How do we do it? What is shared by the (in this case involuntary) sender and by the receiver? Is this mechanism the precursor of willed communications? Our approach is rooted in the observation by Giacomo Rizzolatti and his colleagues in Parma, Italy (Rizzolatti *et al.*, 1995) that the monkey brain contains a set of neurons (*mirror neurons*) each of which is active both when the monkey itself is carrying out a particular action with his hands or when the monkey is observing the same action executed by another. Studies of humans then suggested that the human brain also contains a mirror system for gesture recognition and that this includes Broca's area (Rizzolatti *et al.* 1996). But Broca's area in the left hemisphere of the human brain is crucial for language production (Arbib and Rizzolatti, 1997; Rizzolatti and Arbib 1998) thus argued that "Language is Within our Grasp", and introduced the *Mirror System Hypothesis of Language Evolution*, namely that speech derived from an ancient gestural system based on the mirror mechanism – the link between observer and actor became, in speech, a link between the sender and the receiver of messages.

The second part of this paper will offer a fuller exposition of the mirror system hypothesis and then go "beyond the mirror" to offer more detailed hypotheses about the evolution of the language-ready brain. However, to ground our research project on the evolution of the human brain, we need a detailed understanding of the monkey brain to ground our notions about the brain of the common ancestor of human and monkey. For this reason, the first part of the article introduces computational models of monkey mechanisms for the control of grasping and for the mirror system for grasping.

Computational Models of Monkey Mechanisms for the Control of Grasping

The FARS model of visually-directed grasping

Let me start by presenting concepts in modeling the control of grasping movements of the hand that go back to the 1980s. The key observation that piqued my interest in the subject was the work of Marc Jeannerod and his colleagues from Lyon in France (Jeannerod and Biguer, 1982). When you reach out to grasp an object, you not only transport your hand towards the object but also preshape your hand in a way that anticipates – but is a little bigger than – the shape of the object. The movement concludes when you decrease the aperture between thumb and fingers to contact the object and establish a firm grasp. This work inspired me to conceptualize reaching to grasp in terms of *perceptual schemas* recognizing the key properties of the object and thence providing the input for *motor schemas* which control the motion of the arm and hand (Arbib, 1981). The way in which these various schemas are linked is reminiscent in some ways of the passing of activation in a conventional serial computer program and in other ways of the flow of data between simultaneously active blocks in a control diagram. I thus call the result an example of a *coordinated control program*. For this skill of grasping, the perceptual schema for visual location of the object provides key data to the hand-reaching schema, which instructs the arm first to make a quick movement and then to complete that with a slow movement. Meanwhile, perceptual schemas for recognition of size and orientation provide the inputs needed to preshape the hand and rotate it appropriately.

The particular model shown here has been updated by subsequent work. In particular, we developed notions on how to look more carefully at how the hand shapes itself in relation to the object (Iberall, Bingham and Arbib, 1986). The first idea is that of a *virtual finger* – when we grasp things, often several fingers move together as a functional unit applying force to the object. It is that collection of fingers grouped together that I call a virtual finger. The other concept is that of an *opposition space*. A basic grasp is obtained by moving two virtual fingers towards each other in such a way that they end up contacting two sides of the object. We thus have two opposition axes: one which provides the basic degree of freedom of the movement of the hand as the two contact surfaces move together, and one which defines the target within the object. The goal of a successful reach and grasp is to move and preshape the hand so that the opposition axis in the hand is aligned with the opposition axis in the object and then contact is made.

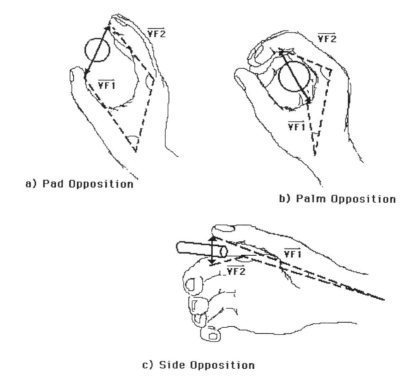

a) Pad Opposition

b) Palm Opposition

c) Side Opposition

Figure 11.1 Each of the three grasp types here is defined by specifying two "virtual fingers" which define two "opposition axes": the *opposition axis in the hand* joining the virtual finger regions to be opposed to each other, and the *opposition axis in the object* joining the regions where the virtual fingers contact the object. (Iberall and Arbib, 1990)

Figure 11.1 shows three types of opposition: the *pad opposition*, which in the particular case when virtual finger two consists only of the index finger is also known as *precision pinch*; palm opposition which is also known as a *power grip*; and finally *side opposition* in which the thumb is opposed to the side of the index finger.

With that background we can move from general functional considerations to some of the data that have guided us in our attempt understand what is actually happening in the brain. These data come from Giacomo Rizzolatti's lab in Parma in Italy and also from Hideo Sakata's lab in Tokyo in Japan (e.g., Sakata *et al.*, 1997). Figure 11.2 shows a side view of the brain of a monkey, highlighting a number of areas including area AIP in parietal cortex and area F5 in frontal cortex. The area shown as F1 is the classical motor cortex whose neurons control both the motor neurons in the spinal cord and other parts of the brain that control motor behavior less directly. F5 is part of premotor cortex, which deserves its name in two ways: Premotor cortex is physically in front of the motor cortex, and it is also premotor in the sense that activity in F5 preceded hand related activity in motor cortex, F1.

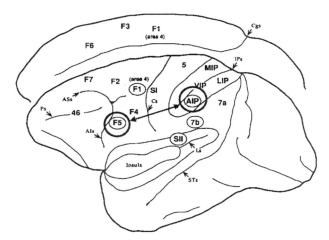

Figure 11.2 A side view of the left hemisphere of the macaque monkey brain dominates the figure, with a glimpse of the medial view of the right hemisphere above it to show certain regions that lie on the inner surface of the hemisphere. (Adapted from Jeannerod, Arbib, Rizzolatti and Sakata, 1995.)

Cells in AIP are excited when the monkey is looking at an object but, interestingly, the firing of a particular neuron does not correlate with the identity of the object but instead with the way in which the object will be grasped (Taira *et al.*, 1990). We say that AIP processes the visual input to come up with *affordances* for grasping. The term "affordance" is due to the American psychologist J.J. Gibson (1966). He distinguishes between processing the sensory input to identify what an object is from processing the sensory input to decide how to interact with the world. Gibson emphasized the notion of optic flow, the way in which the passage of stimuli across the retina can signal an impending collision. For example, when you are walking down a busy street you can avoid collisions without recognizing whom or what you are avoiding. In the same way, here we are talking about affordances for how to grasp objects as distinct from recognizing what the identity of those objects is. AIP is reciprocally connected with F5, which is where Rizzolatti and his colleagues find neuron activity which correlates with the type of grasp that the monkey is about to execute.

To measure activity of a particular cell of the brain, the neurophysiologist uses a microelectrode and places it so that it will track the membrane potential of a single neuron amongst the many, many million neurons of the brain. Consider the right panel of Figure 11.3. At the top are several lines of dots. Each dot corresponds to an action potential, the message that a neuron sends down its output line or axon to all the other neurons to which it sends connections. On ten consecutive trials the monkeys were observed grasping a raisin using a precision pinch. As we can see, on every trial measured from this one particular F5 neuron, a great deal of activity preceded the pinch. This is shown more dramatically by the histogram which sums up all the individual traces. In general we find that different cells in F5 have activity correlated with different types of grasps or with other hand motions such as tearing or twisting back and forth.

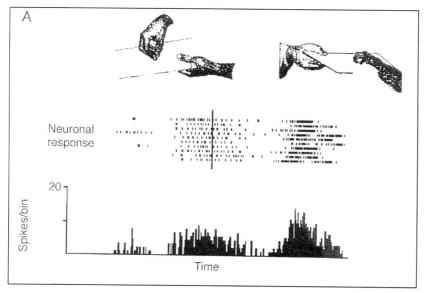

Figure 11.3 Mirror neurons constitute a subset of the grasp-related premotor neurons of F5 which discharge not only when the monkey executes a certain class of actions (right) but also when the monkey observes more or less similar meaningful hand movements made by the experimenter or another monkey (left).

Figure 11.4 provides an overview of how F5 (or premotor cortex more generally) and AIP (or posterior parietal cortex more generally) may work together

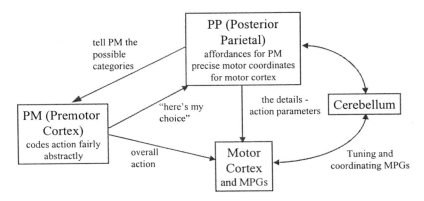

Figure 11.4 An overview of how premotor cortex and posterior parietal cortex may work together to establish the behavior of the human or animal.

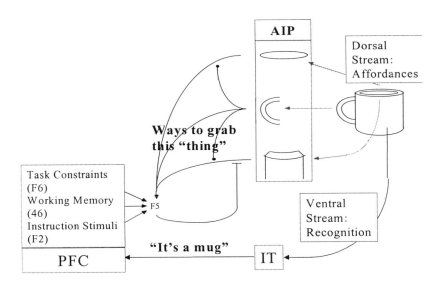

Figure 11.5 Overview of the FARS Model. AIP extracts the affordances and F5 selects the appropriate grasp from the AIP 'menu'. Various biases are sent to F5 by PFC which relies on the recognition of the object by IT.

to establish the behavior of the human or animal. Our view here is that parietal cortex informs premotor cortex of what the possible affordances are. Then, employing information from many other parts of the brain, premotor cortex will make its choice and it will then both prime motor cortex and send a signal back to inform parietal cortex of its choice. Meanwhile, parietal cortex has been gathering precise metric information about the appropriate movement. Thus, once premotor cortex tells it what the chosen action is, it can send the necessary coordinates down to motor cortex, which can then instruct the Motor Program Generators (the MPGs) on how to proceed. Although it is not important for this article, I also note that usually we need the cerebellum to fine tune the activity of the motor cortex and compensate for any interaction forces in arm and hand dynamics to achieve a smooth and graceful movement.

We turn from this general view to focus on the FARS (Fagg-Arbib-Rizzolatti-Sakata) model (Fagg and Arbib, 1998) (Figure 11.5). We see how the dorsal stream – in other words, the pathway from primary visual cortex up to parietal cortex – extracts the affordances for how the object can be grasped. The ventral stream – the stream that proceeds downwards towards IT (inferotemporal cortex) – provides neural machinery to help recognize what the object is, in this case a mug. IT sends this classification on to pre-frontal cortex (PFC) which can use things like task constraints, working memory and various instructions to bias F5 to choose which one of the affordances it is receiving from AIP will determine the future course of action. In the second part, we will recall that pre-frontal planning mechanisms influence F5, but for now I will focus on the top pathway whereby visual information about an object passes through AIP to F5 to generate an appropriate action and then address a new question: How can the brain recognize the action performed by somebody else.

The MNS model of the mirror system for action recognition in grasping

This brings us to the discussion of the data that caused Rizzolatti and his colleagues to talk about a mirror system in the brain. At top right of Figure 11.3, we saw a monkey reaching out towards a raisin on a tray held by the experimenter, and we see the firing pattern earlier discussed as typifying F5 activity. What adds a whole new dimension to the investigation is the top left panel in which both hands belong to the experimenter. As the monkey sees the experimenter execute the precision pinch, we again see strong activity, though not quite as strong as before, in the F5 neuron under study. Rizzolatti *et al.* (1995) found that amongst the F5 neurons which are active when the monkey itself reaches for an object in a particular way (as shown at right of Figure 11.3), there is a major *subset* of neurons which are also active when the monkey observes another monkey or a human conducting the same action (as shown at left of Figure 11.3). They call these *mirror neurons* – we now use the term F5 *canonical neurons* for those grasp-related neurons in F5 that do not have the mirror property but are responsive to the sight of a graspable object. Between them, the canonical and mirror neurons form only about 20% of all the "motor neurons" in F5, i.e., those neurons whose activity correlates with motor activity whether or not visual input is involved. Canonical neurons are active when the monkey itself is doing the grasping in response to sight of an object but not when the monkey sees someone else do the grasping. Mirror neurons always show a link between the effective observed movement and the effective executed movement. Thus F5 in the monkey is endowed with an "observation/execution matching system" and we refer to this as the *mirror system for grasping*. We can think of the activity of mirror neurons as providing a code for action – a code which is shared between the execution system that produces the monkey's own actions and the observation system involved in recognizing actions by other primates.

The monkey mirror system, as observed neurophysiologically by Rizzolatti *et al.*, is concerned with observation of a single action that is already in the monkey's repertoire. Possible roles for such a mirror system may include:

(1) Self-correction: based on the discrepancy between intended and observed self action.

(2) Social interaction. By anticipating what action another monkey has begun, a monkey can determine how best to compete or cooperate with the other monkey.

(3) Learning by imitation at the level of a single action

My hypothesis is that it is function (1) that led to the evolution of the mirror system – i.e., by being able to respond to the relation between hand and object, it laid the basis for generalization from self hand-movements to other's hand movements. (2) was then a byproduct, an "exaptation", although it is the property that most people emphasize when discussing mirror neurons. Mirror neurons enable a monkey to see what other monkeys are doing. This relates to the notion of understanding, of not simply seeing a movement as a movement of a hand but rather recognizing it is a goal-directed action. However, I regard true "understanding" as involving the interaction of many brain regions, not just activity within the F5 mirror neurons. Of course, once one has abstract representations of actions as distinct from unanalyzed records of movements, one is in business for a great deal of further sophistication. Function (3) does not seem to be at all developed in monkeys – monkeys seem poor at imitation in any extended sense – but I will argue below that the evolution of the mirror neuron system to support imitation was a major step on the way to developing a language-ready brain in the hominids.

I now want to focus on the modeling goals that the discovery of the mirror system has posed for us. We have already seen that the FARS model has two primary components, the recognition of the object affordances and the selection by F5 of an appropriate grasp from this menu of affordances. As we move on to the new model, the mirror neuron system or MNS model, we not only have to recognize object affordances but we also have to recognize how the hand is moving and preshaping.

Of course, just as we had to have the parietal region AIP provide visual processing appropriate to extracting the grasp-related affordances of an object so we have to find other regions of the brain that provide appropriate data for trajectory and preshape (Figure 11.6). The caudal intraparietal sulcus or cIPS is not part of this new system but rather provides information about the shape of the object that AIP needs to do its job looking at the orientation and so on of the surfaces of the object. Two other brain regions in Figure 11.6 are in the parietal cortex: PG, which seems to be particularly good at spatial coding for objects, including motion during interaction of objects as well as self motion, and PF which seems more related to somatosensory information, touch and so on but again related to mirror-like responses. The last region, STS (Superior Temporal Sulcus) is in temporal cortex rather than parietal cortex but seems to be very important in detecting biologically meaningful stimuli such as hand movements as well as having sub-regions MT and MST that encode motion related activity.

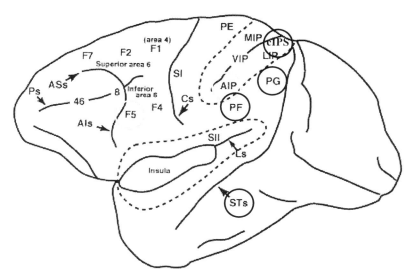

Figure 11.6 cIPS (caudal intraparietal sulcus) extracts axis orientation and surface orientation to provide input for the computation of both affordances and object identity. PF (rostral part of the posterior parietal lobule) and PG (caudal part of the posterior parietal lobule) provide key parietal input for mirror neurons, analyzing motion during interaction of objects and self-motion. STS (Superior Temporal Sulcus) has many subregions, including those involved in detection of biologically meaningful stimuli (e.g., hand actions) and motion-related activity (in areas MT and MST).

The key criteria for activating a mirror neuron are as follows:

(1) The preshape that the monkey is seeing corresponds to the grasp that the mirror neuron encodes.

(2) The preshape that the observed hand is executing is indeed appropriate to the object that the monkey can see (or remember).

(3) The hand must be moving on a trajectory that will indeed bring it to grasp the object.

In modeling this, we could just have tried to explicitly program the various portions of our overall system to yield appropriate mirror activity. However, the

MNS model of Oztop and Arbib (2001) starts with a "brain" in which the F5 canonical neurons are already controlling an interesting set of grasps and then has the mirror neurons learn to recognize how motion of the hand relative to an object correlates with F5 canonical neuron activity during self-generated movements. We will then see how this can be the basis for the recognition of movements executed by others. An interesting sub-issue is that it will be adaptive for the monkey if it can recognize the other monkey's actions as soon as possible. Thus, the ability to activate mirror neurons from smaller and smaller samples of a trajectory will be an important criterion for further developments in the modeling.

Figure 11.7 diagrams the MNS model of the mirror system. It highlights the components that Oztop and Arbib (2001) see as crucial for operation of the mirror system in the monkey. They also posit that such mechanisms in the brain of the human infant underlie the infant's learning to reach and grasp. Humans can perceive variation and assemblages of familiar movements. Our future research will extend the model to imitation, i.e., cases in which the perception of a novel movement helps establish the ability to perform some approximation thereof.

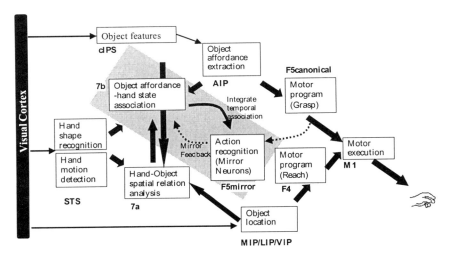

Figure 11.7 The MNS (Mirror Neuron System) model. (i) Top diagonal: a portion of the FARS model. Object features are processed by AIP to extract grasp affordances, these are sent on to the canonical neurons of F5 that choose a particular grasp. (ii) Bottom right. Recognizing the location of the object provides parameters to the motor programming area F4 which computes the reach. The information about the reach and the grasp is taken by the motor cortex M1 to control the hand and the arm. (iii) New elements of the MNS model: Bottom left are two schemas, one to recognize the shape of the hand of the actor being observed by the monkey whose brain we are interested in, and the other to recognize how that hand is moving. Just to the right of these is the schema for hand-object spatial relation analysis. It takes information about object features, the motion of the hand and the location of the object to infer the relation between hand and object. Just above this is the schema for object associating affordances and hand state. Together with F5 canonical neurons, this last schema provides the input to the F5 mirror neurons.

Along the top diagonal there is a portion of the FARS model (cf. Figure 11.5). Object features are processed by AIP to extract grasp affordances, these are sent on to the canonical neurons of F5 that choose a particular grasp. The bottom right of the figure shows how recognizing the location of the object can provide appropriate parameters to another motor programming area – in this case F4, which is adjacent to F5 in premotor cortex – which computes the reach. The information about the reach and the grasp is taken by the motor cortex M1 to actually control the hand and the arm.

The other schemas add the new functionality that allows the mirror neuron system to do its job. At the bottom left we have two schemas which may be localized in area STS of the monkey brain, one to recognize the shape of the hand of the actor being observed by the monkey whose brain we are interested in, and the other to recognize how that hand is moving. Just to the right of these is the schema for hand-object spatial relation analysis. It takes information about object features, the motion of the hand and the location of the object to infer the relation between hand and object. Just above this is the schema for object associating affordances and hand state, which may be in area 7b of the monkey brain. We want to understand how information coming from the F5 canonical neurons during the monkey's own movements can be used to enable the F5 mirror neurons to learn how to recognize actions. We also want to understand how the mirror neurons are activated not only as an accompaniment to the monkey's own movement but also when the monkey observes a similar action by someone else.

We recognize an action when we see that the way in which the hand is moving and shaping is indeed appropriate to one of the affordances of an object. Clearly, in the initial stage of someone else's grasping movement you may not yet be sure as to whether they will be engaged in a precision pinch or a power grip or whatever. Thus it becomes necessary to monitor the possible movement of a variety of portions of the hand along what may turn out to be the appropriate opposition axis for the movement that will be finally recognized. For this reason, Oztop and Arbib (2001) have defined the *hand state* to characterize the hand in relation to the object to be grasped. The definition of the hand state was influenced by our prior analysis of opposition space (Figure 11.1). These parameters provide the data for judging to what extent the hand is being positioned to grasp a particular object. Our current representation of hand state defines a 7-dimensional trajectory *F(t)* with the following components:

$F(t) = (d(t), v(t), a(t), o_1(t), o_2(t), o_3(t), o_4(t))$ where

$d(t)$: distance to target at time t

$v(t)$: tangential velocity of the wrist

$a(t)$: aperture of the virtual fingers involved in grasping

$o_1(t)$: angle between the object axis and the (*index finger tip – thumb tip*) vector

$o_2(t)$: angle between the object axis and the (*index finger knuckle – thumb tip*) vector

$o_3(t)$, $o_4(t)$: the two angles defining how close the thumb is to the hand as measured relative to the side of the hand and to the inner surface of the palm.

Notice that we have carefully defined the hand state in terms of relationships between hand and object. This has the beauty that it will work just as well for measuring how the monkey's own hand is moving to grasp an object as for observing how well another monkey's hand is moving to grasp the object. This is the secret which allows self-observation by the monkey to train a system that can be used for observing the actions of others and recognizing just what those actions are.

Oztop and Arbib (2001) decomposes the mirror neuron system model into three large schemas, for generation of the reach and grasp, analysis of visual input, and the core of the mirror circuit. They use artificial neural nets or other computer programs to capture the high-level function of each schema An artificial neural network is not meant to model actual neural circuitry in the brain but is rather a way of determining whether the input/output behavior needed for a functional schema can be obtained by applying a training method to match some desired function. The disadvantage of using artificial neural nets is that the results may not tell us much about the biology, but these simulations have already yielded predictions for further work on the neurophysiology of the mirror system.

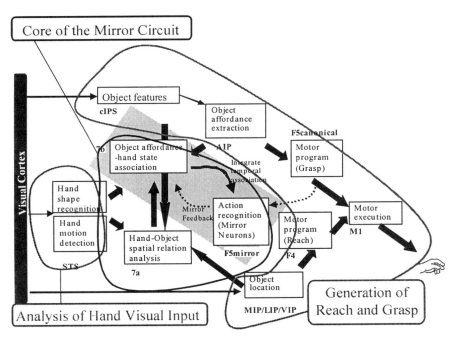

Figure 11.8 The MNS system viewed as involving the interaction of three "grand schemas", for generation of the reach and grasp, analysis of hand visual input, and the core of the mirror circuit.

The details of Oztop and Arbib (2001) are beyond the scope of this chapter, but here I briefly review the work of Erhan Oztop in simulating the three "grand schemas" of Figure 11.8:

1. He built a system which can go from real visual input showing a hand (with colored patches on various joints as well as the wrist to simplify processing) to a representation of the key parameters describing that hand. Two back-propagation networks are used to extract key information about the different features that characterize the view of the hand. The output of the feature extraction process is then used to drive an error minimization process which will find the orientation and joint angles for a 3-D model of the hand (a simple model of linked cylinders based on the actual bones of a human or monkey hand) which provides a two-dimensional view that best matches the given hand image. It is from this parametric representation of the shape of the hand that we can drive our grasp classification process. This, then, implements the schema for Analysis of Hand Visual Input.

2. He designed a reach/grasp simulator that takes as input the representation of the shape and position of an object and the initial position of the arm and hand and yields as output a trajectory which successfully results in a simulated grasping of the object. Figure 11.9 (left) shows the state reached by the hand/arm simulator after executing a power grasp. Figure 11.9 (right) shows the time series for the hand state associated with the power grasp trajectory. For example, the squares show the distance from the hand to the object decreases until the grasp is completed. Similarly, the pluses show how the aperture of the hand first increases to yield a safety margin larger than the size of the object and then decreases until the hand .contacts the object with the aperture corresponding to the width of the object along the axis on which it is grasped. In the brain of the monkey, these hand state trajectories would be extracted by analysis of the visual input. But Oztop has already designed algorithms that can extract the hand state from the visual input (item 1 above). So in the rest of this study we are going to use these simulated hand state trajectories so we can concentrate on the action recognition system without keeping track of details of visual processing.

3. The Core of the Mirror Circuit has to recognize whether the motion of the fingers of the hand is directed towards forming a preshape appropriate to the observed object and whether the motion of the wrist is on a trajectory appropriate to bringing the hand towards the object. This circuit takes as input the object affordance and the hand state generated by the simulator for a particular trajectory. The association neurons combine information on object affordance with information on hand state. The mirror neurons, in the middle of Figure 11.8, must learn to respond appropriately. Initially, the mirror neurons will be receiving coherent input only from the F5 canonical neurons so that for each particular grasp encoded by the activity of appropriate canonical neurons there will be a sub-set of mirror neurons that are active. But at first they are not yet mirror neurons! They are mirror neurons in the making because they do not yet respond alone to object affordance and hand state information as combined by the association neurons. However, time and time again the appropriate hand state and affordance information input coming into the mirror neurons will be paired with their activation by the motor program and eventually the plastic

synopses will become strong enough that the mirror neuron can be activated by a sufficient sample of hand state and affordance information even without input from the canonical neurons. The beauty of this is that, given the ingenious construction of the hand state, we now have a system that can respond to the actions of others as well as to self-generated actions.

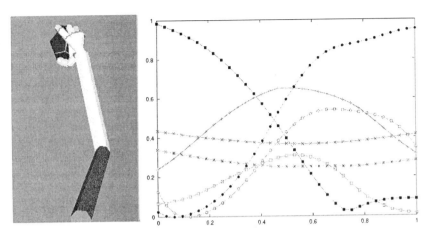

Figure 11.9 (Left) The final state of arm and hand achieved by the reach/grasp simulator in executing a power grasp on the object shown. (Right) The hand state trajectory read off from the simulated arm and hand during the movement whose end-state is shown at left. **Legend:** +: aperture; *: angle 1; **x**: angle 2; •: 1-axisdisp1; • :1-axisdisp2; • : speed; ■: distance.

The result of training is a system that processes successive hand states and builds up an output that expresses a confidence level that the behavior that is being observed corresponds to the particular grasp type correlated with each particular mirror neuron. Figure 11.10 examines the interesting case of a trajectory reaching towards the small object shown at top right in which the original movement of the thumb in relation to the other fingers is somewhat ambiguous. It looks like the hand is shaping up for a power grasp but in the end a precision pinch is actually executed. The lower panel shows the activity of two mirror neurons. At first the mirror neuron for the power grasp is far more active, but as the hand gets closer to the object it becomes clear that only the thumb and the index fingers are moving to contact the object. At this stage the power grip loses out because the initial hypothesis is destroyed. We finally see the system converging with the high activation of the mirror neuron for the precision grip signaling success in classifying the action.

A very important goal for my future work on the mirror system is in neuroscience, actually understanding how the brains of monkeys and humans do their job. Instead of looking at artificial networks that yield the desired input/output behavior of a schema, we will study networks which include neurons which behave in the way that neurophysiologists observe. To complete the model, I will have to hypothesize neurons which complement those that have already been observed. This will allow us to suggest new experiments for our neural scientist colleagues to gain more understanding into the way the brain works. For example, Figure 11.10 shows initial activity of a power grasp mirror neuron losing out to the eventual triumph of a precision pinch mirror neuron. Unfortunately, no neurophysiologist has done this sort of experiment to see if there is indeed neural behavior of this kind and I think we can now encourage new experiments which yield more insight into computational mechanisms of the brain.

With this background, we now turn from the modeling of the mirror system for grasping in the monkey to a grand speculative sweep on the evolution of human language-readiness.

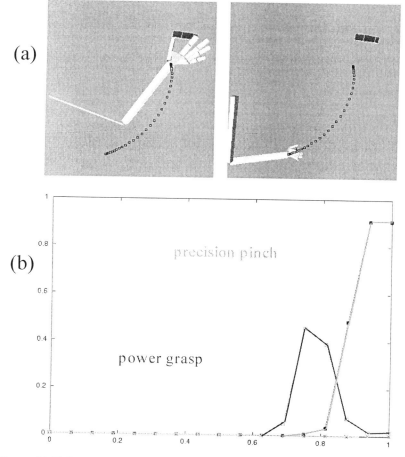

Figure 11.10 Power and precision grasp resolution. The conventions used are as in the previous figure. (a) The right shows the initial configuration and the left shows the final configuration of the hand. (b) The curves for power and precision cross towards the end of the action showing the change of decision of the network.

The Mirror System Hypothesis for the Evolution of Language, and Beyond

Behavior in general is not restricted to repeating actions we already know. Much of our behavior expands our repertoire by making variations on a familiar skill. We also learn complex skills by coming to understand how to put together various simpler actions that we already know how to perform. Once you get this far, you have a system – whether it is an artificial brain or an actual biological brain – that is well equipped for imitation. I thus see one of the major challenges for research on the mirror system to be able to understand how imitation comes about through the ability to both generate and recognize variations on assemblages of familiar actions. Monkeys do not seem to have the ability to imitate at all freely. Presumably, our common ancestors with the monkeys of 20 million years ago could not imitate either. So another biological challenge is to understand how a biological network could allow a basic range of mirror neuron activity and yet not support imitation. Then we can ask what sort of things were added in the evolution of human neural architecture over the last 20 million years that allows us to be so skillful at imitation.

The first part of this chapter has presented the monkey brain's mirror neuron system for grasping. As noted in the Introduction, Rizzolatti *et al.* (1996) used PET imaging to show highly significant activation in the rostral part of Broca's area for both the execution and observation of grasping as against the control task of object observation. Broca's area is thus not just a speech area, but is also involved in complex motor skills involving the hands as well as vocalization. Many authors have hypothesized that, in the evolution of the human capacity for language, some sort of prototypical form of communication using hand gestures provided a crucial link from our early monkey-like ancestors to the language-speaking creatures we are today (cf. Hewes, 1973; Corballis 1991; 1992; Kimura 1993; Armstrong *et al* 1995). On this view, the "generativity" which some see as the hallmark of language (i.e., its openness to new constructions, as distinct from having a fixed repertoire like that of monkey vocalizations) is present in motor behavior which can thus supply the evolutionary substrate for its appearance in language. What Rizzolatti and Arbib (1998) added to the manual origins hypothesis is that the mirror system provides a possible neural "missing link" in the evolution of human language-readiness:

The Mirror-System Hypothesis: Broca's area in humans evolved from a basic mechanism *not* originally related to communication – the *mirror system for grasping in the common ancestor of monkey and human.* The mirror system's capacity to generate *and* recognize a set of actions provides the evolutionary basis for *language parity,* in which an utterance means roughly the same for both speaker and hearer.

Note that the Mirror System Hypothesis does *not* say that having a mirror system is equivalent to having language. Monkeys have mirror systems but do not have language, and we expect that many species have mirror systems for varied socially relevant behaviors. The challenge here is to go "Beyond the Mirror" to understand the human ability to rapidly acquire a vast array of flexible strategies for pragmatic and communicative action; and the ability to generate and comprehend hierarchical structures "on the fly." For want of better data, we will assume that the common ancestor of humans and monkeys shared with monkeys a primate call system (a limited set of species-specific calls) and an oro-facial gesture system (a limited set of gestures expressive of emotion and related social indicators). However, the following two observations suggest that the primate call system does not provide the evolutionary core for human speech:

(i)   combinatorial properties for the openness of communication are virtually absent in basic primate calls and oro-facial communication, even though individual calls may be graded;

(ii)   the neural substrate for primate calls is in a region of cortex distinct from F5, which we have seen to be the monkey homologue of human Broca's area.

Our challenge in charting the evolution of human language, which for most humans is so heavily intertwined with speech, is thus to understand why it is F5, rather than the area already involved in vocalization, which provided the evolutionary substrate for language. The evolution of language-readiness depends not only on the evolution of the F5 mirror system of the common ancestor, but also on the co-evolution of much other circuitry to which F5 is related. Arbib (2001) has postulated – and thus set many of the goals for the proposed research – that the progression from grasp to language through primate and hominid evolution proceeded via seven stages: (1) grasping, (2) a mirror system for grasping, (3) a "simple" imitation system, (4) a "complex" imitation system, (5) a manual-based communication system, (6) protospeech, and (7) language. Stages 1 and 2 follow from the work of Rizzolatti and his colleagues and have been described earlier. The transition from Stages 1 and 2 to Stages 5 and 6 was hypothesized by Rizzolatti and Arbib (1998). The insertion of Stages 3 and 4 is due to Arbib (2001) –arguing that "imitation" takes us beyond the "basic" mirror system for grasping, and that the ability to acquire novel sequences if the sequences are not too long and the components are relatively familiar took the hominid line beyond the level of imitation that humans share with other extant primates. Arbib (2001) also introduces the notion that stages 6 and 7 are separate – characterizing *protospeech* as being the open-ended production and perception of sequences of vocal gestures, without implying that these sequences constitute a language. In the rest of this chapter, I briefly review the argument for these stages, while stressing that I am *not* suggesting that sudden evolutionary leaps supported sharp transitions from one stage to another. Instead, the stages summarize what may have been the cumulative effect of a multitude of biological and biosocial innovations.

Stage 3: A "Simple" Imitation System. Twenty million years separate monkeys and humans from their common ancestor, while five million years separate chimps and humans from their common ancestor (cf. Gamble, 1994, Figure 4.2). How have the mirror systems of monkey and human diverged from that of their common ancestor? How have the mirror systems of chimp and human diverged from that of their common ancestor?

Since language played no role in the evolution of monkey or chimp or the common ancestors we share with them, any changes we chart prior to the hominid line should be shown to be adaptive in their own right, rather than as precursors of language. It is clear that mirror neurons may well be fundamental to *imitation*, so that the utility of the mirror system in the common ancestor of human and monkey may have resided in simple forms of imitation as well as in the infant's learning how to observe its own motor behavior, and in learning how to relate its own actions to those of others. In any case, I argue that *extension of the mirror system from single actions to compound actions* was a key innovation in the brain's evolution relevant to thee emergence of language-readiness.

Stage 4: A "Complex" Imitation System. I next hypothesize that the transition from australopithecines to early *Homo* coincided with the transition from a mirror system used only for action recognition and "simple" imitation to more elaborate forms of imitation that depended on/provided evolutionary pressure for the elaboration of a whole complex of systems that integrated the F5 mirror system for execution/observation of single actions into a far larger system for the execution/observation of complex behaviors. Putting this another way, I argue that what marks hominids as distinct from their common ancestors with chimpanzees is the ability to rapidly exploit novel sequences as the basis for immediate imitation or for the immediate construction of an appropriate response, as well as contributing to the longer-term enrichment of experience. Of course, even this human ability requires that the sequences be not too long and the components be relatively familiar,

Stage 5: A Manual-Based Communication System. I then hypothesize that this ability for "complex" imitation formed the basis for an increasingly human-like mirror system used for intentional communication. However, I want to stress again the thesis that biological evolution created a "language-ready" brain, and that it took a long period of cultural evolution for humans to extend their basic capability for communication to symbol systems with the complexity that characterizes all known human languages. Our hypothetical sequence (Rizzolatti and Arbib, 1998) for manual gesture is then:

(i)   pragmatic action directed towards a goal object;

(ii)   imitation of such actions;

(iii) pantomime in which similar actions are produced in the absence of a goal object;

(iv)   abstract gestures divorced from their pragmatic origins (if such existed). In pantomime it might be hard to distinguish a grasping movement signifying "grasping" from one meaning "a [graspable] raisin", thus providing an "incentive" for coming up with an arbitrary gesture to distinguish the two meanings;

(v)   the use of such elements for the formation of compounds which can be paired with meanings in more or less arbitrary fashion.

My current hypothesis is that stages (iii) and (iv) were present in pre-human hominids, that (v) was present in a rather limited form in *Homo erectus*, and that the "explosive" development of (v) that we know as language depended on "cultural evolution" well after biological evolution had formed modern *Homo sapiens*. This remains speculative, and one should note that biological evolution may have continued to reshape the human genome and brain even after the skeletal form of *Homo sapiens* was essentially stabilized.

Stage 6: Protospeech. We earlier noted that the neural substrate for primate calls is in a region of cortex distinct from F5, which we have seen to be the monkey of human Broca's area. We thus need to explain why F5, rather than the a priori more likely "primate call area", provided the evolutionary substrate for speech in particular, and language in general. Rizzolatti and Arbib (1998) answer this by suggesting two evolutionary stages going beyond basic vocalization and oro-facial gesture, the first of which is really stage (v) in the above list:

(v) A *distinct* manuo-brachial communication system evolved to complement the primate calls/oro-facial communication system. The "speech-like" areas of early hominids (i.e., not only the areas homologous to monkey F5 and human Broca's area, but the larger system mediating "complex" imitation of which they were part) mediated oro-facial and manuo-brachial communication but not speech.

(vi) The manual-orofacial symbolic system then "recruited" vocalization. Association of vocalization with manual gestures allowed them to assume a more open referential character, and exploit the capacity for imitation of the underlying brachio-manual system. I stress again that the form of speech reached at this stage involved the open-ended production and perception of sequences of vocal gestures. One can have speech in this sense without the constituent sequences constituting a language in the sense of a modern human language like Swahili or Korean.

Thus, we answer the question "Why did F5, rather than the primate call area provide the evolutionary substrate for speech and language?" by saying that the primate call area could not of itself access the combinatorial properties inherent in the manuo-brachial system.

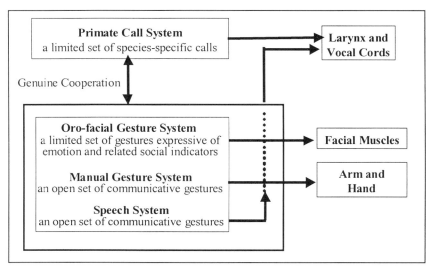

Figure 11.11 The human language system integrates speech with manual and oro-facial gesture, in cooperation with a system akin to that for primate vocalization.

Our use of writing as a record of speech has long since created the mistaken impression that language is a speech-based system. However, McNeill (1992) has analyzed videotapes to show the crucial use that people make of gestures synchronized with speech. Even blind people use manual gestures when speaking. As deaf people have always known, but linguists have only relatively recently discovered (Klima and Bellugi, 1979), sign languages are full human languages, rich in lexicon, syntax, and semantics. Moreover, not only deaf people use sign language. So do some aboriginal Australian tribes, and some native populations in North America. These studies suggest that we locate language-readiness in a *speech-manual-orofacial gesture complex*. I then hypothesize that during language acquisition a normal person shifts the major information load of language – but by no means all of it – into the speech domain, whereas for a deaf person the major information load is removed from speech and taken over by hand and orofacial gestures. On this basis, I show in Figure 11.11 a single communication system but stress that it involves many brain regions, each with its own evolutionary story.

The reader should be warned of the dangerous methodology that I employ here. My work is anchored in a rigorous knowledge of modern neuroscience, but when I assert that *Homo sapiens* was not endowed with language in anything like its modern human richness, I am not appealing to hard data, but rather forwarding a hypothesis based on, but in no sense implied by, a variety of evidence. However, I do not (nor should the reader) accept my hypotheses uncritically. Rather, each new hypothesis is confronted with new data and competing hypotheses as my reading progresses. The hypotheses presented in the second part of this article have thus survived a great deal of "cross-examination", and have been refined in the process. For example, while my reading of historical linguistics impressed me with the rapidity with which languages change (and I view human language as the sum of human languages, not as some abstract entity above and beyond these bio-cultural products) and thus to distinguish the notion of a "language-ready" brain from a "language-equipped" brain, further reading and reflection leads me to accept that the dichotomy here is not as sharp as I may have believed earlier.

From Language-readiness to Language: I have argued that the *biological* evolution of *Homo sapiens* yielded a mirror system embedded in a far larger system for execution, observation and imitation of compound behaviors composed from oro-facial, manual, and vocal gestures. I also accept that this system supported communication in *Homo erectus*, since otherwise it is hard to see what selective pressure could have brought about the lowering of the larynx which, as Lieberman (1991) observes, makes humans able to articulate more precisely than other primates, but at the cost of an increased likelihood of choking. However, I do not accept that this means that the earliest *Homo sapiens* was endowed with language in anything like its modern human richness. Rather, I argue that biological evolution equipped early humans with "language-ready brains" which proved rich enough to support the *cultural evolution* of human languages in all their commonalities and diversities. The divergence of the Romance languages took about one thousand years. The divergence of the Indo-European languages to form the immense diversity of Hindi, German, Italian, English, etc., took about 6 000 years. How can we imagine what has changed since the emergence of *Homo* *sapiens* some 200000 years ago? Or in 5000000 years of prior hominid evolution? Here I need to make two crucial points:

1. There is a danger, typified by Chomsky's notion of Universal Grammar (e.g., Chomsky, 1980), of thinking that any basic characteristics typical of all languages or of broad families of languages must be "hard-wired" into the human genome and thus the human infant's brain. I disagree, arguing that we can each know one or a few languages, but there is no such thing as Language-with-a-big-L that we all know, let alone "know in our genes". It is difficult to learn a foreign language because languages can be inherently different, but more-or-less adequate translations between languages are possible because languages have been shaped both by the need to express a range of basic human experiences, and by cultural diffusion.

2. Bickerton (1995) has spoken of *proto-language* as being language restricted, basically, to two-word utterances (generally comprising a verb and a noun in some order), and suggests that it is common to chimps trained to exhibit some form of language-like use of symbols, two-year-old humans, and people speaking pidgins. He then argues that proto-language was possessed by the hominid precursors of humans. However, I note that proto-language is only observed in creatures (chimps or humans) exposed to human language. I thus argue (with equally little proof, but as a counter-hypothesis to Bickerton's) that early *Homo sapiens* did not have proto-language in this sense. Rather, I would argue that what they possessed was the ability to name events with novel sequences of (manual and/or vocal) gestures, but that this capability does not imply the ability to separately name the objects and actions that comprised those events. I then claim that the latter ability was a momentous discovery made by humans perhaps 100 000 to 50 000 years ago, rather than a biological heritage from earlier hominids.

The ability for visual scene perception that must underlie the ability to employ verb-argument structures – the perception of *Action-object Frames* in which an actor, an action, and related role players can be perceived in relationship – was well established in the primate line, supporting a variety of complex behaviors and social relations. I thus hypothesize that the ability to communicate a fair number of such frames was established in the hominid line prior to the emergence of *Homo sapiens.*

We need here to make a careful distinction. The *action-object frame* is non-linguistic: the representation of an action involving one or more objects and agents. By contrast, a *verb-argument structure* is an overt linguistic representation – in modern human languages, the action is generally named by a *verb* (or verb phrase) and the objects are named by *nouns* (or *noun phrases*). A *grammar* for a language is then a specific mechanism (whether explicit or implicit) for converting verb-argument structures in particular, and more complex structures based on marked compounds of verb-argument structures more generally, into strings of words. Thus

John hit Mary with his hand

is an English sentence for the structure:

Hit (John, Mary, his hand).

With this background, I view the biological basis of language-readiness as "Knowing that there are things and events", one may thus trace a plausible evolutionary sequence:

(1) acting on objects,

(2) recognizing action-object frames,

(3) naming action-object frames,

with (1) and (2) being properties presumably possessed by most animals, although a particular species may have a very limited repertoire of objects and actions for its action-object frames. For example, a frog might be primarily responsive to prey, predators, mates, barriers, ponds, and resting places. Thus much crucial cognitive evolution may have occurred in the primate line to extend the repertoire of action-object frames which could possibly affect the behavior of each animal. It is (3) that marks the transition, in the hominid line, to symbolizing or naming. Monkeys have a fixed set of species-specific vocalizations, but these cannot be combined to yield novel messages – if a "genius monkey" makes the leopard vocalization, the other monkeys will scurry up nearby trees long before the first monkey would have a chance to create a sequence of vocalizations which made clear that the first vocalization was not really the leopard call at all, but rather the first, meaningless, syllable of an utterance intended to convey some quite distinct meaning.

The key invention of the hominids, then, was the creation of a "symbol toolkit" of meaningless elements from which an open ended class of symbols can be generated, with these abstract symbols initially grounded in action-oriented perception. The transition to language then involved two crucial steps:

1. separation of symbols for actions and objects, yielding verb-argument structures linked to action-object frames; and

2. abstraction and compounding of more generic verb-argument structure.

Syntax and semantics then developed together as the compounding of utterances, "going recursive", forced the development of morphological and syntactical conventions as the complexity of utterances meant that the actions and roles expressed became ambiguous without such devices. Indeed, there is pressure for such devices even at the simplest level: e.g., we have little trouble inferring the roles in the sequence "Eat John Apple" but we need some convention, such as word order, to determine who was hitter and whom hittee in "Hit John Bill".

Once again, we must be careful to distinguish the way we represent action-object frames and verb-argument structures on paper or the computer screen from their neural representations. Hit (John, Mary, his hand) is a symbolic string which represents two different representations in the brain, neither of which looks like this structure. The action-object frame represents (whether as a result of perception or motor planning or both) the relation between an action, two agents and an "object" without demanding that any names or words or explicit symbols be attached to any of these entities. The verb-argument structure is an abstraction from the action-object frame in that it lacks any graded representation of the specific event, but is enriched by the linkage of each entity to a specific name or symbol.

Moreover, *we* may classify the specific structure Hit (John, Mary, his hand) as an instance of a more general structure Hit (Agent, Recipient, Instrument) but the brain-representations of the constituent entities may or may not entail their recognition as belonging to these structures. As students of brain and language, we must ensure that descriptive categories are not automatically ascribed to the "neural strategies" of the subject. Thus, further challenges to the work begun here include the questions:

- How are action-object frames and verb-argument structures represented in the brain?
- How are action-object frames mapped to and from verb-argument structures, and how are the latter mapped to and from the utterances of (spoken, written, signed) language?

The Transition to *Homo sapiens* may have involved "language amplification" through increased protospeech ability coupled with the ability to name certain actions and objects separately, leading in turn to the ability to create an unlimited set of verb-argument structures, and thence the ability to compound those structures in diverse ways. This is a long way from seeing X-bar theory, or the nature of nouns in relation to adjectives, determiners and noun phrases as being coded within the human genome. Indeed, I suggest that many ways of expressing these relationships were the discovery of *Homo sapiens*, i.e., these might well have been "post-biological" in their origin. Two examples:

(i) Separating verbs from nouns lets one learn m+n+p words (or less if the same noun can fill two roles) to be able to form m*n*p of the most basic utterances.

(ii) Discovery of the one word *ripe* halves the number of fruit names to be learned. Recognition of hierarchical structure rather than mere sequencing could provide the bridge to constituent analysis in language, e.g., relating particular subactions (themselves further decomposable) to achievement of certain subgoals in a complex manipulation.

Consideration of the spatial basis for "prepositions" may help show how visuomotor coordination underlies some aspects of language. However, the basic semantic-syntactic correspondences have been overlaid by a multitude of later innovations and borrowings.

The further power of language comes from breaking away from the here-and-now, not just by hierarchicalization but also by negation and abstraction. We still need to ask how the needs of human biology and the constraints of the human brain shaped these basic "discoveries", extending the repertoire of recognizable and describable actions, e.g., analyzing how the brain can support counterfactual cognitive representations and relate them to language.

Towards a mirror-system based neurolinguistics: Figure 11.7 showed that the monkey needs many brain regions for the mirror system for grasping. We will need many more brain regions for an account of language-readiness that goes "beyond the mirror" to develop a full neurolinguistic model that extends the linkages far beyond the F5 ≈ Broca's area homology. To set the stage for the future development of such a model, we briefly link our view of AIP and F5 in monkey to data on human abilities. Studies of the visual system of monkey led Ungerleider and Mishkin (1982) to distinguish inferotemporal mechanisms for object recognition ("What") from parietal mechanisms for localizing objects ("Where"). Goodale and Milner (1992) extended this to human, studying a patient with damage to the inferotemporal pathway who could grasp and orient objects appropriately for manipulating them but could not report – either verbally or by pantomime – on how big an object was or what the orientation of a slot was. They thus viewed location as just one parameter relevant to how one interacts with an object, re-naming the "Where" pathway as the "How" pathway. Another patient with damage to the parietal pathway could communicate the size of a cylinder but not preshape appropriately. However, she could preshape appropriately if the "semantics" of an object indicated its size – suggesting the path from IT to the controller for grasping shown in Figure 11.12.

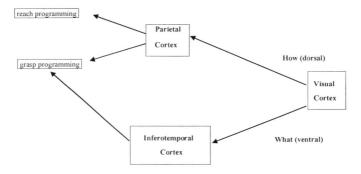

Figure 11.12 "What" versus "How". Lesion of parietal pathway yields inability to verbalize or pantomime size or orientation; lesion of the inferotemporal pathway yields inability to preshape (except for objects with size "in the semantics").

Let us now try to reconcile these observations with our mirror-system based approach to language. Our evolutionary theory suggests a progression from action to action recognition to language as follows:

(1) object •   AIP •   F5$_{canonical}$:   pragmatics

(2) action •   PF •   F5$_{mirror}$:   action understanding

(3) scene •   Wernicke's •   Broca's: utterance

The "zero order" model of the Figure 11.12 data is:

(4) Parietal "affordances" •   preshape

(5) IT "perception of object" •   pantomime or verbally describe size

This implies that one cannot directly pantomime or verbalize an affordance; one needs a "unified view of the object" (IT) before one can express attributes.

The problem with this is that the "language" path as shown in (5) is completely independent of the parietal $\rightarrow$ F5 system, and so the data seem to contradict our view in (3). To resolve this paradox, we recall (Figure 11.5) that although AIP extracts a set of affordances it is IT and PFC that are crucial to F5's selection of the affordance to execute, and then offer the scheme shown in Figure 11.13. This is the merest of sketches. Turning it into a well-formed neurolinguistic model – and developing an integrated view of syntax and semantics to go with it – constitutes a central aim for going "Beyond the Mirror" to establish the mechanisms of the MNS circuitry of Figure 11.7 as the evolutionary heart of a fully articulated model which links neurolinguistics (Arbib and Caplan, 1979) to the basic neural mechanisms for the recognition of the interactions of actors and objects, and for the elaboration of suitable motor plans for interacting with the environment so perceived.

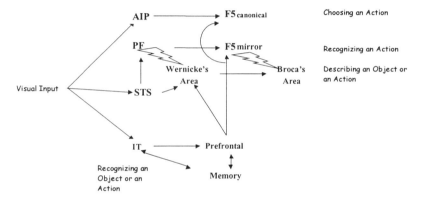

Figure 11.13 An early pass on a mirror-system based neurolinguistics.

References

Arbib MA (1981) Perceptual structures and distributed motor control. In: Brooks VB (ed) *Handbook of physiology (Section 2: The nervous system, Vol. II, Motor control, Part 1)*. American Physiological Society, pp 1449-1480

Arbib MA (2001) The mirror system, imitation, and the evolution of language. In: Nehaniv C, Dautenhahn D (eds) *Imitation in animals and artifacts*. MIT Press

Arbib MA, Caplan D (1979) Neurolinguistics must be computational. *Behavioral and Brain Sciences*, 2: 449-483.

Arbib MA, Rizzolatti G (1997) Neural expectations: A possible evolutionary path from manual skills to language. *Communication and Cognition*, 29: 393-424

Armstrong DF, Stokoe WC, Wilcox SE (1995) *Gesture and the nature of language*. Cambridge University Press, Cambridge

Bickerton D, 1995, *Language and Human Behavior*. University of Washington Press, Seattle

Chomsky N (1980) *Rules and representations*. Columbia University Press, New York

Corballis MC (1991) *The lopsided ape: Evolution of the generative mind*. Oxford University Press, New York

Corballis MC (1992) On the evolution of language and generativity. *Cognition*, 44: 197-226

Fagg AH, Arbib MA (1998) Modeling parietal-premotor interactions in primate control of grasping. *Neural Networks*, 11: 1277-1303

Gamble C (1994), *Timewalkers: The prehistory of global colonization*. Harvard University Press, Cambridge MA

Gibson JJ (1966) *The senses considered as perceptual systems*. Allen & Unwin, London

Goodale MA, Milner AD (1992) Separate visual pathways for perception and action. *Trends in Neuroscience*, 15: 20-25

Hewes G (1973) Primate communication and the gestural origin of language. *Current Anthropology*, 14:5-24

Iberall T, Arbib MA (1990) Schemas for the control of hand movements: An essay on cortical localization. In: Goodale MA (ed) *Vision and action: The control of grasping*. Ablex Publishing Corporation, pp 204-242

Iberall T, Bingham G, Arbib MA (1986) Opposition space as a structuring concept for the analysis of skilled hand movements. *Experimental Brain Research Series*, 15: 158-173

Jeannerod M, Arbib MA, Rizzolatti G, Sakata H (1995) Grasping objects: The cortical mechanisms of visuomotor transformation. *Trends in Neurosciences*, 18: 314-320

Jeannerod M, Biguer B (1982) Visuomotor mechanisms in reaching within extra-personal space. In: Ingle DJ, Mansfield RJW, Goodale MA (eds) *Advances in the analysis of visual behavior*. MIT Press, pp387-409

Kimura D (1993) *Neuromotor mechanisms in human communication*. Oxford University Press, Oxford, New York

Klima ES, Bellugi U (1979) *The signs of language*. Harvard University Press, Cambridge MA

Lieberman P (1991) *Uniquely human: The evolution of speech, thought, and selfless behavior*. Harvard University Press, Cambridge MA

McNeill D (1992) *Hand and Mind: What gestures reveal about thought*. The University of Chicago Press, Chicago

Oztop E, Arbib MA (2001) *Mirror neuron system for grasping: A computational model*. submitted for publication

Rizzolatti G, Arbib MA (1998) Language within our grasp. *Trends in Neurosciences*, 21(5): 188-194

Rizzolatti G, Fadiga L, Gallese V, Fogassi L (1995) Premotor cortex and the recognition of motor actions. *Cognitive Brain Research*, 3: 131-141

Rizzolatti G, Fadiga L, Matelli M, Bettinardi V, Perani D, Fazio F (1996) Localization of grasp representations in humans by positron emission tomography: 1 Observation versus execution. *Experimental Brain Research*, 111:246-252

Sakata H, Taira M, Kusunoki M, Murata A, Tanaka Y (1997) The parietal association cortex in depth perception and visual control of action. *Trends in Neuroscience*, 20: 350-357

Taira M, Mine S, Georgopoulos AP, Murata A, Sakata H (1990) Parietal cortex neurons of the monkey related to the visual guidance of hand movement. *Experimental Brain Research*, 83: 29-36

Ungerleider LG, Mishkin M (1982) Two cortical visual systems. In: Ingle DJ, Goodale MA, Mansfield RJW (eds) *Analysis of visual behavior*. The MIT Press, Cambridge MA

Chapter 12

A Unified Simulation Scenario for Language Development, Evolution and Historical Change

Domenico Parisi and Angelo Cangelosi

Three Aspects of Language Change

When we see a human being who is talking or is listening to and understanding the language produced by someone else, we may ask: What is the origin and past history of this ability? How has this ability taken its present form? Asking and answering questions about the origin and past history of biological, behavioral, and cultural phenomena is a crucial step to an understanding of these phenomena. Biological, behavioral, and cultural phenomena have an intrinsically historical nature. Their current properties are the result of their past history and they retain in themselves the traces of this past history. Language is at the same time a biological, a behavioral, and a cultural phenomenon. Therefore, to understand what language is now requires that we reconstruct how it is has become what it is now.

For language the question "What is the origin and past history of an individual's ability to speak a language?" can be interpreted in three different ways:

1. What is the origin and past history of the ability to speak in the *individual*?

2. What is the origin and past history of the ability to speak in the *species* of which the individual is a member?

3. What is the origin and past history of the particular language spoken by the *social group* of which the individual is a member?

The first question is how language is acquired by new members of the species during the first years of their life, i.e., *language development*. The second question asks how language first emerged in populations of individuals that did not speak a language and what stages it may have gone through before possessing the properties of all known human languages, i.e., *language evolution*. The third question is how particular historical languages originate and how they change across successive generations of speakers, i.e., *language change*. All three are important research questions that science should be able to answer. But we believe that none of these questions can be answered independently from the other two and that all three questions should be investigated within a single, unified, theoretical and methodological framework. This of course poses a formidable problem for research. Each of the three questions is complex and difficult to answer and each is studied by a distinct scientific discipline using a distinct theoretical framework and methodology. How can we study all three questions together using a unified theoretical and methodological framework?

Before we try to answer this challenge, let's try to justify our claim that none of the three questions - language development, language evolution, and language change - can be studied and understood in isolation from the other two.

The first question is: How is language acquired by the child? Language is acquired by the child through cultural transmission. A newborn individual acquires a language by learning the language from others. This is demonstrated by the fact that an isolated newborn individual would acquire no language and that the language acquired by a newborn individual is the particular language spoken by the other individuals in the social group of which the individual is a member. However, learning from others is not sufficient to explain language acquisition in the individual. Language acquisition in humans is made possible by the possession of an appropriate genetic endowment which is specific to the human species, as shown by the fact that only human beings are capable of acquiring a language. In order to understand this genetically inherited basis for language acquisition we have to study the process of biological evolution which has modified the genetic endowment of an initial population of individuals that did not possess a language (our ancestors still lacking a language) and has created in a succession of generations the genetic basis uniquely possessed by the human species that makes the acquisition of language by children possible. Therefore, to answer the first question "What is the origin and past history of the ability to speak in the individual?" (language development) it is necessary to answer the second question "What is the origin and past history of the ability to speak in the species of which the individual is a member?" (language evolution). Only by reconstructing the innate endowment that makes language acquisition possible and which is a result of language evolution can we properly study language development. And vice versa, since any evolutionarily intermediate form of language must have been acquired by children in order to be transmitted to the next generation, one can only study language evolution if one also studies language development.

A newborn human learns from others how to speak and how to understand linguistic signals but these 'others' are not simply other human beings who already speak a language. They are the members of a group of individuals who speak the same language. They speak the same language because linguistic signals are culturally transmitted, that is, each individual tends to imitate the other group's members when it comes to producing and understanding linguistic signals. Imitating the language spoken by others runs through all three channels of cultural transmission, the vertical channel from parents to offspring, the oblique channel from individuals of one generation to unrelated individuals of the next generation, and the horizontal channel among individuals of the same generation (Cavalli-Sforza and Feldman, 1981). If two groups of individuals have few contacts the two groups will speak different languages because the conditions for imitating others that make the existence of a single language possible are not met and the members of one group won't be able to speak and understand the language of the other group.

Any particular language, like all biologically or culturally transmitted entities, is a pool of variants, a set of slightly different ways of talking about things, and the transmission of the language from one generation to the next is selective in the sense that some variants will tend to be more imitated, reproduced, learned by others, than other variants. Furthermore, as in all biologically or culturally transmitted pools of variants, there will be a constant addition of new variants to the common pool, that is, modification of old ways of talking about things and introduction of new ways internally generated by the group or imported from outside, from the languages of other groups. The result is language change. When the language spoken by a group becomes so different from the language previously spoken that an individual who speaks the current language does not any more understand the previous language, the languages have become two and we can answer the third question "What is the origin and past history of the particular language spoken by the social group of which the individual is a member?" (language change). But changed forms of language must be acquired by children, and are in part changed during the course of the acquisition process, and therefore they must be compatible with the genetically inherited basis for language. Therefore, language development, language evolution, and language change cannot be studied in reciprocal isolation. Each of these three classes of phenomena can be understood only by understanding the other two.

A unified approach to language development, evolution, and historical change

Granting that language development, language evolution, and language change should be studied together, how can we study all three using a single, unified, theoretical and methodological framework.

The answer is computer simulations. Computer simulations are a new way of expressing scientific theories. Scientific theories are traditionally expressed either verbally or with the help of mathematical and formal symbols. Simulations are a third way of expressing scientific theories: simulations are scientific theories expressed as computer programs. Like all theories in science, a simulation is a set of hypotheses about the causes, mechanisms, and processes that underlie a given class of phenomena and explain these phenomena. When the program runs in the computer, the results of the simulation are the empirical predictions derived from the theory embodied in the simulation. If the simulation results match the observed phenomena, the theory is confirmed. If not, the theory-simulation must be modified or abandoned. In addition to being theories expressed in a novel way, simulations are virtual experimental laboratories in which, as in the real laboratory, the researcher can observe phenomena in controlled conditions, manipulate the conditions and variables that influence the phenomena, and observe the consequences of these manipulations.

Simulations are the needed tool to study the phenomena of language development, language evolution, and language change using a single, unified framework. These phenomena have properties that make them difficult to investigate using the traditional tools of science: observation, laboratory experiments, verbally or formally expressed theories. The phenomena of language acquisition, language evolution, and language change occur at very different time scales (months and years for language acquisition, centuries for language change, hundreds of thousand or even million years for language evolution) and many of the phenomena cannot be directly observed or experimented with. Furthermore, linguistic phenomena involve very different entities and processes that occur at various spatial scales and at various levels of integration (neural, behavioral, social, cultural) and are the result of complicated interactions among these entities and processes. Finally, linguistic phenomena have many of the properties exhibited by complex systems, i.e., systems made up of very large numbers of elements which by interacting locally give rise to global phenomena that cannot be predicted or deduced even if we know perfectly the elements and the rules that govern their interactions. Examples are large sets of neurons that make up nervous systems capable of linguistic behaviors and large sets of individuals speaking the same language.

Given the enormous memory and computing power of the computer and its great flexibility, computer simulations make it possible to create scenarios in which phenomena may occur at different temporal and spatial scales and be observed and experimented with on the computer screen. The computer can "calculate" the interactions between phenomena at different levels of integration (neural, behavioral, social, cultural) which would be impossible for us to do and, especially important, it allows us to see what phenomena emerge from the local interactions of very large numbers of entities.

In the last fifteen years many computer simulation models for the study of language have been developed. Most of these models have focused on a single aspect of language, that is its development in children, or the evolution of communication and linguistic systems, or historical language change. Simulations of the acquisition of language are mainly based on connectionist models, that is, artificial neural networks trained with a specific language corpus or structure (Broeder and Murre, 2000; Christiansen and Chater, 1999). For example, some models simulate the acquisition of syntax (Elman, 1990), the learning of morphology (Rumelhart and McClelland, 1986; Plunkett and Marchman, 1993), reading (Seidenberg and McClelland, 1989), the acquisition of the lexicon (Plunkett, Sinha, Møller and Strandsby, 1992), and language deficits (Harley, 1993). Simulation models of the origin and evolution of communication and language have used a great variety of methods, such as genetic algorithms, neural networks, rule systems, and robotics (see Chapter 2). Each individual model mainly focuses on a specific aspect of the evolution of language, such as the evolution of honest signaling (Bullock, 1998), the evolution of shared lexicons (Hutchins and Hazlehurst, 1995; Steels and Kaplan, 1998), and the emergence of compositionality (Kirby, 2001). Finally, some simulations focus on the emergence of dialects and linguistic diversity (Livingstone and Fyfe, 1999; Nettle, 1999).

Only a few of these models, however, have simultaneously studied the interaction between the phenomena of language development, evolution, and change. For example, Hare and Elman (1995) looked at the similarities between the learning of the English past tense morphology and some historical changes in old and modern English. Various simulation models have been proposed for studying the interaction between evolution and learning (Hinton and Nowlan, 1987; Belew and Mitchell, 1996; Nolfi and Floreano, 1999) but not with respect to language, although Batali (1994) has studied the evolution of a critical period for grammar acquisition.

In this chapter we propose a unified approach to the study of language development, evolution, and change. The simulation scenario uses (1) genetic algorithms to study the emergence of a biological basis for language, (2) neural network learning to study language acquisition, and (3) both neural network learning (learning from others) and genetic algorithms to study cultural transmission and historical change of language. The simulation scenario adopts the Artificial Life Neural Network framework (Parisi, 1997; Cangelosi, 2001). The simulations simulate evolving populations of autonomous organisms which live in an environment and interact both with the environment and with their conspecifics. Each individual can inherit, either biologically or culturally, various sensorimotor, cognitive, and linguistic abilities. The entire behavior of the organism is controlled by an artificial neural network, which makes it possible to study the effects of behavioral and neural factors in the evolution, learning, and transmission of language.

In the following section we give a detailed description of the simulation scenario. In the second part of the chapter we describe some specific simulations based on the scenario. These model the evolution of a shared language with an emphasis of its adaptive role, the evolution of a simple compositional language, and the first emergence of word classes (verbs and nouns).

A Simulation Scenario for Language Change

We look at the computer screen and we see a certain number of artificial organisms which live in an environment and must capture the resources contained in the environment to survive and reproduce. The behavior of each individual is controlled by the individual's nervous system, which is simulated by an artificial neural network. A neural network is a set of units (neurons) linked by unidirectional connections (synapses between pairs of neurons). Basically, the network includes input units, output units, and one or more layers of internal units. The input units send their connections to the internal units and the internal units send their connections to the output units. At any given time each unit has an activation level (firing rate of neurons) which for the network's input units depends on physico-chemical causes outside the organism and for the remaining units (internal and output units) depends on the activations arriving at the unit from other connected units. The activation sent by a unit to another unit through a connection depends on the first unit's current activation level and on the weight and the plus or minus 'sign' of the connection (respectively, the number of synaptic sites between pairs of neurons and the excitatory or inhibitory character of the connection). In each cycle the network receives some input from outside (pattern of activation values on the network's input units), activation spreads through the network, and the network responds to the input with an output (pattern of activation values on the network's output units) which controls the movements of some portion of the organism's body. The way in which the network responds to the input depends on the input and on the network's connection weights (and, of course, on the network's architecture, i.e., the connection schema that specifies which unit is connected with which unit).

If the organisms on the computer screen are human beings, their neural network is somewhat more complicated in that it includes a rich circuitry of recurrent connections, i.e., connections that do not go "forward" from input to output but go "backward" from units near the output units to units near the input units. These recurrent connections underlie a human being's mental life, i.e., his or her mental images, recollections of past events, predictions of future states, imagined future states, plans. Mental life is a neural network's ability to self-generate its own input and to respond to this self-generated input. Hence, a neural network which is capable of mental life is a network that may respond to external input not with some external behavior but by internally generating some further input and that may produce some external behavior not in response to external input but in response to self-generated input.

The artificial human beings on the computer screen have language. This means that they are able to produce linguistic signals (sounds physically caused by movements of an individual's phono-articulatory organs) in the appropriate circumstances, i.e., in response to the appropriate external or internally generated input, and they are able to respond to linguistic signals (heard sounds) with the appropriate external behavior or with the appropriate internally generated input. Much of mental life consists in internally generated linguistic signals and responses to these internally generated linguistic signals.

Each individual is born, develops, reproduces, and dies. Reproduction means that one or more new individuals are created which inherit the genotype of their parents. The genotype contains information which determines some of the properties of the new individual, in particular the individual neural network's architecture and/or connection weights. Reproduction is selective, that is, some individuals have more offspring than other individuals and some individuals die without leaving offspring. Furthermore, reproduction is accompanied by the constant addition of new variants of genotypes because the inherited genotype may be modified by random genetic mutations and because the recombination of portions of the mother's genotype with portions of the father's genotype in sexual reproduction results in new genotypes. Selective reproduction and the constant addition of new variants of genotypes determine biological evolution, i.e., change in the genetic pool of the population across successive generations of individuals.

Biological evolution means that some traits of the individuals that were absent at some evolutionary stage can emerge in the population later in evolution. These evolutionary changes may be adaptive, i.e., they may result from the fact that individuals inheriting the new trait tend to have more offspring in the particular environment than individuals not inheriting the trait, or they may be due to chance, linkage to other adaptive traits, and other adaptively neutral factors. If the linguistic abilities of our organisms were completely genetically inherited, with learning and experience playing no role in their development in the individual, we might construct a simulation in which a population of artificial organisms starts with genotypes that do not specify these linguistic abilities (or, more precisely, the network's architecture and connection weights that make it possible to produce and understand linguistic signals) and in a succession of generations evolve new genotypes that include the specification for these linguistic abilities. These simulations could be used to test theories on the particular selective pressures that have caused the evolutionary emergence of language, on the consequences for the population of individuals of possessing a language, on how linguistic abilities develop (or, better, mature) in new members of the population, etc.

However, human language is not entirely genetically inherited but it develops in new members of the population as a consequence of a process of learning from others which changes their neural network in the first years of their life. Learning in neural networks may be simulated as change in the network's connection weights as a result of experience. There exist many different learning algorithms for neural networks. The most popular one is the back-propagation algorithm. A neural network is exposed to some input and it responds with some output. At the beginning of learning the network has connection weights that do not allow the network to respond with the appropriate output. However, in each learning cycle the network receives a teaching input from outside, i.e., a specification of the appropriate or desired output. The network compares its own output with the desired output and, on the basis of the discrepancy between the two (error), modifies its connection weights in such a way that after a certain number of learning cycles the network's connection weights are able to generate the appropriate output in response to each input.

Human language is not simply learned but it is learned from others. Learning from others can be simulated in neural networks by having two neural networks, a learner and a model (or teacher) network, that are exposed to the same input. Unlike the learner, the model already knows how to respond appropriately to the input. The learner generates some, initially wrong, output in response to the input and uses the model's output as its own teaching input. After a given number of learning cycles the learner will respond to the input in the same way as the model and any behavioral capacity possessed by the model will have been transmitted to the learner.

Learning a language from others can be simulated in the following way. The neural network of language-using organisms has two sets of input units and two sets of output units. The first set of input units encodes non-linguistic input (e.g., visual input from objects) and the second set of input units encodes linguistic signals (heard sounds). The first set of output units encodes non-linguistic movements (e.g., movements of the arms and hands) and the second set of output units encodes movements of the phono-articulatory organs that produce linguistic signals (sounds). Learning from others how to *produce* linguistic signals requires that both the learner and the models are exposed to the same visual input and both respond with the production of some linguistic signal. At the beginning of learning the linguistic signal produced by the learner will not be appropriate. However, the learner compares the signal it has produced with the signal produced by the model and modifies its connection weights in such a way that after a while that particular visual input is responded to (named, designated) by the learner using the same linguistic signal of the model.

Learning from others how to *understand* linguistic signals requires that the learner is exposed to some linguistic signal (which of course has been produced by some other individual) to which the model is also exposed and the learner compares its own behavioral response to that signal with the response of the model. Using the back-propagation procedure the learner modifies its connection weights so that at the end of learning the learner responds to the linguistic signal in the same way as the model.

We should mention three complications that should be added to this rather simple picture of learning a language from others. First, mental life makes language learning more complicated. A linguistic signal can be produced not in response to some external input but in response to some self-generated internal input. A linguistic signal can be produced internally, not externally. A heard linguistic signal can be responded to not with an external behavior but with an internally self-generated input. In all these cases learning from a model becomes more difficult. However, we can assume that these more sophisticated behaviors (mental life) are more typical of grown-up individuals who have already mastered most of the language spoken around them.

Second, the back-propagation procedure models only one aspect of learning language from others. Other aspects can be modeled using another learning algorithm which can be used with neural networks, the reinforcement procedure. In the back-propagation procedure the learner is provided with a specification of the correct response to some input so that it can compare its own response with the correct response. In the reinforcement procedure the learner is provided only with a global, positive or negative, evaluation of its response to some input, and the learner uses this externally provided information to modify its connection weights in such a way that in future occasions its response to the same input will receive better evaluations. Language is learned through this type of reinforcement learning when the model (or, more appropriately in this case, the teacher) "rewards" or "punishes" the language learner with positive or negative evaluations of its linguistic performances. Or, more generally, the learner receives indirect social reinforcement when linguistic interactions with others lead to success or failure.

Third, as we have said, language development depends both on learning from others and on genetically inherited species-specific information. Hence, although simulations in which language emerges purely biologically and is entirely genetically inherited and simulations in which language emerges purely culturally and has no genetic basis can be informative and useful, the simulation scenario we have described should enable us to simulate a more realistic framework in which language is learned from others but only by individuals that are born with a genetically inherited and biologically evolved specification that makes this learning from others possible. An important research goal is to explore using the simulation scenario different hypotheses on what must be genetically inherited for learning a language from others to be possible and what selective pressures have created this genetic basis for language learning.

Some Examples of Simulations Using the Scenario

Evolutionary emergence of a simple language

Language has evolved and, therefore, unless we assume that it has evolved entirely by chance, language must have some positive influence on the reproductive success of the individuals exhibiting it. One can advance various hypotheses on the selective advantages provided by language and the simulation scenario described in the previous section can be used to explore and test these hypotheses.

One possible selective advantage of language is that with language one can know the environment beyond the limits of one's senses. Organisms without language can only know the environment based on the information directly provided to them by their own senses. In a population of linguistic organisms an individual can be informed about the environment beyond direct sensory access by linguistic signals produced by conspecifics and it can respond adaptively to this linguistically mediated information. This can result in better performance so that linguistic organisms will gradually out-compete and replace non-linguistic organisms.

To study the adaptive advantage of language in populations that evolve a shared signaling system we executed a first simulation using the proposed scenario (Cangelosi and Parisi, 1998). A population of organisms lives in an environment that contains both edible and poisonous mushrooms, each individual mushroom being different from all others. To survive and reproduce the organisms must be able, first, to recognize encountered mushrooms as either edible or poisonous and, second, to respond appropriately by approaching and eating the edible mushrooms and avoiding the poisonous ones. In the simulation we compare two populations, one with language and the other without language. In both populations an organism is informed by its senses about the *location* of the single nearest mushroom whatever the distance of the mushroom from the organism but the organism can perceive the *specific properties* of the mushroom and therefore recognize the mushroom as either edible or poisonous only if the mushroom is very near to the organism. Given this sensory limitation the organisms lacking language can only approach any mushroom they locate in the environment until they are sufficiently close to the mushroom that they can recognize the mushroom as either edible or poisonous. The organisms with language have another possibility. When they locate a mushroom, whatever their distance from the mushroom, they receive from the environment an additional sensory input: a heard linguistic signal. This signal is produced by some conspecific but for the moment we will ignore why and how conspecifics produce linguistic signals. The only properties of signals we will mention are that there are only two linguistic signals and that the two signals are perceptually very different. One signal always accompanies edible mushrooms and the other signal always accompanies poisonous mushrooms. Hence, linguistic organisms can know what type of mushroom they are currently perceiving not only when the mushroom is very close but also when it is more distant. The consequence is that, while non-linguistic organisms must approach every mushroom they encounter to discover if it is edible or poisonous, linguistic organisms can exhibit a more efficient behavior: they can approach perceived mushrooms only when they are edible.

For each population the simulation starts with a certain number of individuals whose behavior is controlled by a neural network. Each neural network has input units encoding the location of the nearest mushroom and input units encoding the mushroom's properties, but the mushroom's properties can only be perceived if the mushroom is sufficiently close to the organism. For the organisms with language there is an additional set of input units encoding heard linguistic signals. All the networks have output units encoding the movements with which the organisms move around in the environment and hidden units intermediate between the input and the output units. All the connection weights of the initial population of neural networks are randomly generated and therefore the behavior of the first generation of organisms is not very efficient. They are not very good at distinguishing edible from poisonous mushrooms and at associating the appropriate behavior (approaching or avoiding) to each type of mushrooms. The organisms reproduce selectively based on their individual fitness, which increases with the number of edible mushrooms eaten and decreases with the number of poisonous mushrooms eaten. Furthermore, the genotype inherited by their offspring, which encodes the network's connection weights, is in part modified by random genetic mutations. The selective reproduction of the best individuals and the genetic mutations which, even if rarely, can result in offspring which are better than their parent (reproduction is non-sexual), progressively increase the average fitness across successive generations in both populations. But here the two populations differ: the average fitness of the population with language is almost twice as great as that of the population without language (Figure 12.1).

In these simulation organisms with language outperform organisms without language because language allows an individual to know more of its environment than is directly provided to the individual by its senses and therefore linguistic organisms behave more efficiently than organisms that in order to know the environment can only rely on their senses. Since language is adaptively useful it will evolve in a population initially lacking language. However, in the simulation just described what evolves is only the ability to *understand* language, not the ability to produce language. The linguistic signals are inputs for the neural network of the linguistic organisms and these organisms evolve the appropriate connection weights for processing the linguistic input in useful ways. These connection weights constitute the ability to understand linguistic signals. Now we should ask: can we simulate the evolutionary emergence of the ability to *produce* linguistic signals?

In the simulation we have described we have assumed that linguistic signals are produced by conspecifics but we have not simulated why and how an individual produces useful linguistic signals for other individuals. The appropriate linguistic signals are "hardwired", i.e., they are provided by us. To simulate the evolutionary emergence of language production we have conducted a second simulation in which we modify the network architecture by adding a new set of output units that encode produced linguistic signals. When an individual encounters a mushroom we assume there is another individual nearby which perceives the same mushroom and produces a linguistic signal as output. It is this linguistic signal which is heard by the first individual and which tells the individual whether the perceived mushroom is edible or poisonous. How are the individuals able to produce useful linguistic signals, that is, one linguistic signal for edible mushroom and a different signal for poisonous mushrooms?

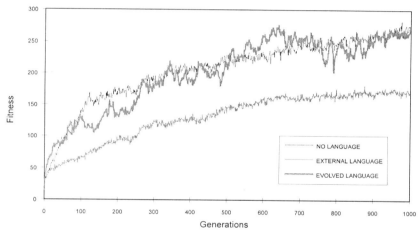

Figure 12.1 Average fitness of three different populations: without language, with external "hardwired" signals, and with evolved language. (From Cangelosi and Parisi, 1998)

The ability to understand and the ability to produce linguistic signals must co-evolve together in a population of organisms. Unless conspecifics are able to produce useful signals, there is nothing useful to understand, and unless conspecifics are able to correctly understand the signals they hear it is no good to produce useful linguistic signals. At the beginning of this second simulation all connection weights are randomly generated, which means that both the ability to understand and to produce linguistic signals are non-existent. After a certain number of generations both abilities have evolved. The average fitness of the population is similar to the fitness of the population with hardwired language in the preceding simulation (Figure 12.1), which means that the population has evolved both an ability to produce useful linguistic signals *and* an ability to understand those signals. In fact, when we look at the actual linguistic signals produced by the organisms we discover that they produce very efficient signals, that is, signals which (a) are distinct for functionally different categories of objects (edible and poisonous mushrooms), (b) are identical for all the members of each category, (c) are the same signals used by the other individuals in the population. (Eve Clark has argued that principles similar to these govern children's acquisition of language. Cf. Clark, 1993).

Our simulation scenario allows us to look at linguistic abilities not in isolation from other behavioral and cognitive skills, but in direct interrelation with them. For example, in the population where foraging behavior depends on language, a positive and statistically significant correlation between average fitness (that reflects the ability to categorize and approach mushrooms) and a measure of the quality of language in the population was observed (Cangelosi and Parisi, 1998). Moreover, it was shown that the ability to appropriately categorize encountered mushrooms positively affects the evolution of good languages also when the communication signals are not directly used by the organisms. These results can be used to support Burling's (1993) language origin hypothesis. Burling hypothesized that human language has emerged from the cognitive capacities of our prelinguistic ancestors rather than from their primate-level communicative behavior. The production of efficient linguistic signals can be viewed as a by-product of a pre-existing ability to categorize the environment on a sensory-motor basis.

But the relation between language and the categorization of experience can be more directly examined in our simulations. In the simulations described so far the selective advantage provided by language is mainly due to the capacity of linguistic signals to "stand for" something which is absent. Since the detailed perceptual properties of mushrooms are inaccessible to their senses at greater distances, non-linguistic organisms have no way of knowing which type of mushroom they are currently perceiving unless the mushroom is very close to them. On the contrary, for linguistic organisms linguistic signals "stand for" the perceptual properties of mushrooms and, therefore, linguistic organisms know from the linguistic signal which is provided to them if the mushroom they have located in the environment is edible or poisonous.

However, the selective advantages of language go well beyond language's capacity to stand for something which is not directly perceived. As we have said, to survive in the mushroom environment our organisms must possess two cognitive abilities, an ability to correctly categorize encountered mushrooms as either edible or poisonous and an ability to respond to each distinct category of mushrooms with the appropriate behavior. Linguistic signals can be evolutionarily advantageous not only because they can "stand for" things we do not perceive but also because they help us to categorize more efficiently what we perceive. This has been shown in another simulation (Parisi, Denaro and Cangelosi, 1998) in which the organisms are able to perceive the properties of encountered mushrooms whatever the distance of the mushroom from the organism and therefore linguistic signals never "stand for" something which is perceptually absent. This notwithstanding, linguistic signals have a positive influence on the behavior of approaching and eating edible mushrooms and avoiding poisonous ones because perceiving a mushroom and at the same time hearing a linguistic signal that distinguish between edible and poisonous mushrooms results in better categorization of the encountered mushroom.

What does it mean for a neural network to categorize experience appropriately? For a neural network an experience is the pattern of activation values observed in the network's input units at some particular time. The connection weights linking the input units to the internal units transform this pattern of activation into another pattern of activation at the level of the internal units. We will call the pattern of activation observed in the internal units the "internal representation" of the input. For a neural network to categorize an experience appropriately is to possess connection weights such that the internal representations of inputs that must be responded to with the same output tend to be similar, while the internal representations of inputs that must be responded to with different outputs tend to be different. Similarity and difference among activation patterns can be measured precisely as the distance between points, each corresponding to one particular activation pattern, in the abstract multidimensional space representing a particular layer of units. A layer of N units is represented by an abstract space of N dimensions. One particular activation pattern on the N units is represented as a point located, for each dimension of the space, in the position specified by the activation value of the relevant unit. All the points that are internal representations of inputs that must be responded to with the same output constitute what we will call a "cloud", whereas points that are internal representations of inputs that must be responded to with different outputs constitute separate 'clouds'. To categorize experience efficiently means that, at the level of internal units, (a) the distance between clouds is great, (b) cloud size is small, and (c) there is little overlap between clouds.

Given this framework, when we say that language results in better categorization of experience, we mean that for our organisms the clouds of points observed at the level of their internal units have more of the properties (a)–(c) if the organisms live in a linguistic world and they experience mushrooms together with the appropriate linguistic signals than if they lack language. This phenomenon has been observed in various simulations (Parisi *et al.*, 1998; Cangelosi and Parisi, 2001; Cangelosi and Harnad, in press) and a further detailed example of it will be discussed in the final section on verb-noun languages.

Evolutionary emergence of a simple compositional language

In the previous simulations the signals used by organisms to communicate are genetically inherited and language is fixed during their entire lifetime. Moreover, the language evolved by the organisms is quite simple since it is constituted by single, isolated signals. To simulate more complex languages and the process of cultural transmission of language a new simulation has been developed (Cangelosi, 1999; 2001). The simulation uses the same general scenario we have already described, with some important changes in (1) the set of foraging stimuli, (2) the ability of the organisms to produce and use compositional signals, i.e., complex signals composed of two simple linguistic signals, and (3) the process of language learning and cultural transmission.

The population of organisms performs a foraging task which consists in avoiding poisonous mushrooms and collecting and manipulating edible mushrooms. The foraging task stimuli are organized into a hierarchy of functional categories inspired by Savage-Rumbaugh and Rumbaugh's (1978) ape language experiments. They consist of two high-level categories (edible and poisonous mushrooms) and three low-level categories (large, medium, and small mushrooms). An individual's fitness depends on its ability to discriminate between the two high-level categories of mushrooms (approaching the edible mushrooms and avoiding the poisonous ones) but also on its ability to behave differently with respect to the three sizes of edible mushrooms (e.g., "cut" the big ones, "smash" the medium size, and "wash" the small ones).

In the input layer of the organism's neural network, in addition to the sensory units that encode the perceptual features of the nearest mushroom and its location, there are eight linguistic units that encode the signals available for communication. In the output layer three units are used to control the organism's behavior (moving around in the environment and producing the action appropriate for each mushroom size), and eight units are used to produce the communication signals. The linguistic output units are organized into two winner-takes-all clusters of competitive units (one cluster of six units, one of two). The two most active units, one for each cluster, represent the pair of linguistic signals chosen to name a mushroom.

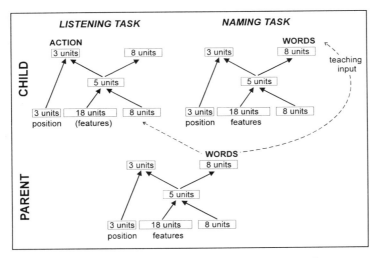

Figure 12.2 Neural network architecture and the interaction between a child organism and its parent (from Cangelosi, 1999).

In the first few hundred generations organisms evolve the ability to discriminate between the six types of mushrooms and no communication is allowed. Only in the second stage of the simulation organisms are they allowed to communicate using the linguistic units and to teach each other the language. Moreover, during this second evolutionary stage a mushroom's visual features are rarely given in input to the organisms (10% of the total time). Communication and cultural transmission is limited to parents and their children. The parent-child interaction consists of two tasks (Figure 12.2). In the Listening task the parent names the mushroom which is close to its child by activating two signals, one for each linguistic output cluster. The child uses this linguistic description to decide what to do with the mushroom. The child's fitness depends on its performance with respect to this mushroom. The parent only acts as speaker and linguistic teacher since it cannot reproduce anymore. In the Language Learning task the child learns to name the mushrooms by imitating its parent's linguistic behavior. The child network compares the linguistic signal it has produced in response to some particular mushroom with the signal produced by its parent in response to the same mushroom (teaching input) and it modifies the connection weights of its neural network using the back-propagation procedure. During these learning trials some noise is added to the teaching input in order to introduce variability in the process of cultural transmission (Denaro and Parisi, 1997).

The results of the simulation show that the majority (60%) of the populations evolve languages that use combinations of symbols, with one symbol designating the more general action of approach or avoiding encountered mushrooms and another symbol designating the more specific action which is appropriate for each size of edible mushrooms.

Overall, this simulation shows that it is possible to evolve compositional languages through processes based on both biological evolution and cultural transmission. Biological evolution supports the emergence of behavior which requires an appropriate internal representation of the environment. During the evolution of foraging behavior, the organisms' neural networks evolve internal representations that discriminate between different functional categories of mushrooms. These categorical representations constitute the basis for the subsequent emergence of a set of linguistic signals which are culturally, not biologically, transmitted. Children learn how to produce the appropriate linguistic signals for naming different types of mushrooms by imitating their parent's naming behavior. In addition to studying the interaction between the phenomena of language learning and evolution, this simulation also permits the study of the evolution of compositional languages based on symbolic communication. As Deacon (1997) explains, symbolic communication is characterized by the existence of two types of relationships: (1) links between the symbols and their referents in the real world, and (2) relationships among the symbols. The compositional languages that emerge in the present simulation possess both types of relationships and therefore can be considered symbolic (see Cangelosi 2001 for a more detailed discussion of the symbolic nature of languages).

Behavioral and neural factors in verb-noun languages

Linguistic signals by themselves are just sounds that an organism hears or produces by executing phono-articulatory movements. To become linguistic signals the organism must associate a meaning to the sounds. Linguistic signals receive their meaning from the fact that the organism experiences the sounds in concomitance with non-linguistic experience. Different sounds co-vary with different non-linguistic experiences and this co-variation endows sounds with their specific meaning and transforms them into linguistic signals.

An important distinction that may have emerged early in the evolution of language is a distinction between signals that in our ancestors' experience co-varied with the particular content of their visual experience and signals that co-varied with the particular action with which they responded to this visual experience. We will call the first type of linguistic signals 'nouns' (or 'arguments') and the second type 'verbs' (or 'predicates'). These signals are not what linguists call nouns and verbs but they may have been precursors of nouns and verbs.

To study the evolution of simple verb-noun languages, and their effects on the organisms' behavior and adaptation, the following simulation was conducted (Cangelosi and Parisi, 2001). Imagine an organism whose behavior is controlled by a neural network with visual input units encoding the content of a 'retina' and motor output units encoding the movements of a two-segment arm. The network has an additional set of proprioceptive input units encoding the current position of the arm. At any given time the retina encodes one of two perceptually different objects, a horizontal object A and a vertical object B. Either object may be located in one of four different locations in the retina and the organism responds by moving its arm and either pushing the object away from itself or pulling the object toward itself. The organism's life is divided up into a certain number of epochs, each made up of a fixed number of input/output cycles. At the beginning of each epoch the arm's endpoint (the hand) is already grasping the object so that all the arm must do is to either push or pull the object. At the end of life the organism is scored for the number of epochs in which the object has been moved correctly. 'Correct' means that A objects must be pushed and B objects must be pulled. (An object is successfully pushed or pulled if the arm moves the appropriate object sufficiently (i.e., beyond a certain threshold) away from or near the organism.)

The connection weights for the neural network controlling the organism's behavior are found using a genetic algorithm. At the beginning of the simulation a population of neural networks is created with randomly assigned connection weights. The individuals with the highest scores reproduce selectively and their offspring inherit the connection weights of their (single) parents, except that some random genetic mutations modify the value of some of the weights. After a certain number of generations the organisms have developed the appropriate behavior. They use their arm to push A objects and to pull B objects, whatever the position of the objects on the retina.

Imagine now another population of organisms which are identical to the preceding ones except that they live in a linguistic environment. In some of the epochs of their life they do not only visually perceive the object and must respond to the object in the appropriate way but they also experience a linguistic signal. Their neural network has an additional set of input units encoding linguistic signals. There are four possible signals. Two of these signals are nouns and two are verbs. One of the two nouns invariably accompanies A objects and the other noun invariably accompanies B objects. On the other hand, when the organism hears one of the two verbs it is scored correct if it responds with the action of pushing the object, whether the object is A or B, while when it hears the other verb the organism is scored correct if it pulls the object, whether the object is A or B. In other words, nouns co-vary with the visually perceived object whereas verbs co-vary with the action with which the organism must respond to visually perceived objects.

During its life an organism is exposed to a variety of situations. In some situations it perceives the object with no accompanying linguistic signal and it must respond to the object with the appropriate action (pushing A objects and pulling B objects). In other situations it is exposed to both an object and a linguistic signal but the linguistic signals can vary. The object can be accompanied by a noun or by a verb. The noun accompanying a given object is necessarily congruent with the object (A objects are always paired with the noun designating A objects and B objects are always paired with the noun designating B objects) but the verb can indicate either the action congruent with the visually perceived object (A objects are accompanied by the verb designating the action of pushing and B objects by the verb designating the action of pulling) or the action incongruent with the object (A objects are accompanied by the verb designating the action of pulling and B objects by the verb designating the action of pushing). And when the object is accompanied by a verb designating an action incongruent with the object, the organism must respond in the manner required by the verb, not by the object. In other words, while nouns always designate what is visually perceived and they elicit actions that the organism is used to perform in response to the object seen, verbs detach the behavior of the organism from what is visually perceived. If the organism is asked to push or pull, it must push or pull independently from the visually perceived object that must be pushed or pulled. Finally, in still other situations an organism experiences only a linguistic signal, without visually perceiving the object, and it must respond in the manner specified by the linguistic signal.

What are the results of these simulations? One first result is that the addition of language enhances the performance of the organisms but only if language is added after the population of organisms has already evolved an ability to respond appropriately to the visually perceived objects without language. We have compared two populations (Cangelosi and Parisi, 2001), one population in which language is added from the beginning of the simulation and it evolves together with the ability to respond to visually perceived objects in the absence of language, and another population which lives for a certain number of generations without language and it reaches a certain level of ability in responding to the visual input without language and then language is added until the end of the simulation. At the end of the simulation the population which has language from the beginning reaches a level of performance which is <u>lower</u> than that of the population which never experiences language. In contrast, adding language after the population has already reached a certain level of ability without language, at first produces a drop in performance due to the novelty of language but then it leads to a higher level of performance at the end of the simulation compared with the population which never experiences language. In other words, language is useful if it can build on an already existing basis of sensory-motor ability.

A second result of the simulations is that verbs appear to have a larger positive effect on performance than nouns. If we score the individuals separately for their performance in the situations in which nouns accompany the visually perceived objects and in the situations in which verbs accompany the objects, we see that performance in the latter case is better than in the former. Verbs tell an organism directly what they have to do whereas nouns accompany objects that in different occasions may need to be responded to in different ways.

A better understanding of the causes of these different effects of nouns and verbs on performance can be obtained if we analyze the internal organization of the organisms' neural networks at the end of evolution. To exhibit some desired performance neural networks organize themselves so that internal representations of inputs that must responded to with the same action tend to be close to each other in the abstract space of internal representations, whereas internal representations of inputs that must be responded to with different actions tend to be more distant. As will be recalled, we use the term 'cloud' to refer to set of points (internal representations) in the abstract space of internal representations that must be responded to with the same action. In the present simulations there are only two actions, pulling or pushing the object, and therefore there are only two 'clouds'. As we have seen in a preceding section, a neural network has a good internal organization if the different 'clouds' of points in the abstract space of its internal units have reduced size and are widely separated one from the other. If we measure the size of the 'clouds' and the distance between 'clouds' at the end of the present simulation we observe the following results (Figure 12.3). 'Cloud' size is smaller in the situations with language (all taken together) than in the situations without language. When we compare the two linguistic conditions, with nouns and with verbs, the cloud is smaller in the situations with verbs than in those with nouns. The results with respect to the distance between the two 'clouds' are similar. The distance between 'clouds' is greater in the situations with language than in the situations without language, and greater in the situations with verbs than in those with nouns. These results confirm what we already saw in the section on the emergence of simple languages. Language has a positive effect on categorization in that it increases the distance between 'clouds' and reduces their size, i.e., it makes members of different categories more different and members of the same category more similar with respect to each other.

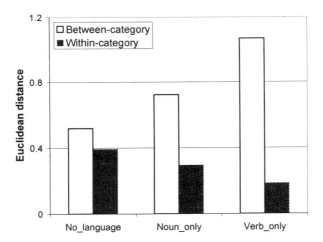

Figure 12.3 Average Euclidean distances in the 'clouds' of points for the input without language, with nouns, and with verbs.

What can we conclude from these simulations? Verbs (or predicates or perhaps proto-verbs) tend to be distinguished from nouns (or arguments or proto-nouns) in that verbs tend to co-vary with the action with which the organism must respond to the sensory input independently from the sensory input whereas nouns are linguistic signals that tend to co-vary with the sensory input. Verbs are important because they allow behavior to become detached from the sensory input. If an organism responds to a verb it executes the action designated by the verb whatever the sensory input in response to which the action is executed. If this is the function of verbs, what is the function of nouns? In our simulations nouns are redundant (unless they appear alone, without the object), since they always require the organism to execute the same action the organism is already used to executing with respect to a given object. At this primitive level of language evolution linguistic messages could be constituted by a single linguistic signal: a verb. The specific function of nouns can only emerge in somewhat more complex situations in which the organism simultaneously perceives two or more different objects. In these situations nouns can have a selective attention function in that they tell the organism which of the two or more objects is the one with respect to which the action specified by the verb must be executed. At this somewhat more advanced level of language evolution linguistic messages would be composed of two signals: a noun and a verb.

Conclusion

The simulation scenario that has been proposed in this chapter aims at studying language through a unified approach in which all aspects of language change (development, evolution, historical change) are simultaneously included in the model. This permits a better understanding of the way biological, behavioral, and cultural phenomena interact and of their role in the origin and evolution of language. The proposed approach is based on the simulation of populations of organisms that live in an environment and interact with both the environment and their conspecifics. The behavior of each organism is controlled by an artificial neural network, which makes it possible to study the effects of both behavioral and neural factors in the evolution, learning, and cultural transmission of language.

In this chapter we have described the simulation scenario and we have presented some examples of simulations based on the scenario. We have simulated how a very simple language can emerge spontaneously in a population of organisms because it helps the organisms to respond appropriately even in the absence of direct experience of objects, and to better categorize the objects when they encounter them. In a second simulation we have shown how a simple compositional language with two-signal messages can emerge and help the organisms to organize their experience in terms of a hierarchy of categories. In the third simulations we have simulated the emergence of a distinction between (proto-)verbs and (proto-)nouns, respectively defined as linguistic signals that co-vary with the organism's actions (verbs) or with the organism's visual input (nouns) and we have examined their different influences on the organisms' performance and internal categorization of experience.

Of course with the simulations we have described we have only scratched the surface of the many phenomena of language change and we have only exploited a very limited number of the many possibilities offered by the simulation scenario.

For example, with respect to our first simulation, many other possible evolutionary advantages of language could be explored such as its role in social coordination or in making predictions of future events or of possible consequences of planned but not executed actions. The potential for compositionality present in human languages of course is much greater than the very beginning of compositionality which has been explored in our second simulation but our simulation scenario can be used to further explore the emergence of more complex linguistic signals. The same for word classes. Our third simulation explores the emergence of a distinction between verbs and nouns but our simulation scenario can be used to further explore the emergence of, say, adjectives and proper nouns. By putting together word classes which are defined in our simulations in purely cognitive terms and compositional syntactical messages, our simulations scenario can tell us something about the mixed cognitive/formal nature of word classes in human languages.

References

Batali J (1994) Innate biases and critical periods: combining evolution and learning in the acquisition of syntax. In: Brooks R, Maes P (eds) *Proceedings of the Fourth Artificial Life Workshop*. MIT Press, Cambridge MA

Belew RK, Mitchell M (eds) (1996) *Adaptive individuals in evolving populations*. Addison-Wesley

Broeder P, Murre JMJ (eds) (2000) *Models of language acquisition: Inductive and deductive approaches*. Oxford University Press

Bullock S (1998) A continuous evolutionary simulation model of the attainability of honest signaling equilibria. In: Adami C, Belew R, Kitano H, Taylor C (eds) *Artificial Life VI*. MIT Press, Cambridge MA, pp 339-348

Burling R (1993) Primate calls, human language, and nonverbal communication. *Current Anthropology*, 34(1): 25-53

Cangelosi A (1999) Modeling the evolution of communication: From stimulus associations to grounded symbolic associations. In: Floreano D, Nicoud JD, Mondada F (eds) *Proceedings of ECAL99 the Fifth European Conference on Artificial Life (Lecture Notes in Artificial Intelligence)*. Springer-Verlag, Berlin

Cangelosi A (2001) Evolution of communication and language using signals, symbols and words. *IEEE Transactions in Evolutionary Computation*, 5(2): 93-101

Cangelosi A, Harnad S (in press) The adaptive advantage of symbolic theft over sensorimotor toil: Grounding language in perceptual categories. *Evolution of Communication*

Cangelosi A, Parisi D (1998) The emergence of a 'language' in an evolving population of neural networks. *Connection Science*, 10: 83-97, http://www.tandf.co.uk

Cangelosi A, Parisi D (2001). How nouns and verbs differentially affect the behavior of artificial organisms. In Moore JD, Stenning K (eds), *Proceedings of the 23rd Annual Conference of the Cognitive Science Society*, Lawrence Erlbaum Associates, pp 170-175

Cavalli-Sforza LL, Feldmann MW (1981) *Cultural transmission and evolution. A quantitative approach*. Princeton University Press

Christiansen MH, Chater N (1999) Connectionist natural language processing: The state of the art. *Cognitive Science*, 23(4): 417-437

Clark E (1993) *The lexicon in acquisition*. Cambridge University Press, Cambridge MA

Deacon TW (1997) *The symbolic species: The coevolution of language and human brain.* Penguin, London

Denaro D, Parisi D (1997) Cultural evolution in a population of neural networks. In: Marinaro M, Tagliaferri R (eds), *Neural Nets. Wirn-96.* Springer Verlag, New York

Elman JL (1990) Finding structure in time. *Cognitive Science,* 14: 179-211

Hare M, Elman JL (1995) Learning and morphological change. *Cognition,* 56: 61-98

Harley TA (1993) Connectionist approaches to language disorders. *Aphasiology,* 7: 221-249

Hinton GE, Nowlan SJ (1987) How learning can guide evolution. *Complex Systems,* 1: 495-502

Hutchins E, Hazelhurst B (1995) How to invent a lexicon: the development of shared symbols in interaction. In: Gilbert N, Conte R (eds) *Artificial societies: The computer simulation of social life.* UCL Press

Kirby S (2001) Spontaneous evolution of linguistic structure: An iterated learning model of the emergence of regularity and irregularity. *IEEE Transactions in Evolutionary Computation,* 5(2): 102-110

Livingstone D, Fyfe C (1999) Modeling the evolution of linguistic diversity. In: Floreano D, Nicoud J-D, Mondada F (eds) *Proceedings of the Fifth European Conference on Artificial Life, ECAL 99 (Lecture Notes in Artificial Intelligence, Volume 1674)* Springer-Verlag, Berlin

Nettle D (1999) *Linguistic diversity,* Oxford University Press

Nolfi S, Floreano D (1999) Learning and evolution. *Autonomous Robots,* 7(1): 89-113

Parisi D (1997) An Artificial Life approach to language. *Mind and Language,* 59: 121-146

Parisi D, Denaro D, Cangelosi A (1998) Language as an aid to categorisation. Poster presented at the *Second International Conference on the Evolution of Language,* London

Plunkett K, Marchman V (1993) From rote learning to system building: acquiring verb morphology in children and connectionist nets. *Cognition,* 48: 21-69

Plunkett K, Sinha C, Møller MF, Strandsby O (1992) Symbol grounding or the emergence of symbols? Vocabulary growth in children and a connectionist net. *Connection Science,* 4: 293-312

Rumelhart DE, McClelland JL (1986) On learning the past tense of English verbs. In: Rumelhart DE, McClelland JL, PDP group (eds), *Parallel distributed processing: Exploration in the microstructure of cognition.* MIT Press, Cambridge MA , pp 318-362

Savage-Rumbaugh S, Rumbaugh DM (1978) Symbolization, language, and Chimpanzees: A theoretical reevaluation on initial language acquisition processes in four Young Pan troglodytes. *Brain and Language,* 6: 265-300

Seidenberg MS, McClelland JL (1989) A distributed, developmental model of word recognition and naming. *Psychological Review,* 96: 523-568

Steels L, Kaplan F (1998) Situated grounded word semantics. In: *Proceedings of IJCAI-99,* Morgan Kauffman Publishing, Los Angeles, pp 862-867

Part VI

AUTO-ORGANIZATION AND DYNAMIC FACTORS

Chapter 13

Auto-organization and Emergence of Shared Language Structure

Edwin Hutchins and Brian Hazlehurst

The principal goal of attempts to construct computational models of the emergence of language is to shed light on the kinds of processes that may have led to the development of such phenomena as shared lexicons and grammars in the history of the human species. Researchers who attempt to model the emergence of lexicons make a set of shared assumptions about the nature of the problem to be solved. First, there are constraints on what counts as a shared lexicon. A *lexicon* is a systematic set of associations (a mapping) between *forms* and *meanings*. Forms are patterns. *Tokens* of a form are physical structures that bear the pattern of a particular form. For example, words are forms in this sense. Each instance of a particular word is a token of that word because it bears the pattern (sequence of sounds or letters) of that word. Forms must be discriminable from one another. *Meanings* are generally taken to be mental structures which, on the one hand, shape agents' interactions with a world of objects and, on the other hand, also shape agents' interactions with forms.

A lexicon is said to be *shared* when the members of a community adopt similar forms, meanings, and the mapping between these two elements. This is a requirement for the communication of meanings via forms. A shared lexicon is thus a systematic set of form-meaning mappings in which the forms are discriminable, the mappings are (roughly) one-to-one, and the set of associations between forms and meanings is shared by members of a community. The mappings are roughly one-to-one because synonyms (two or more forms for a single meaning) and homonyms (two or more meanings for a single form) are possible but do not dominate the mappings. The lexicons of natural languages can be described by these properties (among others).

The emergence of a shared grammar presents a more complex problem. Grammar refers to properties of language involving sequences of lexical forms.

These sequences are called expressions or sentences. A grammar implies constraints on the internal organizations of expressions. Grammatical expressions of a given language constitute a subset of the universe of possible sequences of forms drawn from the lexicon. This property of grammar is called *systematicity* and is often characterized by reference to the structure of expressions. For example, in English, structure is evidenced in a class of words called "noun" which includes the member forms "John" and "Mary." Language is systematic because the members of the set of allowable sequences share patterns (e.g., NVN → John Loves Mary, Mary Loves John, Marry Hates John, etc.). Looked at from the standpoint of language production, a grammar enables complex expressions to be easily built from simpler parts. The meaning of a grammatical expression depends on both the meanings of the lexical forms from which the expression is composed, and on the relations among the forms in the expression. For example, the sentences "John loves Mary." and "Mary loves John." contain the same lexical forms, but have different meanings because the forms bear different relations to one another in the two sentences. This property of grammar is called *compositionality*. Finally, it should be possible to create novel meanings by composing new expressions from the available set of forms. This property of grammar is called *generativity*. A grammar is said to be shared when the members of a community adopt similar systems for composing expressions, including similar mappings between expressions and meanings.

Using definitions of shared lexicon and shared grammar such as these, researchers ask, "What sort of process could lead to the development of a shared language?" Clearly, some historical process led human ancestors from the condition in which there was no shared language to the condition in which a shared language exists. It is assumed that language, and many other aspects of culture, develop without any central control. That is, there could not have been a "teacher" who knew the language first and then taught it to others. Rather, a shared language should be expected to emerge somehow from the interactions among the members of a community who must communicate. Since we have no direct access to the historical events that led to the development of language, a common strategy for addressing this question is to construct a computational simulation model. Such a model begins in a state in which a shared language clearly does not exist. The model is then run and eventually reaches a state in which a shared language does exist.

In our discussion of models addressing the emergence of language, we will attempt to clarify the different stances that modelers take regarding the elements of language and the relations among those elements. Figure 13.1 depicts the elements of language (in boxes), and their relations (connections among boxes), as exemplified in the simulation models that address the emergence of lexicon. The question marks (?) indicate relations that are treated differently by different simulation models. In addition to these components, each model also specifies processes that bring agents into interaction with one another. In a successful simulation, language-like structures emerge from these interactions.

In a computational simulation of the emergence of a shared lexicon, a community of virtual agents is created. Each agent is capable of implementing a form-meaning mapping. In the initial condition of the simulation, no systematic form-meaning pattern exists. The simulation will include a model of the interactions among agents, such that agents can be changed by the experience of interaction. This interaction protocol determines the organization of the interactions among agents and between agents and their simulated world. As the simulation runs, and as virtual agents are changed by repeated interactions with meanings and forms, and possibly with objects in the world and other agents, a shared lexicon *emerges*. The lexicon is not explicitly specified in advance and there is no teacher in the system telling the agents how to construct the mapping that will become shared. The term *auto-organization* refers to this process of emergence in which the community of agents organizes itself over time.

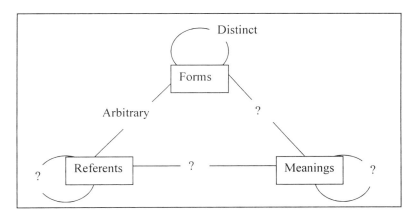

Figure 13.1 Elements of language (in boxes) and their relations (connections among boxes) in the simulation models on the emergence of lexicon.

Auto-organization is not magical. New patterns result from the organized interaction of patterns that are present in the initial conditions of the model or are present in the history of interaction itself. All of the models discussed in this chapter operate on a principle of modulated positive feedback. Modulated positive feedback is positive feedback with a resonant filter that favors some signals in the loop and causes others to dissipate. This principle underlies many kinds of auto-organizing processes including those that produce bio-convection and many other animal built structures (Turner, 2000). In the models discussed here some kinds of structure are simply given or assumed by fiat, some random process interacts with the assumed structure to produce small initial differences and similarities, and then a modulated positive feedback loop operates to amplify the initial differences and similarities. When we consider the significance of the models with respect to the goal of contributing to our understanding of the processes that might have led to language-like phenomena, it is essential to understand two kinds of question. The structure question concerns the kinds of structure the emergence of language might plausibly have been built upon. In our review of the models we ask, What has been given by fiat? What produces the initial patterns that are later amplified? The process question asks, What sort of process implements the modulated positive feedback that leads to the emergence of new structure? Is this process a plausible candidate given our expectations about the conditions under which language-like behaviors arose?

In all of the models considered in this chapter, the initial condition of the model includes structure in the architecture of the agents, and structure in the interaction protocol. Many of the models begin with structure in a set of candidate meanings. Some of the models also begin with structure in a pre-determined set of possible forms. Other models begin without structure in the forms, but include structure in the world with which agents interact. Auto-organization is the transformation and propagation of these structures into new, sometimes surprising, structures inside or between the individual agents. The most compelling models are those in which the emergent structures cannot be anticipated from an inspection of the initial conditions.

In all models of the emergence of lexicon, each agent takes the behaviors of other agents as the target for its own behavior. The idea that humans are, and that their ancestors were, prodigious imitators is an important theme of contemporary studies of primate behavior (Tomasello, 1996). This mutual and reciprocal targeting of behavior creates a positive feedback loop. Once a behavior enters the repertoire of one agent, for whatever reason, it is likely to enter the repertoires of others, which makes it even more likely to enter the repertoires of others, and so on. In order to produce a shared lexicon, the positive feedback loop created by mutual and reciprocal behavioral targeting must be modulated or filtered in some way. The solutions to the modulation problem vary depending on the assumptions on which the model is built and on the choices made regarding the representation of the various elements of the model. We turn now to the details of the models.

Three Frameworks for Modeling the Emergence of Shared Lexicons

Three major frameworks have been employed to simulate the emergence of shared lexicons. Each framework makes a different set of explicit and implicit theoretical assumptions. The implicit assumptions are often revealed by choices concerning the representation of meanings, forms, and referents, the mappings among these, and choices of algorithms that constitute the interaction protocols.

The three frameworks are:

1.   Expression/Induction (E/I)
2.   Form-tuning
3.   Embodied guessing game.

Expression/Induction models (Hurford, 1999; Oliphant, 1997)

In these models, forms are provided in a large pre-defined closed set. The forms have no relevant internal structure and there are no relevant relations among the forms. Typically, the set of forms are constructed by the researcher through random selection of characters from the English alphabet.

There is also a relatively small closed, pre-defined set of structured meanings. The meanings are unrelated and atomic. Meanings are assumed to reside inside individual agents. The agents are completely disembodied, and the emergence of lexicon is taken to be a formal problem. Typically, meanings are represented in the model as a set of strings that invoke a simple "world" for the reader of the research. This world has no independent representation and plays no role in the simulation.

The constraints on forms and meanings require that the form-meaning mappings must be learned as an unstructured list. The agents cannot learn or exploit any higher-level regularity in the set of forms or in the set of meanings, or in the set of possible form-meaning mappings. This is done in part to ensure arbitrary relations between forms and meanings. Because the distinctiveness of forms is built into the architecture of the model, this approach cannot address the processes by which distinctiveness of forms might arise. Similarly, because meanings are pre-defined and arbitrary the model cannot address any role that relationships among meanings might have upon the emergence of a lexicon. (See Kaplan 1999, for a model which explicitly addresses this question. See also the discussion below of Hutchins and Hazlehurst, 1995.)

A simulation begins with the creation of a community of agents. Agents then participate in pair-wise interactions that provide a mechanism for the transmission of form-meaning associations. A speaker, a listener, and a meaning are chosen for an interaction. From the outset, each agent is capable of representing all of the meanings internally. To express a meaning, the speaker chooses a form from the set of forms it has experienced other agents use for that meaning. If it has not yet experienced other agents expressing that meaning, it chooses a form at random (from a very large set). The listener knows which meaning the speaker is trying to express (the model forces this to happen) and simply adds the form-meaning pair expressed by the speaker to its list of observations. Note that since the meaning to be represented is given to both agents, the interaction is not a model of the communication of meanings via forms.

A system with only these parts will produce a lexicon in which each meaning is associated with many forms (as many forms as there were interaction events in which the speaker had not previously experienced a form for that meaning). To produce a lexicon in which each meaning has a single form, a production bottleneck is introduced. A *production bottleneck* exists if the method the speaker uses to choose a form to express a meaning results in some forms never being used by that speaker for that meaning, even though these form-meaning pairs may have been acquired by the agent. For example, a production bottleneck can be implemented by biasing agents to utilize those form-meaning mappings that they have experienced most frequently in the past. When a production bottleneck is in operation, form-meaning pairs will gradually drop out of the system, and the community will eventually converge on a single form for each meaning.

There is a positive feedback loop here because the contracting set of form-meaning pairs in use by speakers limits the range of experience of listeners, which further limits the range of production when those listeners become speakers, and so on. The positive feedback amplifies small differences in the frequencies of form-meaning associations in the "aboriginal" lexicon. The positive feedback is modulated or filtered by the rule that implements the production bottleneck.

As long as the production selection rules lead agents to fail to produce some observed form-meaning pairs, the production bottleneck will act as a weak filter, favoring some forms over others. Sharing among agents is produced by the elimination of some forms and the inability of an individual to eliminate a form that is frequently used by others (even if dropped by an agent, it will force itself back into the agent's repertoire). E/I models assume the distinctiveness of forms and the a priori sharing of distinct meanings. They use modulated positive feedback to produce the emergence of shared mappings by amplifying differences in the frequencies of form-meaning mappings. They adjust the distribution of intact form-meaning mappings in a population of agents. As Oliphant points out, "Innately specifying the set of signals and meanings, however, negates what is perhaps the primary benefit of a learned system – extendibility" (Oliphant, 1997:117).

Form-tuning model (Hutchins and Hazlehurst, 1995)

In this model, there is no pre-determined set of forms. In some sense there are no internal meanings either. Instead, there is a set of structurally related "visual experiences" imagined to result from stimulation of a visual sensory surface caused by objects within the simulation world. Visual experiences can stimulate production of agent behaviors without any explicitly represented meanings. Every agent produces a behavior (taken to be a verbal production) in response to each visual experience. Early in the simulation, the behaviors are undifferentiated. On each cycle of the model, a pair of individuals encounters a particular visual experience and each agent produces a behavior in response to that experience. Each agent then tries to shape its own behavior in two ways. First, it tries to shape its behavior so that it matches the behavior produced by the other agent. Second, its internal organization (implemented as a connectionist autoassociator network) leads it to change the structure of the behavior evoked by an experience so that it is different from the behaviors evoked by the other experiences. In this way, the behaviors are gradually tuned so that they (a) match the behaviors produced by other agents in the presence of the various visual experiences, and (b) discriminate among the visual experiences. The model will produce as many distinct behaviors as there are distinguishable visual experiences.

Each individual is an autoassociator network consisting of 36 visual input units, four hidden units, four verbal input-output units and 36 visual output units, as shown in Figure 13.2. The simulation proceeds via interactions – one interaction is one time step in the simulation. An interaction consists of the presentation of a scene (drawn from the set of 12 scenes, see Figure 13.3) to two chosen individuals, a *speaker* and a *listener* (drawn from the set of four individuals).

One of the individuals chosen (say A) responds to the scene by producing a pattern of activation on its verbal output layer (A speaks). The other individual, (say B) also generates a representation of what it would say in this context.

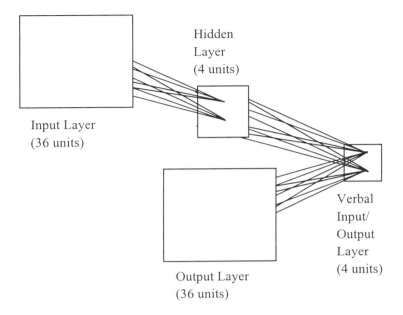

Figure 13.2 The autoassociator network used in the simulation.

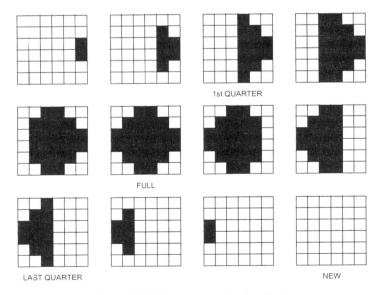

Figure 13.3 The scenes to be classified.

As listener, B uses what A said as a target to correct its own verbal representation. This moves B's behavior toward a match with the behavior produced by A. The listener, B, is also engaged in a standard autoassociator learning trial on the current scene, which means its own verbal representation – in addition to being a token for comparison with A's verbal representation – is *also* being used to produce a visual output by feeding activation forward to the visual output layer. The pattern of activation on the visual output is compared to the input signal to establish an error pattern, which is back-propagated through the network. This changes the structure of the behavior evoked by an experience so that it is different from the behaviors evoked by the other experiences. These two sorts of learning together produce the matching of discriminable verbal behaviors. Over time, by choosing interactants and scenes randomly, every individual has the opportunity to interact with all the others in both speaking and listening roles in all visual contexts.

At the outset, the behaviors that are evoked by the visual experiences are not really forms that can play a role in a form-meaning mapping. All visual experiences give rise to approximately the same verbal behavior in all agents as shown in Figure 13.4. This means that responses to visual stimuli initially carry no information. As the simulation progresses, however, the behaviors come to have the properties of forms and play the role of forms in form-referent mappings. These properties emerge because the networks develop internal structure that solves the dual problem of producing forms that distinguish visual experiences and producing shared form-referent maps among members of the community.

Contrary to what is generally expected of the relationships between forms and meanings, the mappings produced by this model are not completely arbitrary. Because the architecture of the agent requires the forms to produce an efficient or condensed encoding of the structure of the set of visual experiences, parts of the form (individual unit activations) encode features of the visual scene.

The patterns that arise on each agent's verbal medium come to discriminate among the patterns on that agent's visual medium and come to agree with the patterns that arise on the verbal medium of the other agents in the presence of each visual pattern as shown in Figure 13.5. It is important to note that the individual agents become functionally equivalent, but not structurally identical. Agent internal organization provides functional equivalence yet is variable among members of the community.

In this model mutual reciprocal targeting of behavior supports a feedback loop as described for the E/I models. Instead of simply choosing a discrete form that matches the discrete form chosen by the other agent (as was done in the E/I models), agents in this model tune the structure of the forms they produce. What is shared in the "tuning" model is a skill to produce particular behaviors, rather than choices for pre-formed tokens. As a shared solution begins to emerge among a few agents, the experience of other agents is changed so that they are drawn toward that solution, which then affects more agents, and so on.

In fact, this model involves positive feedback organized by two kinds of tuning simultaneously. Consider the constraints on forms as forces applied to the weight agent in response to each interaction. The constraint that *forms be shared* applies a

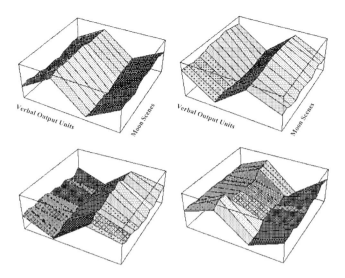

Figure 13.4 Patterns of activation of the verbal output units in four individuals early in the simulation. None of the individuals can distinguish among the twelve moon scenes and there is no shared pattern of activation across individuals for any particular moon scene.

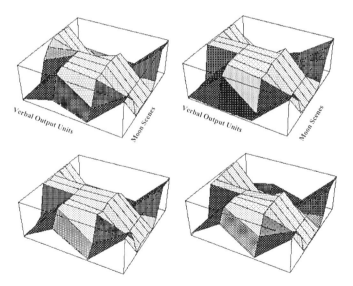

Figure 13.5 Patterns of activation of the verbal output units in four individuals after an average of 2000 interactions for each pair of individuals. All of the individuals distingu:sh among the twelve moon scenes by producing a distinctive pattern of activation on the verbal output units. A shared set of form-referent mappings has developed.

force to shift the weights of the agent. This shifting has the effect that the next time the same experience is encountered the weights will produce a form more like the form that was produced by the other agent. The constraint that *forms be discriminable* one from the other applies a second force. The effect of this force is to shift the weights so that the form produced for each experience is maximally different from the forms produced for other experiences. Imagine both forces pushing the value of each weight. At the outset, there is no reason for these forces to be aligned. In fact, they often act in opposite directions. If two agents do not agree on how to represent an experience, the sharing force may act opposite the discrimination force on every weight. However, if two agents agree, even partially, then the sharing force may work in concert with and add to the discrimination force.

The ease of finding a solution that satisfies both constraints depends on the initial conditions of the model, especially the initial distribution of weight values. One positive feedback loop, call it the distinctiveness loop, amplifies initial random differences in the forms produced within agents across visual experiences. Another positive feedback loop, the sharing loop, amplifies initial random similarities in the forms produced among agents for each particular visual experience. As learning proceeds, the two loops may become mutually reinforcing. It is crucial to note that the structure of the forms is a product of the interaction of structure that is present in the visual experiences and emergent structure inside the agents. The two positive feedback loops amplify differences and similarities in that interaction between external structure and developing internal structure. Form-tuning models assume a structure of visual experiences and small random differences among the internal structures of individual agents. The distinctiveness of forms and the sharing of form-meaning mappings emerges from the operation of two, interacting, positive feedback loops.

It is easy to see that E/I models have given forms and given meanings and no referents at all. It is more difficult to say what the elements of the form-tuning model represent. One might take the visual experiences to be meanings analogous to the meanings of the E/I models. After all, it is only the modeler's assertion that these things are external to the agents that makes them so. In that case, meanings are given a priori. The process could then be said to produce distinctive forms and shared form-meaning mappings. Alternatively, the visual experiences could be taken to be referents outside the agents. In that case, the behaviors that are produced in response to the referents might be either meanings or forms. The interpretation of the behaviors as meanings is supported by the fact that they entail feature decompositions of the visual experiences. Feature theories of meaning have a long and venerable history. However, if the behaviors are meanings, then where are the forms and what is meant by the direct sharing of meanings? The interpretation under which the visual experiences are referents and the behaviors are forms seems at first glance somewhat anomalous because then the relation of form to referent appears not to be mediated by meaning. A third alternative is that the behaviors are forms and the meanings are represented implicitly in the structure of the weight vectors of the agent, rather than explicitly as an autonomous interpretable representation. Under this interpretation, the meaning of a visual experience is the process that is required to produce its representation as a form.

This last interpretation is important because under it, the models can be said to model the emergence of referent-meaning-form mappings with emergent forms and meanings, but without making a commitment to symbolic representations of meanings.

Embodied guessing game (Steels, 1996; Steels and Kaplan, 1999)

As was the case in the E/I models, forms in these models are arbitrary strings of syllables drawn from a fixed and given alphabet.

These models have the most complex representation of meanings among the three frameworks considered in this section. Meanings are symbolic representations of context-sensitive distinctions over a set of complex reference objects. As was the case in the form-tuning model, objects have properties that impinge upon agents' visual sensory surfaces. This process produces agent perceptions of features that are represented by continuous ranges along the properties sensed by the sensory surface. The task facing an agent is to discriminate topic (foreground) objects in the context of ground (background) objects within a scene using sets of features. In order to solve this problem agents employ an error signal generated by a built-in need for producing unique feature sets among all topic/ground possibilities.

In these models, there is an explicit distinction between objects in the world (referents) and symbolic representations of properties of those objects (meanings). This is the only framework of the three that includes an explicit representation of form, meaning, referent, and the relations among the three terms. In interaction, speaker and listener share awareness of the reference object. The speaker creates an utterance to encode a feature of the object. The listener tries to anticipate which feature the speaker will encode. When the listener guesses correctly, the association of the form (utterance) to the meaning (feature description) is strengthened for both speaker and listener. When the listener guesses incorrectly, the form meaning association is weakened. Importantly, the feature sets held by different agents may not be identical. However, as the universe of objects about which agreement is required becomes large, convergence toward shared perceptual distinctions develop, and shared form-meaning-referent mappings emerge.

The entire set of all possible form-meaning pairs is implicitly present in the model. Interactions strengthen some pairs and weaken others. A modulated positive feedback loop drives the process because stronger associations are more likely to lead to successful communication, which will make them even stronger. Meanings are grounded in the sensed properties of referents. The structure that is amplified by the positive feedback is produced by fortuitous agreements in the application of the meaning-making process. Forms have arbitrary relations to meanings and to referents.

Modeling the Emergence of Grammar

One goal of models that simulate the emergence of lexicon is to demonstrate possible origins for the "denotation" function of language. Denotation is made possible through development of shared and coherent relationships among forms, meanings, and referents. In general, forms within these models are atomic units. Models of simple lexical denotation make no effort to account for combinations of forms. Instead, the single relation of "distinctiveness" holds among the forms of the lexicon.

Simulations of the emergence of grammar attempt to explain the origins of a more complex but related phenomenon, namely the systematic nature of language. Every language organizes words of the lexicon into sequences that express complex meanings. The grammar of a language describes the internal structure of these sequences and accounts for the mapping from this structure to the complex meanings that are expressed. The fact that sequences of forms have internal structure implies new classes of relations among forms. The goal of models simulating the emergence of grammar is to describe the origins of complex form-form relations and the manner in which these constructs map onto complex meanings.

The computational requirements of a language with syntax entail:

1. **Compositionality**
   The capacity to construct composite forms (representing complex meanings) from simpler parts. Because human language involves a serial production device, complex forms (e.g., sentences) are composed through sequential concatenation of atomic forms (e.g., words).

2. **Systematicity**
   The capacity for complex constructs to entail "roles" for elements in the construct (e.g., noun) such that these elements retain their individual meanings (e.g., as a word) yet also serve the meaning defined by the roles they fill within the construct (e.g., as the subject of sentence).

Combining the properties of compositionality and systematicity accounts for the open-ended yet structured nature of language. The notion that language is inherently "generative" follows directly from these properties. With such a system, novel meanings can be expressed with novel forms and yet these sentences are easily understood by virtue of conforming to the grammar, the form-meaning mapping of the language.

A great challenge in the study of language is capturing the structural properties of such an arrangement, modeled as an abstract formal system, while addressing what we know about human evolution and history as well as what we know that language accomplishes in the world as a vehicle for situated communication. Attempts to model the emergence of grammar highlight certain relations among the elements of language while disregarding others. We now turn to an examination of a set of models addressing the emergence of grammar. For each model we try to illuminate the consequences of choices made in the representation of language elements and relations among the elements, as well as the processes which employ these elements and relations.

The capacity to learn systematic form-meaning mappings (Batali 1998, Kirby 1999)

In this group of models, which all employ the E/I framework (Hurford, 1999), each simulation includes a finite set of discrete tokens and a process that can assemble the tokens into complex forms or sequences. The entire set of such sequences is open-ended or potentially infinite. The simulation world also includes a set of structured symbolic meanings which agents share. The structure of meanings is given in some type of propositional language (e.g., a simplified predicate logic). There are no referents included in the simulation world.

These simulations demonstrate the development, through learning, of shared mappings between the structure of meanings and sequential patterns of forms. At the start of the simulation, there is no patterning within or among the forms (i.e., the relationships among forms is unspecified). A primary goal of these simulations is to demonstrate that learning devices employed in the service of inter-agent communication can produce the specific type of mappings characteristic of language. If such mappings can emerge from sharing propositional meanings through verbal encounters, then the nature of these encounters may provide a non-genetic generator of language structure.

As with models simulating the emergence of lexicon, agents in these models come together in interaction in the roles of speaker and listener. The speaker produces a sequence from a meaning, and the listener is challenged to produce the inverse mapping from form back to meaning. Importantly, the meanings in the encounter are unproblematically shared by the interactants. In all models of this group, the error signal that promotes consensus is made available from the a priori sharing of structured meanings together with the experience of specific forms produced by the speaker. These models assume that the capacity to construct complex propositional meanings arose prior to and independent of the capacity to express meaning, and that the sharing of meaning is accomplished through some non-linguistic means.

In Kirby (1999), the agents' abilities to map between forms and meanings are provided by symbolic rules. The set of such rules used by an agent constitutes that agent's grammar. Rules are induced and maintained through an algorithm that strives to accommodate all experienced sentence examples in the simplest form possible. Given a complex form and the meaning that this form expresses, the induction algorithm of the learning agent first checks to see if the form conforms to the agent's grammar. If that check succeeds, then nothing else is required. Otherwise, a new rule, which produces this sentence, is entered into the agent's grammar. The new rule uses the given meaning to create the given form. This new rule produces the sentence as a singleton. At this point a generalization algorithm attempts to decompose the new rule in a fashion which exploits parts that are already available within other rules of the grammar. This is possible because of the well-formed symbol structures representing meanings. Over time, the set of rules becomes compact and coherent, making the grammar efficient while accommodating all experienced sequences.

At first, agent productions are simple one-to-one mappings from complex meanings to random sequences of tokens. This state of affairs results from the fact that agents have minimal linguistic knowledge and thus speakers engage in many instances of "invention", modeled as the random selection of forms for gaps in production knowledge. Over time, after many instances of language production, rule creation, and generalization, agents converge upon systematic and shared sets of rules constituting a single grammar with properties of compositionality and recursion. Convergence in this process results from the properties of learning, which strives to generalize linguistic knowledge. Each new form-meaning mapping acquired by a learner is "chunked" so as to fit within the existing rule structure of the grammar, whenever possible. As a consequence, early in the simulation experienced forms are one-to-one with rules while later on the number of form instances dominates the number of rules. The more comprehensive the rule set becomes, the more likely it is that the rule set provides the mechanism for production (thus, "invention" is not required). This process creates positive feedback, which builds the structure of the rule set. The rule set, in turn, acts as a filter on the kinds of forms that can be produced.

In the general multi-agent case, the distribution of rule sets among agents produces the observed distribution of forms[1]. At the same time, the observed distribution of forms sets the targets for the agents' rule sets. If fortuitous independent invention increases the representation of a particular form for a particular meaning, that form-meaning pair will be a more frequent target for the rule sets of other agents. If it is the sort of mapping that the induction algorithm can learn, it will become more likely that the rule sets of other agents will come to produce that form for that meaning, which will further increase the representation of that form-meaning pair. The positive feedback loop amplifies the effects of fortuitous coincidental form-meaning pairs, and is modulated by the learning processes that govern the modifications of rule sets. The induction algorithm is a sort of resonant filter on the positive feedback loop, reinforcing some signals and causing others to dissipate.

In Batali (1998), a very different representational mechanism is employed. However, the methodology is similar. The objective of the model is a shared mapping between structured meanings and patterned sequences. In this model, an agent's ability to map between forms and meanings is implemented by a recurrent neural network. The weights of the network propagate structure from input units to output units. Network weights develop in such a way as to support a systematic mapping from sequences of basic tokens to complex meaning structures. This mapping is taken to be a grammar, and it emerges from the interactions among agents.

As Elman (1991) showed, sentence processing by an agent with this type of architecture can be understood in terms of trajectories through the network's internal state space. The representational space of the network can simultaneously encode information about individual tokens (inputs to the network) and the positional or context-sensitive information given by the sequence of tokens constituting a sentence. The learning algorithm applied to a recurrent neural network partitions this space so as to accommodate all of the examples experienced, while simultaneously enabling systematic generalization, capturing information about the structure inherent in the set of examples.

In Batali's model, as with all of the others examined in this paper, agents interact via the roles of speaker and listener. More accurately, the model explicitly treats the problem as that of a "learner" sampling the productions of "teachers" in

---

[1] In Kirby (1999) the population is composed of a single speaker/listener pair.

each iteration of the simulation. Each teacher produces sequences through a protocol that selects the sequence of tokens which "best" invokes the given meaning. This meaning and the produced sequence then provide the learning instance for the learner. Over time, preferred mappings emerge because learners are also teachers. These preferences result from (1) the distribution of initial conditions inherent in the starting weights of the networks, (2) biases which may be associated with the ordering of interactions, and (3) the constraints of emerging agent representational spaces which may promote some mappings and inhibit others.

Despite very different choices of representational frameworks, the Kirby model and the Batali model share high-level assumptions. Objects in the world are assumed to be irrelevant to the problem of language systematicity. Abstract structures are declared to be meanings by stipulation alone.

At a finer level of description, the shared assumptions made by the two models entail four important features: (1) the structure of meanings (i.e., the meaning-meaning relation) is given rather than emergent, (2) communication has no role because meanings are always shared perfectly before interactions take place. Interaction is present only as a context for the production and comparison of form-meaning mappings, (3) the distribution of understandings held by agents plays no role in the models because all agents understand the world in exactly the same way and, (4) the framework assumes that the origin of grammar is explained by the propagation of a pre-existing internal language of thought to an external language in the world.

We will return to this set of shared features in the discussion section. First, we review a very different simulation that nonetheless has the shared goal of elucidating the nature and possible origins of language systematicity.

The emergence of propositional descriptions about the world (Hazlehurst and Hutchins, 1998)

In Hutchins & Hazlehurs (1995 - H&H'95) and Steels (1996), the emergence of "lexicon" is modeled as the development of *shared descriptions of perceptual distinctions*. Agents in these models inhabit shared worlds containing a variety of objects. Meanings are emergent perceptual structures that mediate relations between forms and aspects of objects in the simulated world. Forms are descriptions of the perceptual structures. The simulations model the development of mappings between emergent forms, emergent meanings, and given environmental structure. These meanings are, thus, not structural entities created in advance by the researchers but rather are themselves developed in the course of the simulated interaction processes. The agents engage the objects as part of an explicit communication task.

In modeling the emergence of grammar, Hazlehurst and Hutchins (1998 - H&H'98) add the social and cognitive problems of *sharing attention* to their earlier model. Sharing of attention introduces a time-based sequential coordination constraint on interactions, providing a temporal scaffold for the simultaneous emergence of patterned forms and systematic form-meaning mappings.

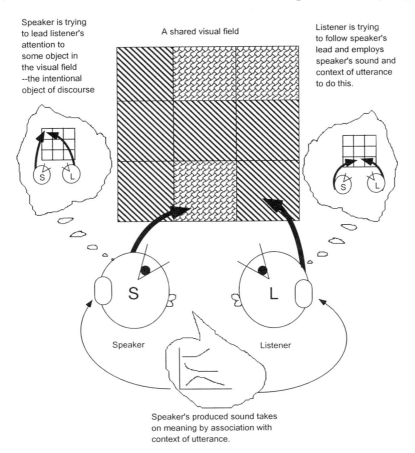

Figure 13.6 An Interaction. A simulation is composed of a sequence of interactions involving two agents chosen at random from the population, one in the role of speaker and the other in the role of listener. These agents engage each other in a discourse structured by a specific communication task. The speaker has in mind an object (the "intentional object") that is located within the shared visual field but unknown to the listener. The speaker employs sounds and gaze to direct listener's attention to the intentional object. The listener employs the speaker's sounds and gaze to coordinate in attempt to accomplish the communication task of identifying the intentional object.

Hutchins and Hazlehurs assume that language is a set of resources that have been shaped in response to the problem of coordinating action. Agents have some built-in cognitive properties. For example, agents have the ability to perceive the world via a visual modality. But even basic cognitive properties may be shaped by their use. For example, perceptual structures are tuned by constraints that are imposed by the need to represent that which is perceived. Acts of communication are both shaped by the organization of cognitive processes and put constraints on the organization of cognitive processes. Meanings and forms arise together. The requirement to produce forms that coordinate with meanings shapes the nature of meaning.

In H&H '95, agents that engage each other as discourse participants share the scene about which they are speaking. The sharing of a scene in interaction is given as a property of the simulation and both participants attend to the shared scene in its entirety. In H&H '98 discourse participants also share a scene, but rather than a scene being a single object, a scene is composed of multiple objects located on a spatial grid which is mapped onto the visual sensory surfaces of each agent in interaction. Now, the agents must negotiate a shared focus of attention in order to communicate successfully. The problem of communication now includes the creation of shared understanding about which referent is being discussed.

The speaker in each interaction has some specific object in mind that is present within the current scene. We refer to this as the "intentional object" of an interaction. For the purposes of the interaction, this chosen (but privately held) object can be thought of as foreground against a background comprised of a scene of spatially arranged objects. The listener sees the entire scene but at the start of the interaction only attends to the foregrounded object by chance. When taking the role of speaker, agents produce non-linguistic structure that may lead the listener (in concert with produced linguistic structure) to attend to the referent held in the speaker's mind. When taking the role of listener, agents may make use of this structure to determine what the speaker has in mind. Communication is successful when the listener identifies what is on the speaker's mind by achieving shared focus of attention upon the intentional object (see Figure 13.6).

The agents' abilities to accomplish this communication task depend on the speaker's ability to guide the listener's attention to the object held in the speaker's private mind. It is also dependent on the listener's ability to follow the speaker's lead. Agent focus of attention is materially available to interactants as direction of gaze, which identifies a location within the grid of objects that make up the visual scene. In other words, coordination is made possible through (non-identical) visual and verbal inputs to each agent. Coordination is only accomplished through success at the respective (and asymmetric) tasks of the two participants, and failure to coordinate within an interaction terminates the interaction.

Agents are composed of complex, modular, connectionist networks which map from aural and visual sensory surfaces (inputs) to motor controls effecting gaze and verbal actions (outputs). In interaction, motor productions by the speaker create sounds (forms) that impinge upon the aural sensory surfaces (the "ears") of both speaker and listener. In addition, each agent perceives the location of gaze (the "focus of attention") of both agents on each time step of the interaction. Actions are produced on each time step of the interaction, first by the speaker and then by the listener. Actions are realized in the world as shifts in the location of gaze of each agent and a public sound (form) produced by the speaker.

As with H&H'95, the forms that might do communicative work in this process are of no use early in a simulation run. Tokens are represented by continuous values along dimensions in agent verbal "articulatory space," so the set of possible tokens is infinite. At the beginning of the simulation, there is no useful structure in agent verbal productions (see Figure 13.7). Stringing such tokens together in sequential constructs also carries no information at this point.

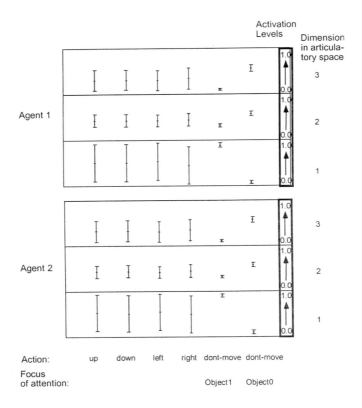

Figure 13.7 A lexicon after 10000 interactions of a simulation run. In the simulation run detailed in Figures 13.7-13.10, there are two agents in the population, two objects in the environment, and each object can occupy a single location within a 3x3 scene. There are, thus, 512 possible scenes. We limit the simulation to 462 scenes, one of which is selected to be the "shared visual field" for each interaction, and saved 50 scenes (randomly chosen) aside for testing of the population on novel scenes. Sounds are represented by a layer of three output units, each unit representing a feature or dimension of agent verbal articulatory space. Thus a sound is represented by three real-valued numbers, each within the range [0.0,1.0]. The figure shows the distribution of sounds that each agent produces in all possible contexts. Distributions are represented by the value, along with one standard deviation error bars. The figure lumps together many different specific contexts (there are thousands, depending on how much history is considered that leads up to the agent's current situation) and differentiates only by the agent's concurrent choice of gaze motor action (shift up, down, left, right, stay-focused-on-Object1, stay-focused-on-Object0). At the beginning of the simulation, all sounds are nearly identical – in fact, all articulatory features take on mid-range values for all contexts (not shown here). After 10000 interactions (shown here), we see that there is emerging consensus for using the third articulatory feature to enable a contrast in denotation of Object0 and Object1. The lexicon is beginning to emerge.

Early in a simulation run the agents often terminate interactions quickly, because they are unable to sustain the coordination required for the "follow the leader" communication task, and therefore the verbal sequences they produce are short. In fact, early in a simulation run, the communication task is accomplished by chance alone and verbal forms are meaningless entities – just noise. At this point, there is no artifactual structure (forms or behaviors) capable of mediating agent accomplishment of the task. Later in the simulation run, structure in support of the task (both internal to and among agents) does develop and success at the communication task rises.

One way to measure this success is by accounting for how interactions terminate. There are five possible conditions for interaction termination in a simulation run:

1. *Invalid Shift.* Speaker attempts to shift gaze outside of the visual field.

2. *Disagree.* Speaker and listener gaze become uncoordinated.

3. *Halt.* Speaker and listener successfully conclude the interaction by halting at the intentional object held in the speakers mind and located within the shared visual field.

4. *Cycle.* Speaker revisits a location with gaze that was attended to on the previous time step.

5. *Max.* Some maximum number of time steps have occurred within the interaction.

Figure 13.8 shows the evolution of a simulation run in terms of the frequency of each type of termination condition as the run proceeds. As shown, the simulation evolves to a point where the communication task is reliably accomplished. This is seen by the rise in occurrence of the *Halt* condition and the extinction of all other conditions for terminating interactions.

Over time within a single simulation run, several kinds of structure begin to emerge that mediate the organization of behavior which accomplishes the communication task and which provide the foundations for a simple language. First, agents produce a class of forms and meanings, and the mapping between them, which denote objects in the world of the simulation (see Figure 13.7). This occurs because despite the problem of coordinating attention, agents nonetheless attend (by chance) to the same object on the first turn of some interactions. When this happens, the agents can tune their verbal productions for the object to match that produced by the other. This emergence of lexicon is a replication of the result obtained in H&H'95. Structure which builds internally in service of perceptual distinctions among referents propagates to the verbal form-producing and form-interpreting mechanisms of agents. These properties enable simple denotation of the current focus of attention on the first turns of interactions, precisely as happened in H&H'95.

Second, agents develop internal structure that permits them to use the gaze of the other as a guide to attention. This produces the joint management of attention. This attention management function is structurally similar for speakers and listeners although it entails distinct tasks for each of them.

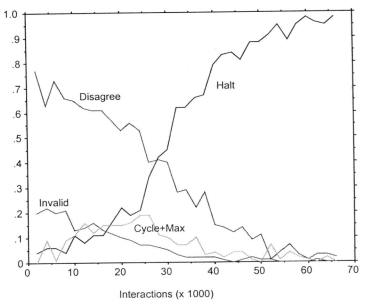

Interactions (x 1000)

Figure 13.8 The evolution of a simulation run. Every interaction terminates under one of five conditions (*Disagree, Halt, Invalid, Cycle, Max*). See main text for explanations of these conditions. Over time, interactions nearly all terminate under the successful Halt condition, indicating that the communication task is being solved in almost every instance. The y-axis represents the fraction of interactions which terminate under each condition. (See Figure 13.7 for a description of the parameters of this simulation run.)

As discussed above, the speaker learns to produce shifts in gaze that can direct the listener's gaze to the object held in mind, while the listener learns to follow the speaker's gaze. In the process of producing coordinated attention management, agents develop another (second) class of verbal forms that refer not to objects in the visual scene but rather to *the actions required to maintain coordination in visual attention*. In other words, a new class of internal structure, involving the motor control of attention, propagates to the verbal form-producing and form-interpreting mechanisms of agents (see Figure 13.9). These forms (when fed back to agents via an auditory sensory surface) produce internal meanings which, when allowed to activate motor control portions of agent architecture, produce actions such as "shift gaze leftward". Agents develop this second class of forms (and the sharing of these forms) through the process of agreement in motor control structure associated with controlling gaze.

This development can be seen as analogous to the externalization of perceptual structure that serves as the basis for development of the lexicon. In each case there is the requirement for coordination between something internal (motor or perceptual structure), something external (an object or an action), and something verbalized (a language form).

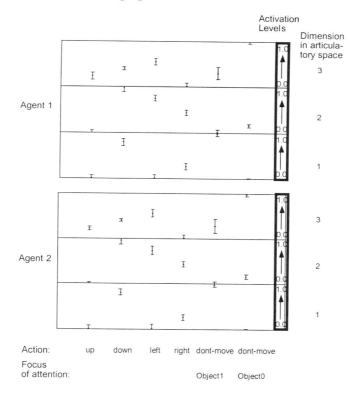

Figure 13.9 A lexicon after 60 000 interactions (for the same simulation run as shown in 7a). While the early denotation of objects in the simulation still holds (see Figure 13.7), the lexicon is now elaborated to include forms which coincide with actions associated with gaze and the spatial shifting of a focus of attention. This elaboration is emergent and shared by the agents of the population.

As the attention management task is sequential, so is the production of language forms that come to reliably represent the coordination task. In fact, after a period of time agents can use the language constructs alone to simulate the appropriate motor events, without actually manifesting any of the entailed actions in the world (see Figure 13.10).

This development provides the agents with a system for predicating and communicating the relative arrangement of objects in visual space. Clearly, this system has emerged from the interactions of the agents and has not been specified by any central designer. In this complex system, multiple interacting modulated positive feedback loops, operating simultaneously on several aspects of agent/environment coupling, amplify emergent patterns that link perceptual processes to words and features of objects, motor processes to words and shifts in attention, and attention management processes to gestures and direction of gaze.

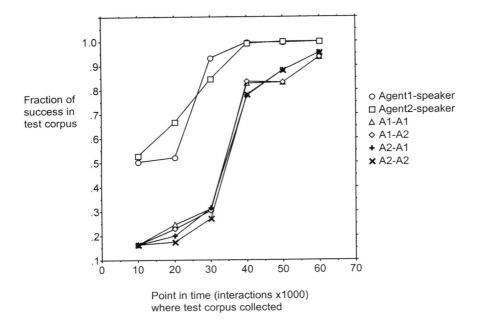

Point in time (interactions x1000)
where test corpus collected

Figure 13.10 Emergence of the ability of language structure alone to guide attention. Every 10 000 interactions a "test corpus" of language constructs was elicited from agents using the following method. Each agent, in role as speaker and in a complete set of contexts created from 10 novel scenes, produced actions precisely as they would in a normal interaction except that there was no listener with which to negotiate the interaction. In particular, only two types of interaction termination applied: *Halt*, and *Max* (see Figure 13.8). The fraction of successful termination by speakers (i.e., *Halt*) is shown by the two upper curves in the diagram. These two conditions represent reference curves for the others in the figure, and the actions produced by these speakers (gaze and sounds) constitute the test corpus. In the four conditions represented by the four lower curves, we took the test corpus of one speaker and had each agent process the language constructs in the role of "blind listener". In particular, the listening agent received no visual inputs and had to rely on language inputs alone to produce the appropriate actions. Appropriateness of actions (and thus "success" as identified in the figure) was determined by comparison with speaker's actions. As can be seen, once the speakers had organized coherent language structure which solved the communication task (by about 40 000 interactions in the simulation run) then agent's abilities to perform the same task solely with access to the language constructs follows (by about 60,000 interactions in the simulation run).

At this point a system of language with some very interesting properties has been bootstrapped. H&H show that this language exhibits both systematicity and compositionality, and that agent facility in producing valid strings of the language (and rejecting all others) demonstrates agent grammatical competence in the language. In particular, there are an infinite number of valid strings that are possible, yet they all conform to a compact structural description that demonstrates word classes (objects and actions), context-free composition from the lexicon (e.g., the words representing objects take the same form regardless of position in a complex construction) and recursion (i.e., complex constructions involve the systematic use of component sequences).

Discussion

There are some important similarities and differences between the assumptions made by the two classes of models reviewed in this section. Both classes of models deal with the emergence of grammar, taken to be a mapping between forms and meanings. Both classes of models address how this mapping takes on specific properties entailing the emergence of certain complex constructions or form-form relations. Of particular interest to all of the models are constructions that demonstrate compositionality and systematicity in language. Finally, in none of the models is the mapping built into the "genetic material" of the agents. Neither is there a centralized designer that defines that grammar. Rather, the mapping emerges in the course of exchanges that produce a feedback effect encouraging convergence upon a shared grammar.

The classes of model differ, however, in the processes and component structures of the simulations that put the mapping (the grammar) in place. Although both classes of models employ a framework of "communication" as the mechanism by which shared grammar emerges, the models take very different stances on what this actually means. In the E/I models of Kirby and Batali, agents come together in interaction *already* sharing the meaning of what is being talked about *without* requiring a language to accomplish this. The goal of these simulations is to demonstrate that the structure inherent in complex proposition-like meanings that are already available and shared can propagate into form-meaning mappings (the grammar) simply through the process of agents learning from each other's productions. Thus, interacting agents are imitating each other's speech behavior but not really communicating, per se. Hurford (1999) explains that the emergence of systematic form-meaning mappings in these models results from the imposition of "bottlenecks" which prune the choices (all implicitly present in the meaning and form repertoire of the agents) down to a compact, shared, and explicit set. He speculates that these models would work equally well by eliminating the multi-agent component and simply imposing constraints upon an agent's production or semantic repertoire.

By contrast, in the model of H&H'98 the task presented to agents is built upon the premise that agents engage each other in order to resolve the communication problem that arises when one agent (the speaker) has something in mind, an attentional relationship to a perceived object, that is not directly accessible to the other agent (the listener). In this case, the structure of meanings and forms must co-emerge. There is no "language of thought" available to agents prior to engaging the world and negotiating its meaning. Rather, agents learn about the world through perceiving it, and coordinating action within it, and talking about it. Thus, the constraints of language production are a resource for developing cognitive structure involved in perception and motor tasks. Simultaneously, as structure associated with solving these tasks begins to emerge, that structure can become a resource for language production. H&H'98 claim that the origins of language systematicity and structured meanings reside in the processes that produce this coordination among structures.

In contrast to Hurford's (apparently correct) observations about the E/I framework employed by Batali and Kirby, isolated individual agents in H&H'98 *cannot* produce a systematic language. This is so, even though an individual agent is fully equipped in principle to produce a private language. In the case of an individual learner, the cognitive task of tracking a target location in visual space (without requiring coordination with a listener) does not produce a systematic language. For individual learners, this motor control task is solved by treating the visual field as a gradient encoding the distance to target. In this case, any path that minimizes distance to target is a good path, and there is no need to adopt a convention for parsing the visual field. Such a representation of the visual problem (unencumbered by the constraints of communication) does not yield the cognitive structure necessary to produce a systematic language. It is only in the course of coordinating action among agents in interaction that a systematic language emerges from this model. When agents must coordinate their actions with others, but not when they act alone, the preferred pathways through visual space build upon conventions that yield the compositional and systematic construction of sequences.

General Discussion of the Models

Framing the issue in terms of the modulation of positive feedback permits us to apply the process and structure questions to the modeling attempts. What is the process in which modulated positive feedback could operate to produce language-like structure? What is the relation of the process by which structure develops to the processes by which structure is thought to change in historical time? What is the substrate on which the growth of structure takes place? The answers to these questions clarify the contribution of the modeling efforts to the high-level goals of understanding the sorts of processes that may have led to the development of language.

We do not expect language to emerge from a process that does not include the challenge of communication. We find it implausible to look for the origins of language in interactions where fully composed meanings are injected into the mind of the listener before a public expression of that meaning is encountered. E/I models may be interesting engineering solutions, but this aspect of the process violates our understanding of the conditions under which language-like behavior might have arisen.

The structure question is more difficult to handle. The field of cognitive science is nowhere near a consensus on the nature of linguistic representations for contemporary humans. There seems little hope that we could resolve this question for proto-humans at some unknown point in the evolution of intelligence. If the argument cannot proceed from evidence, what else is there?

Expression/Induction models have nothing to say about the development of the language of thought or the representation of meanings because they simply assume that complexly structured meanings exist prior to the language phenomena that emerge. The problem they address is the development of shared mappings between pre-existing meanings and pre-existing forms (in the case of lexicon) or pre-existing complex meanings and emergent complex forms (in the case of grammar). These models fit a view of language in which the language of thought somehow arises prior to and independent of the development of public language. As the assumptions of the E/I models make clear, if confronted with the children's riddle, "Which came first, the chicken or the egg?", E/I modelers would blurt out "Egg!". The language chicken simply emerges from the well-formed meaning egg. This view is grounded in the Physical Symbol System Hypothesis (Newell and Simon, 1990; Newell *et al.*, 1989), which assumes that all intelligent processes are characterized by the manipulation of symbolic structures.

The positive feedback loop in E/I models of the emergence of lexicon amplifies small differences in the frequencies of form-meaning associations in the "aboriginal" lexicon and is modulated by the rule that implements the production bottleneck. E/I models have no effect on the structure of forms, the structure of meanings, or the structure of form-meaning mappings because these are all given. Rather, the models simply adjust the initially random distribution of intact abstract form-meaning mappings in a population of agents.

Hutchins and Hazlehurst, and we suspect Steels as well, have a different image of the historical processes that led to the development of language and the cognitive abilities that go with it. Their response to the chicken-and-egg riddle is "neither". Imagine a history in which structural regularities arise in the interactions among individuals. Imagine further that some of these regularities are true emergent properties in the sense that they cannot be fully constrained by the behaviors of either interactant alone. In such a situation, individual agents might begin to develop internal structure that permits them to maintain coordination with the emergent regularities. Eventually, they might even develop the ability to produce the formerly emergent structure alone. In such a world, symbols could arise between agents first, and become internalized later. Meaning structures could be embodied in forms that are assembled in interaction long before those meaning structures are internally represented by any agent. Language of thought need not precede public language. Structure and process always go hand in hand. If the process through which language emerged involved complex acts of embodied communication, then public propositional expressions could be the source rather than the result of the development of propositional mental representations. The news here is not just that chicken and egg develop together. The bigger point is that this viewpoint introduces another place to look for the origins of cognitive complexity. It has long been argued that the ontogeny of high-level cognition depends on social interaction (Vygotsky, 1978; Wertsch, 1985). We propose that the same may be true of the phylogeny of high-level cognition.

The form-tuning models of Hutchins and Hazlehurst (1991; 1995; 1998) attempt to demonstrate the dialectical development of internal and external structure. Form-tuning models assume a structure of visual experiences and small random differences among the internal structures of individual agents. The distinctiveness of forms and the sharing of form-meaning mappings emerges from the operation of multiple, mutually reinforcing, positive feedback loops that amplify differences and similarities in the interaction among stable external structure (referents), emergent external structure (forms), and emergent internal structure (meanings). These models have weaknesses. For example, the lexicon model lacks arbitrariness in the form-meaning relation, and the grammar model derives its sequentiality from a course of shared action (which cannot account for much of the sequential structure of real languages). Nevertheless, the models do suggest, and are informed by, a wider view of the nature of the issues that should be addressed by models of the emergence of language. The coordination of emergence of internal and external structure suggests that we should look for the origins of cognitive complexity in embodied interactions among agents.

The structure of language must come from somewhere. All of the models we have looked at try to show how it emerges in a process of auto-organization. All of these models work by constructing positive feedback loops that amplify certain, nearly invisible, initial differences. But what are the differences and what are the processes that amplify them? We have tried to use the positive-feedback-loop framework to identify the signals and the processes that modulate the propagation of those signals in the emergence of novel structure. In E/I models of the emergence of lexicon, the original signal is simply irrelevant noise introduced by random correspondences of arbitrary patterns. It may indeed be possible to get a community of agents to discover regularities in those patterns and to conventionalize those regularities. One must ask, however, could real language come from a process like that? Could language be a structure that has been extracted entirely from random fluctuations in arbitrary patterns? We think not. We expect language to be a structure that highlights and focuses patterns of embodied experience. E/I models of the emergence of grammar assume a complex propositional representation of meaning. Is the problem of the emergence of the grammar of public language simply a matter of propagating fully formed, complex, internal representations into public representations? Inserting meaning structure by fiat in these models means it must be accounted for some other way. Surely some kinds of meaning precede the advent of language. The physical symbol system hypothesis assumes that propositional representations of meaning precede the advent of language. We offer an alternative; a cultural symbol systems hypothesis, according to which symbols arise in interactions among agents concurrently with, or even before, the internal structures with which they are coordinated. Under this hypothesis, the ability to give meanings a propositional representation is a consequence rather than a cause of the ability to create grammatical external forms.

References

Batali J (1998) Computational simulations of the emergence of grammar. In: Hurford J, Studdert-Kennedy M, Knight C. (eds) *Approaches to the evolution of language: Social and cognitive bases.* Cambridge University Press, Cambridge UK pp 405-426

Elman JL (1991) Finding structure in time. *Cognitive Science,* 14: 179-211

Hazlehurst B, Hutchins E (1998) The emergence of propositions from the co-ordination of talk and action in a shared world. *Language and Cognitive Processes*, 13(2/3): 373-424.

Hurford J (1999) Expression/Induction models of language evolution: Dimensions and issues. In: Briscoe EJ (ed) *Linguistic evolution through language acquisition*. Cambridge University Press, Cambridge UK

Hutchins E, Hazlehurst B (1991) Learning in the cultural process. In: Langton C, Taylor C, Farmer D, Rasmussen S (eds) *Artificial Life (Vol. II. Santa Fe Institute Studies in the Sciences of Complexity, Proc. Vol. X)*. Addison-Wesley, Redwood City CA, pp 689-706

Hutchins E, Hazlehurst B (1995) How to invent a lexicon: The development of shared symbols in interaction. In: Gilbert N, Conte R (eds) *Artificial societies: the computer simulation of social life*. UCL Press, London, pp 157-189

Kaplan F (1999) A new approach to class formation in multi-agent simulations of language evolution. In: *Proceedings of the Third International Conference on Multi Agent Systems ICMAS'98*. IEEE Computer Society Press

Kirby S (1999) Syntax out of learning: The cultural evolution of structured communication in a population of induction algorithms. In: Floreano D, Nicoud J-D, Mondada F (eds) *Advances in Artificial Life, Proc. of the 5th European Conference, ECAL'99*. Springer, Berlin, pp 694-703

Newell A, Simon H (1990) Computer science as empirical enquiry: Symbols and search. In: Garfield JL (ed) *Foundations of cognitive science: The essential readings*. Paragon House

Newell A, Rosenbloom PS, Laird JE (1989) Symbolic architectures for cognition. In: Posner M (ed) *Foundations of cognitive science*. MIT Press.

Oliphant M (1997) *Formal approaches to innate and learned communication: Laying the foundation of language*. Unpublished doctoral dissertation, University of California, San Diego

Steels L (1996) Self-organizing vocabularies. In Langton C, Shimohara K (eds) *Proceedings of Alife V*. MIT Press, Cambridge MA, pp 177-184

Steels L, Kaplan F. (1999) Collective learning and semiotic dynamics. In: Floreano D, Nicoud J-D, Mondada F (eds) *Advances in Artificial Life, Proc. of the 5th European Conference, ECAL'99*. Springer, Berlin, pp 679-688)

Tomasello M (1996) Do apes ape? In: Heyes C, Galef B (eds) *Social learning in animals: The roots of culture*. Academic Press, San Diego, pp 319-436

Turner JS (2000) *The extended organism: The physiology of animal-built structures*. Harvard University Press

Vygotsky LS (1978) *Mind in society: The development of higher psychological processes*. Harvard University Press

Wertsch J (1985) *Vygotsky and the social formation of mind*. Harvard University Press

Chapter 14

The Constructive Approach to the Dynamic View of Language

Takashi Hashimoto

Introduction

Human languages have a variety of characteristics. To study the evolution and dynamics of language using simulation models, it is important to consider which characteristics should be adopted and abstracted when constructing the model. The features that are adopted and the way in which they are modeled represent what researchers of language evolution recognize as the essence of language[1].

In many cases, the role of language as a tool for communication is abstracted as its essence. That is, the evolution and dynamics of language are formalized in terms of how people come to use the same lexicon and grammar for communication. Although communication is one of the most important aspects of the evolution of proto-language from animal communication systems, human language is not only a tool for conveying one's mind to others. The activities of communication induce various effects on the speakers and listeners in addition to the exact "transmission of messages". For example, the act of communicating extends and reshapes the cognitive structures of language users. This function may separate human language from animal communication systems.

---

[1] This refers not only to the study of the evolution and dynamics of language but to the whole of linguistic studies. Tokieda (1941) has insisted upon the relationship between the essence of language and the study of language as follows, "The mission of the study of language should be not to arrange particular linguistic data and organize them into linguistic laws but to clarify the profile of language as the subject of study of language (p.iv)."

Let us consider the system of symbols in the light of this point. The establishment of shared connections between symbols and their referents in human language does not in itself demonstrate a qualitative difference between human language and animal communication systems, but it does show a quantitative difference in the amount of information transmitted. A crucial dissimilarity between human and animal communication systems is that new symbols and referents can be created in human language, and these can bring about changes in the system of symbols. Language users do not learn the system of symbols as an existing structure of the world, but create the system through the subjective activity of perpetual interaction with the external world and through communication with other individuals. The system of symbols changes when the relationship of individuals to the world changes through this creative process. This kind of change can induce modifications in the structure language. To study the evolution of language from a dynamic viewpoint is to consider such dynamics brought about by individuals' creative processes to be a fundamental feature of language use.

Language as a complex system

Models of language that take account of these kinds of dynamics consider language to be a complex system characterized by emergence, subjectivity and dynamics.

It is said that "emergence" is the phenomenon through which the functions or global orders of a system arise spontaneously from the local interactions of elements of the system. Socially shared linguistic components such as grammar and lexicon do not exist a priori but emerge from interactions among language users without peculiar central controls and powers. Thus it is natural to look on the evolution of language as a typical phenomenon of emergence.

In a linguistic system the agents who participate in the local interactions are the language users. Language comes into existence when subjective users of the language exist. The subjective user embodies a cognitive system that makes its internal structure by embedding the behavior of others and of the external world within him/herself. He/she engages in cognitive activities by acting toward others and the external world, and changes its structure and relationships with others and the external world (Kaneko and Tsuda, 1994; Kaneko and Ikegami, 1998).

Evolutionary linguistics focuses not on the linguistic structure at a temporal point, but on change in language. Language is regarded as essentially dynamic. The dynamics of language, hereafter DOL, can be divided into the following four levels in terms of time:

I.   Origin,

II.   Evolution,

III. Development,

IV.   Sense-making and Conceptualization.

In the study of the origin and evolution of language, we turn our attention to the long-term change of linguistic structure. The origin of language involves the problem of how a change from a situation without a linguistic system to one with such a system can occur. The studies on the evolution of language try to understand how the structural changes of language systems are established.

Development in language treats dynamics on a shorter time scale, and involves language acquisition, in which we study how infants can learn the existing language structure, and second language acquisition, in which we study how users of one language learn a new language.

'Sense-making' is the process of giving words meaning (Fukaya and Tanaka, 1996), and also involves dynamics on a much shorter time scale. Words that are exchanged between subjects do not have a priori meanings but are given subjective meanings in the context of communication. In the next subsection, we discuss the sense-making activity of individual language users as the basis of language dynamics.

Dynamic view of language

The complex systems' viewpoint on language puts a premium on the emergence, subjectivity, and dynamics of language as discussed above. Its fundamental tenet can be summarized as follows: language is carried on the shoulders of subjective language users who embed the whole of a system within which they behave and act according to the structure embedded. This means that they both form a linguistic system and are reflexively subject to the imposed restrictions of the system.

The dynamic view of language envisages language as it is altered with use. If a linguistic system did nothing more than conform to the subjects or users of the system, then the system would be static. In actuality, however, language systems continually change. The subjects not only confirm the linguistic system, but can also break linguistic 'rules' in their linguistic activities. The conceptualizing and sense-making acts of the language users as well as external forces such as environmental changes cause the linguistic system to change over time.

For instance, let us consider creative metaphors. Although metaphorical expressions that are already common are often a part of a linguistic system, we can also manufacture novel expressions. Such expressions are only 'invalid' if they are viewed from within the existing linguistic rules. When we listen to or read such expressions, we can subjectively understand them to some extent. How should we make sense of this phenomenon? If we consider subjectivity to be indispensable to language, the subjects make sense by themselves.

The production of creative metaphors may seem too unusual an occurrence to provide a basis for the proposal that language users create meaning. The subjective activity of making sense, however, is involved in all linguistic processes such as speaking and recognizing. Cognitive linguistics, in which metaphors are not mere rhetoric, but rather are important instruments for structuring our cognition, indicate that many expressions which are not used as metaphors can be considered originally metaphorical (Lakoff and Johnson, 1980; Taylor, 1995). Since such expressions have deeply permeated our minds, we are not conscious of their metaphorical nature. This means that there is no strict distinction between metaphorical and non-metaphorical expressions, and the assertions about metaphors are more or less applicable to daily expressions. Let us broaden our interpretation of the function of metaphor as the (re)conceptualization (Nogales, 1999) of daily linguistic expressions. Even a declarative sentence is not a mere description of an objective phenomenon but a manifestation of how a subject recognizes the phenomenon. The subject who accepts a declarative sentence (re)structures his/her own conceptualization through being constrained by situations, contexts, experiences and the linguistic system.

Constructive approach

When the object of study, like the origin and evolution of language, is a difficult one on which to carry out experiments or to observe, we can rely on the constructive modeling that is often used in the field of complex systems (Kaneko and Tsuda, 1994; 2001; Kaneko and Ikegami, 1998), and especially in the study of artificial life. Models constructed with computers are regarded as laboratories in which we can show various emergent phenomena and clarify their information structure (Casti, 1999). Here we consider three major advantages of this approach.

The main merit of the constructive approach is to make subjects and their dynamics, both of which are likely to be stripped from the study of human-related phenomena like language, economics and social systems, into an object of mathematical scientific study. Although subjective activities are indispensable to language, "linguistic rules" are often regarded as objective entities detached from language users when language is treated descriptively. The subjective individuals who use language are eliminated. In our constructive approach, we base the model on the activities of speaking, listening and understanding language and describe "linguistic rules" as the result of the development and interactions between individuals.

A second merit of this approach is in its usefulness for understanding dynamic systems. To describe language, we can assume static structures and describe them. To study the temporal changes in language, one may use the diachronic approach (Saussure, 1959). However, this approach is more or less a way of comparing static models and is not well-suited for capturing language as an essentially dynamic phenomenon, since it shows differences in language by comparing structures that are statically described at different points of time. Adopting the constructive approach makes it possible to thoroughly investigate the state changes of such essentially dynamic systems.

A third merit of the constructive approach comes to light when we try to understand a system that is complex in its nature, such as life, society or language. Reproducing a complex system is often such an intractable difficulty that we, in this approach, do not try to duplicate the complex state of the system as it is. We construct a simpler system which is considered to be its ancestor, together with a mechanism to evolve it. By observing the increasingly complex processes and results, we will be able to understand the dynamics and features of the complex systems under scrutiny.

Development of Categorical Structure

In this section, we show an example of constructive research on language dynamics[2]. Focusing on sense-making activity in language use (level IV in the DOL), we construct a model in which a web of relations among words, which represents the internal structure of a language user, develops in the course of conversation. We call the model 'The Developing Word-Web Model'. We analyze how categorical structures develop as internal structure through conversation (level III in the DOL), and how the spreading of structures in a population and in individuals are related (level II in the DOL).

Modeling the sense-making process

In this model, the internal structure of a subject is expressed by a web of relations between words. We employ usage-based modeling (Wittgenstein, 1953; Langacker, 1999; Barlow and Kemmer, 1999; Hashimoto, 1997), in which language structures are organized according to the use of language in conversation. Therefore, the relations among words are not a copy of the relations among the objects indicated or referenced by the words.

The sense-making activities in using language are modeled as processes in which the words used are situated in a web of relations. Concretely, when a subject accepts a sentence, the relations between the words used in the sentence and all words which the subject knows are computed and renewed. Through this computation, the subjects change their internal structure dynamically in the course of a conversation. This renewal process is schematically indicated in Figure 14.1.

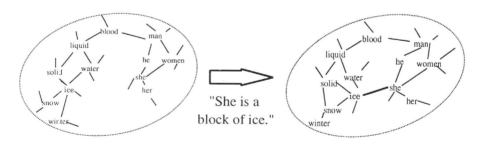

Figure 14.1 A sketch of the change that occurs in a word-web by the subject's accepting of a sentence. When a web representing the internal structure of a language user is in the state depicted on the left, in which there is no connection between the words 'she' and 'ice', he/she accepts the sentence "She is a block of ice" and changes his/her internal structure to the state depicted on the right, in which there comes to be a connection between the words 'she' and 'ice', and the whole shape is modified.

---

[2] Refer to Hashimoto (1998; 1999) for details.

The algorithm used to update the relations between words is based on the method of calculating similarity among words in a corpus proposed by Karov and Edelman (1998), with two modifications. First, we changed the method so as to calculate iteratively with conversation, since we are interested in the dynamics of categorization, not the final state of the structure. Second, we introduced 'texts' as a higher level category of elements of language than words and took the correlation of word appearance frequency in texts into consideration. In the present case, a text is a sequence of sentences spoken and listened to by subjects in conversation.

A relation between words i and j at the acceptance of the nth sentence in the tth text is a linear combination of 'word-similarity' and 'word-correlation'.

$$R_{t,n}(w_i, w_j) = \alpha^w \text{(word-similarity)} + (1-\alpha^w)\text{(word-correlation)} \qquad (1)$$

where α^w is a parameter that signifies the weight of word-similarity in the relation. The first term, word-similarity, is used to calculate the similarity between words in terms of how they are used in sentences. Two words that are used in one sentence come to have a strong word-similarity. The second term, word-correlation, concerns the pattern of appearance of words in texts. If the patterns of appearance of two words are similar, this term in the formula has a positive value, and vice versa. This quantity is determined by calculating the correlation of appearance probabilities of words in texts. The value of this quantity is calculated on the basis of the proximate sentence, $R_{t,n-1}$; if $n=1$, it is calculated on the basis of the last sentence at the proximate text $t-1$. Thus, if a subject continuously utters and accepts sentences, the relations between words change in succession.

Conversations among agents are modeled as exchanges of sentences. Conversation topics are introduced to two agents who are chosen from a given population. A conversation starts when one agent (the speaker) utters a sentence about one topic. If the other (the listener) accepts the sentence, he/she replies to the sentence. To qualify as reply, the sentence need not address the first topic precisely, but should contain any one word from the accepted sentence. The sentences have a chance to be modified before utterance so that new sentences and new words are introduced.

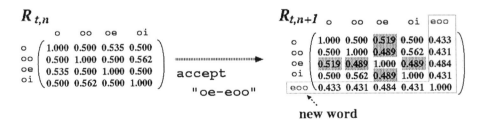

Figure 14.2 An example of how the matrix representing a word-web is renewed upon the acceptance of sentence. When an agent's internal state is represented by the matrix on the left and he/she accepts the sentence "oe-eoo", the elements in the row and column of the word 'oe', i.e., shaded elements in the matrix on the right, are updated. The matrix is then enlarged to incorporate the unknown word 'eoo'.

The word-web is represented by a matrix. When an agent utters or accepts a sentence including a word w, he/she calculates $R(w,w')$ of all known words w' and renews the value in the matrix. If there is only one unknown word in the sentence to which he/she listened, the listener accepts the sentence, enlarges the matrix to incorporate the new word, and calculates its relation with other words (Figure 14.2).

Stability and adaptability of categorical structure

We summarize the results of simulations in which the agents, who initially have no knowledge of words, communicate with each other and develop their internal structures.

The agents create clusters in the word-web based on the strength of the relations among words. There are two major shapes of clusters, flat clusters, in which words have a strong relation to each other, and gradual clusters, in which relations between words gradually change. In actual simulations, these two types of clusters are interlaced.

This cluster formation can be interpreted as a means of categorization. Since the boundary of a flat cluster is sharp, it is quite clear whether an entity is a member of the cluster. This type of cluster is like a category in which the members are rigidly determined by necessary and sufficient conditions. In contrast to flat clusters, gradual clusters show a graded change in the relations among words from strong to weak. These two types of clusters can be combined, with some flat clusters being connected through gradual clusters. This structure is like a prototype category (Lakoff, 1987; Taylor, 1995). The extent to which words are included in a category is a matter of gradient. Part of the flat cluster corresponds to the central member of a category. Words having weak relations to the central member are peripheral members of the category.

When a new word is used or a word is used in an unusual way, word relations can change dramatically. The dynamics of the internal structure of an agent is exemplified in Figure 14.3. The change in the relations between a given word and other words in the course of conversations is superimposed on this graph. In the 21st text ($t=21$), a word is used in a new way, and the relative strength of relations among words is turned over.

The change in the word-web that occurs with the new usage of a word is depicted in Figure 14.4. The corresponding words before and after the change are connected by arrows. While words move coherently with other words in the same cluster, the word that has been used in a new way, which is connected with the corresponding word by a broken arrow, moves in a different direction from the others in its cluster and becomes a member of a different cluster.

The dynamics with which the word-web preserves the whole structure and accommodates itself to new experience indicates the coexistence of global stability and local adaptability. This dynamic quality is the fundamental feature with which languages should be equipped. If a language is too rigid, its users will not be able to formulate new expressions to describe diverse experiences, and if it is too unstable, no structuralization will be possible either at the individual and/or global levels; hence, no communication will take place (Geeraerts, 1985).

Using a word in a new way induces remarkable change in a subject and connects two previously unconnected clusters. If, again, we think of clustering as categorization, the function of the dynamics resembles that of metaphorical expressions. Such expressions reconceptualize our experiences (Nogales, 1999) by connecting two different domains of categories.

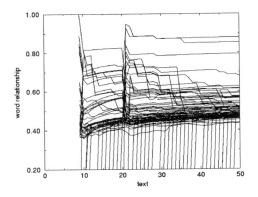

Figure 14.3 The dynamics of word relations in conversation. The x-axis represents the texts and the y-axis represents the relations of a given word with other words. In the 21st text, we observe a large, rapid change in the relations among words. (From Hashimoto, 1998)

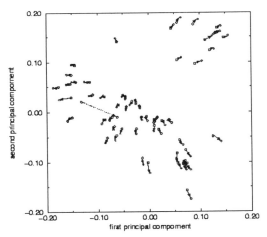

Figure 14.4 Change in the word-web resulting from the turnover of word relations. This diagram is the result of a principal component analysis of the matrices expressing the internal structure. The words that correspond to each other before and after the listener accepted the sentence containing the new usage are connected by arrows. (From Hashimoto, 1998)

Development of commonality and individuality

The population of agents develops both the shared and individual parts of the cluster structure through conversation. Establishing the shared structure is the basis of society, and is necessary for language to serve mutual understanding. The degree of sharing does not necessarily grow with conversation, and sometimes it diminishes. This is because the agents each experience a separate conversation history, and therefore can supply different senses for the same sentence and thus develop individuality. All of the agents within the system do not come to identical conclusions, but instead retain their individuality. Thus the relationships between agents change constantly, and the structure at the global level, namely the language structure, is dynamic. This reconciliation between commonality and individuality is also a fundamental feature of language systems.

Evolution of Grammar

This section describes a study of the evolution of grammar[3] involving longer time scale dynamics (level II in the DOL) than in the previous section. The agents engaging in conversations are assumed to have some internal rules that they use to produce and recognize sentences. The rules are expressed by generative grammar. We observe in a conversation network of agents the process of complexification and structuralization of grammars and the emergence of social rules that evolve from the shared usage of words in the population.

Conversation game between grammar systems

An agent is defined as having a generative grammar,

$$G_i = (\{S,A,B\}, \{0,1\}, F_i, S) \tag{2}$$

where $\{S,A,B\}$ and $\{0,1\}$ are the sets of non-terminal and terminal symbols, respectively, F_i is a set of rewriting rules and S is the initial symbol. This grammar is used both for the production and acceptance of sequences. At the level of production, the agent begins rewriting from the initial symbol, S, applies the rewriting rules to the non-terminal symbols, and stops rewriting when the rewritten sequence consists of the terminal symbols. For example, an agent with a rewriting rule list,

$$S \rightarrow A0B, A \rightarrow 10, B \rightarrow 11 \tag{3}$$

produces a sentence "10011" as

---

[3] Only the summary of the model and the results are explained here. The detailed description is in Hashimoto and Ikegami (1995; 1996)

$$S \overset{S \to A0B}{\Rightarrow} A0B \overset{B \to 11}{\Rightarrow} A011 \overset{A \to 10}{\Rightarrow} 10011 \tag{4}$$

He/she then utters the sequence to all of the agents, who try to recognize it in terms of their own grammars. The acceptance process is the reverse of production. If a sequence that is heard can be rewritten to the initial symbol, it is recognized.

The agents, in a conversation network of P individuals, engage in a game: they score according to the utterance and recognition of sequences. Each agent is able to utter R times in one time unit. The score an agent receives in a time unit is a weighted average of the scores based on the uttering and recognition of sequences as well as having his/her own utterances recognized by others[4],

$$p^{tot} = \frac{1}{R}\{r_{sp}(s^{sp} - f^{sp}) + r_{rec}(s^{rec}\sum_{recog}\frac{1}{step} - f^{rec}) + r_{br}\frac{s^{br} - f^{br}}{P}\} \tag{5}$$

where s^{sp}, s^{rec}, and s^{br} are the times of uttering, recognizing and being recognized, respectively; f^{sp}, f^{rec}, and f^{br} are the times of not uttering, not recognizing and not being recognized, respectively; and r^{sp}, r^{rec}, and r^{br} are the parameters for the weight of each term. The variable *step* is the number of rewriting steps taken before recognition occurs. The sum Σ_{recog} is taken only when an agent can recognize the sentence.

Here we introduce the evolutionary dynamics: some of the original agents are replaced with new agents according to their scores. The new agents are variations of the agents with higher scores. Since an agent is considered here to be a family or a population sharing the same grammar, the grammar of the higher-scoring agents has a higher chance to be inherited and spread with differentiation.

Emergence of common language and evolution of grammar

The simulation starts with the agents having the simplest, non-syntactic rules, namely $S \to 0$ or $S \to 1$, as the initial state. These agents can utter only a word. The product of the length of the sequences uttered and recognized measures the information handled by an agent. The information flow in the network is the average of the amount of information handled by all of the agents.

At the beginning of the evolution, the variety of sequences uttered develops with the growth of the information flow. At some point the growth stops. After that point, the information flow in the network shows a punctuated equilibrium evolution, as depicted in Figure 14.5, in which the flow rapidly grows and stops in turn.

At equilibrium, a group is formed in which agents utter and recognize the common sequences. The agents who can understand such sequences get higher scores. The formation of such a group suggests the emergence of a social rule in which it is preferable to use some particular sequences in communication. Agents

---

[4] The definition of the score (5) is simplified with respect to that of Hashimoto and Ikegami (1995; 1996).

who cannot recognize such sequences do not earn higher scores, even if they have a better ability to understand sequences on average. Such agents are often not able to survive in the network. This means that the formation of groups sharing common sequences suppresses the evolution toward a language with a large variety of sequences.

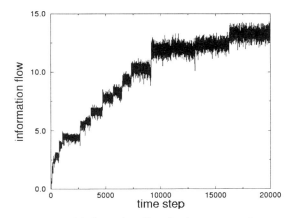

Figure 14.5 Transition of information flow in the conversation network. $r_{sp\_} = 3$, $r_{rec\_} = 1$, $r_{br\_} = 1$.

During the growth of a group with a common language, agents who can recognize a large variety of sequences emerge from the group. They can understand both common and new sequences. The development of a grammar with a high variety of sequences is effected by module-type evolution and the emergence of a loop structure in the grammar. Module-type evolution is to get a rewriting rule that can be attached to other rules to produce many sequences. Such rules resemble the type of word formation known as suffixation. The emergence of a loop structure enables an agent to use a loop in the production process. This allows the agent to produce an infinite number of sequences, in principle, through recursive production processes. Such looping corresponds to the embedding structure of phrases in natural language. These two types of evolution of grammar are often realized by obtaining only one new rewriting rule, which spreads rapidly through the network so that the information flow shows steep growth.

Double articulation

The grammars that agents arrive at are structured by reflecting the characteristics of the language used in a society. The agents, in general, have three types of rewriting rules:

$$N \rightarrow \text{sequence of } Ts, \tag{6}$$

$$N \rightarrow \text{sequence of } Ns, \tag{7}$$

$$N \rightarrow \text{sequence of } Ns \text{ and } Ts, \tag{8}$$

where N and T are a non-terminal and a terminal symbol, respectively. Rules of type (6) often code sequences that are shared in a group. Since being able to understand frequently used sequences quickly carries an advantage, the agents develop grammars in which the common sequences can be recognized in short rewriting steps. That is, they understand common and frequent sequences with rules of the following type:

$$S \rightarrow \text{frequent sequence of } Ts \qquad (9)$$

which is a special version of the type (6) rules. Infrequent sequences are recognized by the joining of the sequences of terminal symbols coded by rule (6) to each other using rules (7) and (8). The type (6) rules can be seen to code 'words' and the type (7) and (8) rules 'sentences'.

This structuring of grammar can be seen as a correspondence of 'double articulation' in natural language. This mechanism enables languages with a finite number of symbols to produce infinitely diverse sentences. By combining symbols to make words, and combining words to make sentences, we can produce an infinite number of structured sentences.

As we can understand from observing the evolution of agents in the model, double articulation brings stability and adaptability to language. Utilizing double articulation, the agents who emerge concurrently with the rapid increase in the information flow get new sequences without large changes in rewriting rules and already known sequences. In the simulation, such characteristics appear as the punctuated equilibrium in the development of the information flow.

Developing Word-webs with Grammar

In the second section, we treated the dynamics of levels IV, III and part of II in the DOL, and in the third section we also discussed level II. In aiming to form a coherent understanding of the dynamics of language, here we integrate the approaches taken in the previous sections into one model. The present section discusses an attempt at such integration by presenting a model of the conversation network of agents having word-webs and grammars[5].

Articulation by grammar

Agents have an algorithm for inferring the relations among words. This algorithm was introduced in the second section, and a generative grammar was defined in the subsequent section. The grammar is used to parse sentences as sequences of words. For example, an agent with a grammar,

$$S \rightarrow A0B, A \rightarrow 10, B \rightarrow 11 \qquad (10)$$

parses a sentence "10011" as

---

[5] This work was first reported in Hashimoto (1997).

$$10011 \overset{A\leftarrow 10}{\Rightarrow} A011 \overset{B\leftarrow 11}{\Rightarrow} A0B \overset{S\leftarrow A0B}{\Rightarrow} S \tag{11}$$

and recognizes it as a sequence of words, "10·0·11". The parsing tree is shown in Figure 14.6(a). According to the results of the parsing, the agent updates his/her matrix of word relations.

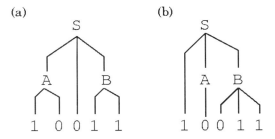

Figure 14.6 Examples of the parsing tree and articulation. (From Hashimoto, 1997, in P. Husbands and I. Harvey, Proceedings of the Fourth European Conference on Artificial Life © Massachussets Institute of Technology)

The incorporation of the grammar into the analysis introduces another level of subjectivity. The parsing of a sentence differs in different grammars. For example, an agent with the grammar

$$S \to 1AB, A \to 0, B \to 011 \tag{10}$$

parses the sentence "10011" as shown in Figure 14.6(b) and accepts it as a sequence of words: "1·0·011".

Structure and dynamics of categories

Words are clustered in a word-web as shown in the results in the second section above. The amount of variation in the shape, which depends on the initial grammar, is greater than the amount of variation in the Word-web without grammar described in the second section above. We classify the cluster structures into six types according to their shapes. Examples of simple structures of each type are shown in Figure 14.7.

The features of these shapes can be summarized as follows: (a) solitary word: a word that has no connection to other words; (b) flat cluster: the words in the cluster have almost identical relations with each other; (c) gradual cluster: the relations between words in the cluster vary with the words and gradually change; (d) two-peak cluster: the words are in a cluster with two peaks of close relations within it; (e) subclustering: the cluster has a stepwise structure, and words are thought to be divided into subclusters; (f) plural clusters: the words form many clusters without relations.

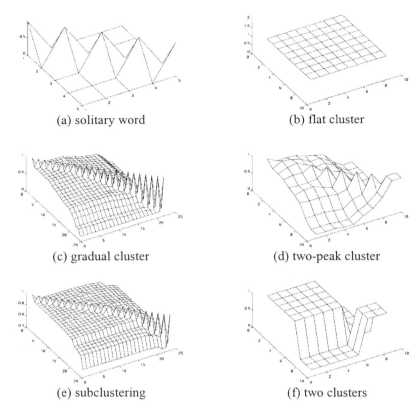

Figure 14.7 Examples of typical structures of word relations. Each number on the X- and Y-axes indicates one word. The Z-axis shows the relation between the words. A word is selected as a standard word and put at the origin of the X–Y plane. The other words are arranged in descending order of the closeness of their relations to the standard word. The standard word is selected as that for which the decline is as smooth as possible. (From Hashimoto, 1997, in P. Husbands and I. Harvey, Proceedings of the Fourth European Conference on Artificial Life © Massachussets Institute of Technology)

A solitary word is a word without any relation to other words (Figure 14.7(a)). Strictly speaking, we cannot say it is a cluster and also a category, since we do not have any categories with only one member in our knowledge system. There might be, however, such a simple structure at the very beginning of our developmental process.

The flat and gradual clusters (Figures 14.7(b) and 14.7(c), respectively) have the same structure and meaning as those in the second section. The two-peak cluster (Figure 14.7(d)) is an analogue of a category with two central members. It can be regarded as a polysemous category. All of the words in a single-peak structure are characterized by how strong relations they have with a central member in the category. Whereas in a plural peak structure, as in a case where there are two peaks, there are words which have strong relation with one central member but not with the other central member, and words which have some degree of relations with both central members of the category.

We can see the subclustering structure in two groups of words in a cluster (Figure 14.7(e)). One group consists of words with strong relations to each other and the other of words with rather weaker relations. The two groups can be regarded as two subcategories within the category. This is the simplest case of a hierarchy of categories.

A general scenario for the development of a cluster structure is the following. At first an agent can recognize only a single one-word sentence. The agent then develops the ability to recognize several sentences, but these are all one-word sentences, and therefore, altogether, consist of several solitary words. The agent becomes able to articulate plural words in a sentence, which constitutes the forming of relations between words. Eventually words form gradual clusters. Following such initial development, the clusters change their structure through such processes as expanding their boundaries, making connections with each other, and incorporating solitary words.

When clusters expand their boundaries, the structure of the original cluster does not undergo a great change, thus satisfying the requirements for adaptability and stability in language. In terms of its flexibility, then, the original cluster resembles, in the context of dynamics, the prototype category.

Parallel to this development in word-space, the syntactic structure also develops from a sequential structure to a branch structure, and finally, to a loop structure.

Discussion

We formulated the evolution and dynamics of language on the basis of sense-making and conceptualization in language use. To base language on subjects' conceptualization is consistent with Tokieda's conception of language as process (Tokieda, 1941). He put forth "a theory in which the essence of language is considered to be a set of mental processes" (Tokieda, 1941: p.i). Bakhtin has also stated that language is understood as an "activity, an unceasing process of creation realized in individual speech acts" in the individualistic subjectivist view of language (Volosinov, 1986).

We have argued that the merit of the constructive approach is to take account of individual subjects in an objective system of scientific study. Thus, we now have an actual apparatus with which to deepen and broaden the discussions begun by Tokieda and Bakhtin.

There are various strands of constructive approaches in the study of the evolution of language (Hashimoto and Ikegami, 1995; 1996; Steels, 1997; 2000; Arita and Koyama, 1998; Kirby, 1998; Batali, 1998). They investigate the process of the emergence and evolution of, for example, a shared lexicon or grammar. The emergence of global order as language-like behavior is observed by modeling individuals in terms of their linguistic interactions. Note that carrying out computer simulations alone is not the same as taking the constructive approach. For example, the evolutionary game theoretical studies of the evolution of language by Nowak *et al.* (Nowak and Krakauer, 1999; Nowak, Plotkin and Jansen, 2000; Nowak, Komarova and Niyogi, 2001) are rather top-down and descriptive analyses.

In this paper, we have attempted to construct and understand the linguistic structures and the ceaseless dynamics within them that are induced by using language, with a special focus on sense-making by subjective individuals. Since the models and mode of analysis introduced in this paper assume that language users are equipped with categorical structures and grammars, they cannot, strictly speaking, treat the problem of the origin of language (level I in the DOL). However, we began with simulations involving agents with minimal abilities, namely, having no knowledge of words and no syntactic structure, and investigated the process of the development of categories and syntax as internal and social structures. We can conclude that the constructive approach is an effective tool for studying the origin as well as the evolution of language.

Conclusion

Language is not a static and objective entity, but an ever-changing system encompassing the subjective activities of language users. In this paper, we propose that the dynamic view of language is a foundation from which we can understand the dynamic aspects of language, especially the evolution of language. The dynamic view sees language as involving subjective processes of sense-making, and sees its dynamics as induced by its use. The constructive approach is advantageous as a methodology for analyzing language according to this dynamic view. In this approach, both subjective individuals as language users and conversations among them are modeled, and the development of the system is observed using computer simulations.

According to the viewpoint that we have articulated, then, we introduced three models that facilitated our inquiry into the evolutionary processes of language categories and grammar. In this study, it was shown that the coexistence of stability and adaptability as well as of commonality and individuality are prerequisites for the existence of ever-changing languages, and these characteristics are actualized, in part, by the proto-typical categorical structure and by double articulation.

The study of evolutionary linguistics should go beyond regarding the establishment of a shared lexicon or complex syntactic structure as the emergence of language. We must aim to construct and comprehend dynamic communication systems in which the linguistic structure can always change in response to the development, through acts of communication, of the internal structures of subjective language users.

Acknowledgment

I am grateful to Takashi Ikegami for his fruitful discussions and collaboration. I give my thanks for Yuzuru Sato and Yukito Iba for their helpful comments. I wish to express my gratitude to Sarah Knutson for her editorial revisions. I thank the

Sony Computer Science Laboratory – Paris and the Santa Fe Institute for supporting me as a visiting researcher. This work was partly supported by a Grant-in-Aid for Scientific Research (No.12780269) from the Ministry of Education, Science, and Culture of Japan and partly by the Special Postdoctoral Researchers Program at the Institute of Physical and Chemical Research (RIKEN), Saitama, Japan.

References

Arita T, Koyama Y (1998) Evolution of linguistic diversity in a simple communication system. *Artificial Life*, 4: 109-124

Barlow M, Kemmer S (1999) Introduction: A usage-based conception of language. In: Barlow M, Kemmer S (eds) *Usage based models of language*. CSLI Publications, Stanford CA, pp vii-xxviii

Batali J (1998) Computational simulations of the emergence of grammar. In: Hurford J, Knight C, Studdert-Kennedy M (eds) *Approaches to the evolution of human language: Social and cognitive basis*. Cambridge University Press, Cambridge UK

Casti JL (1999) The computer as a laboratory. *Complexity*, 4(5): 12-14

Cruse DA (1986) *Lexical semantics*. Cambridge University Press, Cambridge

Fukaya M, Tanaka S (1996) *Kotoba no <Imiduke-ron>* (*A theory of "making sense" of language*. In Japanese), Kinokuniya Shoten, Tokyo

Geeraerts D (1985) Cognitive restrictions on the structure of semantic change. In: Fisiak J (ed) *Historical semantics*. Mouton de Gruyter, Berlin, pp 127-153

Hashimoto T (1997) Usage-based structuralization of relationships between words, In: Husbands P, Harvey I (eds) *Proceedings of the Fourth European Conference on Artificial Life*. MIT Press, Cambridge MA, pp 483-492

Hashimoto T (1998) Dynamics of internal and global structure through linguistic interactions. In: Sichman JS, Conte R, Gilbert N (eds) *Multi-agent systems and agent-based simulation*, Springer-Verlag, Berlin, pp 124-139

Hashimoto T (1999) Modeling categorization dynamics through conversation by constructive approach. In: Floreano D, Nicoud JD, Mondada F (eds) *Proceedings of ECAL99 the Fifth European Conference on Artificial Life (Lecture Notes in Artificial Intelligence)*. Springer-Verlag, Berlin, pp 730-734

Hashimoto T, Ikegami T (1995) Evolution of symbolic grammar systems. In: Morán et al. (eds) *Advances in Artificial Life*. Springer, Berlin, pp 812-823

Hashimoto T, Ikegami T (1996) Emergence of net-grammar in communicating agents, *BioSystems*, 38: 1-14

Kaneko K, Ikegami T (1998) *Fukuzatsu-kei no Shinka-teki Shinario (Evolutionary scenario of complex systems*. In Japanese). Asakura Shoten, Tokyo

Kaneko K, Tsuda I (1994) Constructive complexity and artificial reality: An introduction. *Physica*, 75: 1-10

Kaneko K, Tsuda I (2001) *Complex systems: Chaos and beyond*. Springer-Verlag, Berlin

Karov Y, Edelman S (1998) Similarity-based word sense disambiguation. *Journal of Computational Linguistics*, 24: 41-59

Kirby S (1998) Fitness and the selective adaptation of language. In: Hurford J, Knight C, Studdert-Kennedy M (eds) *Approaches to the evolution of human language*. Cambridge University Press, Cambridge UK, pp 359-383

Lakoff G (1987) *Women, fire, and dangerous things.* The University of Chicago Press, Chicago

Lakoff G, Johnson M (1980) *Metaphors we live by.* The University of Chicago Press, Chicago

Langacker RW (1999) A dynamic usage-based model. In: Barlow M, Kemmer S (eds) *Usage based models of language.* CSLI Publications, Stanford CA, pp 1-63

Nogales PD (1999) *Metaphorically speaking.* CSLI Publications, Stanford CA

Nowak MA, Komarova NL, Niyogi P (2001). Evolution of universal grammar. *Science,* 291: 114-118

Nowak MA, Krakauer D (1999) The evolution of language. *Proceedings of the National Academy of Science, USA,* 96: 8028-8033

Nowak MA, Plotkin JB, Jansen VAA (2000) The evolution of syntactic communication. *Nature,* 404: 495-498

Saussure FD (1959) *Course in general linguistics.* Philosophical Library, New York

Steels L (1997). The synthetic modeling of language origin. *Evolution of Communication,* 1(1): 1-34

Steels L (2000) Language as a complex adaptive system, In: Schoenauer M *et al.* (eds) *Parallel problem solving from nature (PPSN VI).* Springer, Berlin, pp 17-26

Taylor JR (1995) *Linguistic categorization: Prototypes in linguistic theory.* Oxford University Press, Oxford

Tokieda M (1941) *Kokugo-gaku Genron (The Principles of Japanese Linguistics.* In Japanese). Iwanami Shoten, Tokyo

Volosinov VN (1986) *Marxism and philosophy of language.* Harvard University Press, Cambridge MA

Wittgenstein L (1953) *Philosophische untersuchungen.* Basil Blackwell

Part VII

CONCLUSION

Chapter 15

Some Facts about Primate (including Human) Communication and Social Learning

Michael Tomasello

These are exciting times for all those who have refused to listen to the *Société Linguistique de Paris*. Investigation of the origins and evolution of language is back, and investigators have a whole new range of knowledge and methodologies to bring to the task. Among the most exciting of the new scientific tools – for all of the reasons that Cangelosi and Parisi outline in their introduction – are computer models designed to simulate aspects of some biological, psychological, and sociological processes that may have been involved in the origins and evolution of language. The problem, of course, is that a simulation is a simulation of something, and it is very useful if that something is available for inspection and comparison – which is not exactly the case for things that have occurred in the ancient past. Cangelosi and Parisi bemoan "a very limited, in a sense nonexistent, empirical base".

But when the focus is on process, as it is for many modelers, the situation is actually a bit different. There are a number of processes currently occurring in the natural world that were very likely involved in the origins and evolution of language. These can be studied empirically, and simulations can be compared to them. For example, many of the people engaged in computer simulations have attempted to benefit from recent research on such things as non-human primate communication; child language acquisition; imitation and cultural learning; and processes of language change, especially those involved in pidginization and creolization.

My research involves the first three of these sets of processes, broadly conceived as processes of primate communication and social learning, including as a special case child language acquisition. I use observational and experimental techniques, not modeling, and indeed I often fail to comprehend in detail how many models work. Therefore, the most important contribution I can make to the simulation enterprise – which I think is an extremely important one – is to help it to get its facts straight in those areas in which I have some expertise. I focus on what I consider to be seven important facts.

Non-human Primate Communication

1. Vervet monkey alarm calls are not direct precursors for language

Many scientists outside the field take as the paradigm case of non-human primate communication the alarm calls of vervet monkeys. The basic facts are these (Cheney and Seyfarth, 1990). In their natural habitats in east Africa vervet monkeys use three different types of alarm calls to indicate the presence of three different types of predator: leopards, eagles, and snakes. A loud, barking call is given to leopards and other cat species, a short cough-like call is given to two species of eagle, and a "chutter" call is given to a variety of dangerous snake species. Each call elicits a different escape response on the part of vervets who hear the call: to a leopard alarm they run for the trees; to an eagle alarm they look up in the air and sometimes run into the bushes; and to a snake alarm they look down at the ground, sometimes from a bipedal stance. These responses are just as distinct and frequent when researchers play back previously-recorded alarm calls over a loudspeaker, indicating that the responses of the vervets are not dependent on their actually seeing the predator. On the surface, it seems as if the caller is directing the attention of others to something they do not perceive or something they do not know is present. These alarm calls would thus seem to be referential and therefore good candidates for precursors to human language.

But several additional facts argue against this interpretation. First, no ape species has such specific alarm calls (Cheney and Wrangham, 1987). Since human beings are most closely related to apes, it is not possible that vervet monkey alarm calls could be the direct precursor of human language unless apes at some point used them also. Indeed, the fact is that predator-specific alarm calls are used by a number of non-primate species who must deal with multiple predators requiring different types of escape responses – from ground squirrels to domestic chickens (Marler, Evans and Hauser, 1992) – although no one considers these as direct precursors to human language. Second, vervet monkeys do not seem to use any of their other vocalizations referentially. They use "grunts" in various social situations (and some ape species have similar "close" calls as well; Cheney and Seyfarth, 1990), but these mainly serve to regulate dyadic social interactions such as grooming, play, fighting, and sex, not to draw attention to outside entities. Alarm calls thus are not representative of other monkey calls and so they do not embody a generalized form of reference. Third, primate vocalizations are almost certainly not learned, as monkeys and apes raised outside of their normal social environments still call in much the same way as those who grow up in normal social environments (although some aspects of call comprehension and use may be learned; Tomasello and Zuberbühler, in press). One would think that language could only have evolved from socially learned and flexibly used communicative signals.

2. Apes use gestures more flexibly than vocalizations – but still not referentially

Our nearest primate relatives, the chimpanzees, actually communicate in more flexible and interesting ways with gestures than with vocalizations. Although they have a number of more or less involuntary postural and facial displays that express their mood (e.g., piloerection indicating an aggressive mood and 'play-face' indicating a playful mood), they also use a number of gestures intentionally, that is, in flexible ways tailored for particular communicative circumstances. What marks these gestures as different from involuntary displays and most vocalizations is: (i) they are clearly learned as different individuals use different sets of them; (ii) they are used flexibly both in the sense that a single gesture may be used in different contexts and in the sense that different gestures may be used in the same context; and (iii) they are clearly sensitive to audience as the signaler typically waits expectantly for a response from the recipient after the gesture has been produced (Tomasello, George, Kruger, Farrar and Evans, 1985).

In their natural communication with conspecifics chimpanzees employ basically two types of intentional gesture. First, "attractors" are imperative gestures aimed at getting others to look at the self. For example, when youngsters want to initiate play they often attract the attention of a partner to themselves by slapping the ground in front of, poking at, or throwing things at them (Tomasello, Gust, and Frost, 1989). Because their function is limited to attracting attention, attractors most often attain their specific communicative goal from their combination with involuntary displays; for example, the specific desire to play is communicated by the involuntary 'play-face', with the attractor serving only to gain attention to it. The second type of intentional gestures are "incipient actions" that have become ritualized into gestures (see Tinbergen, 1951, on "intention-movements"). These gestures are also imperative in function, but they communicate more directly what specifically is desired. For example, play hitting is an important part of the rough-and-tumble play of chimpanzees, and so many individuals come to use a stylized 'arm-raise' to indicate that they are about to hit the other and thus initiate play. Many youngsters also ritualize signals for asking their mother to lower her back so they can climb on, for example, a brief touch on the top of the rear end, ritualized from occasions on which they attempt to push her rear end down mechanically.

Interestingly, in using their gestures chimpanzees demonstrate an understanding that the bodily and perceptual orientation of the recipient is an important precondition for the gesture to achieve its desired goal. Thus, young chimpanzees use their visually based gestures only when the recipient is looking at them, whereas they use more contact-based gestures when the recipient is oriented in another direction (Tomasello, Call, Nagell, Olguin and Carpenter, 1994). Nevertheless, they still do not use their gestures referentially. This is clear because (1) they almost invariably use them in dyadic contexts – either to attract the attention of others to the self or to request some behavior of another toward the self (e.g., play, grooming, sex) – not triadically, to attract the attention of others to some outside entity; and (2) they use them exclusively for imperative purposes to request actions from others, not for declarative purposes to direct the attention of others to something simply for the sake of sharing interest in it or commenting on it. Thus, perhaps surprisingly, chimpanzees do not point to outside objects or events for others, they do not hold up objects to show them to others, and they do not even hold out objects to offer them to others (Tomasello and Call, 1997).

Following from the initial excitement of the discovery of referentially specific monkey alarm calls, a number of scholars have recently cautioned against using human language as an interpretive framework for nonhuman primate communication (Owings and Morton, 1998; Owren and Rendell, 2001). According to these theorists, non-human primate communicative signals are not used to convey meaning or to convey information or to refer to things or to direct the attention of others, but rather to affect the behavior of others directly. If this interpretation is correct – and it is certainly consistent with the facts outlined above – then the evolutionary foundations of human language lie in the attempts of individuals to influence the behavior of conspecifics not their mental states. Attempting to influence the attention and mental states of others is a uniquely human activity, and so must have arisen only after humans and chimpanzees split from one another some 6 million years ago.

Relevant to this, from their earliest communicative attempts before language, young human children demonstrate this ability by pointing for others declaratively, holding up things to show them to others, offering objects to others, and in general engaging in a number of joint intentional activities by means of which they follow into, direct, and share attention with other people (Tomasello, 1995). This suggests that the social-cognitive prerequisites for language actually concern a much deeper and wider set of joint attentional and social learning activities.

Non-human Primate 'Culture' and Imitation

3. Nonhuman primate culture is different from human culture in important ways

Since language involves cultural transmission, it is important to see whether human cultural transmission has any unique qualities. There currently many reports in the popular press that some non-human primates, especially chimpanzees, live in cultures – and indeed they do in the most general sense of that term. But there are differences between human and non-human primate culture, especially with regard to the transmission mechanisms involved.

The best known case of non-human primate culture is the potato washing of Japanese macaques. Although the "spread" of potato washing as a novel behavior in one group of human provisioned Japanese macaques is well known, it turns out that the most likely explanation for that behavior is individual learning, with some influence of the social environment. It is likely that one individual invented the behavior by walking into the water with the potatoes thrown to her by humans, and her relatives and friends followed her into the water with their potatoes and invented the behavior for themselves (Galef, 1992). Supporting this interpretation is the fact is that potato washing is much less unusual a behavior for monkeys than was originally thought. Many monkeys brush sand off food naturally, and indeed potato washing has also been observed in four other troops of human-provisioned Japanese macaques – implying at least four individuals who learned on their own. Also, in captivity individuals of other monkey species learn quite rapidly to wash their food when provided with sandy fruits and bowls of water (Visalberghi and Fragaszy, 1990).

A much better species for investigating possible cultural processes in is chimpanzees, who have many population-specific behaviors (and in the absence of human provisioning). The best known example is chimpanzee tool use. For example, chimpanzees in the Gombe National Park (as well as several other groups elsewhere) fish for termites by probing termite mounds with small, thin sticks. In other parts of Africa chimpanzees simply destroy termite mounds with large sticks and attempt to scoop up the insects by the handful. Field researchers such as Boesch (1993) and McGrew (1992) have claimed that specific tool use practices such as these are "culturally transmitted" among the individuals of the various communities. The problem is that it is possible that chimpanzees in some localities destroy termite mounds with large sticks because the mounds are soft from much rain, whereas in other localities there is less rain, so the mounds are harder, and thus the chimpanzees there cannot use this strategy. In such a case there would be group differences of behavior – superficially resembling human cultural differences – but with no type of social learning involved at all. In such cases the "culture" is simply a result of individual learning driven by the different local ecologies of the different populations (and so it is sometimes called 'environmental shaping').

The human case is very different. Human beings have species-unique modes of cultural transmission involving imitation and teaching and this leads to some different types of cultural traditions and artifacts. Most importantly, the cultural traditions and artifacts of human beings accumulate modifications over time – so-called cumulative cultural evolution. Basically none of the most complex human artifacts or social practices – including tool industries, symbolic communication, and social institutions – were invented once and for all at a single moment by any one individual or group of individuals. Rather, what happened was that some individual or group of individuals first invented a primitive version of the artifact or practice, and then some later user or users made a modification, an "improvement", that others then adopted perhaps without change for many generations, at which point some other individual or group of individuals made another modification, which was then learned and used by others, and so on over historical time in what has sometimes been dubbed "the ratchet effect" (Tomasello, Kruger and Ratner, 1993).

The process of cumulative cultural evolution thus requires not only creative invention but also, and just as importantly, faithful social transmission that can work as a ratchet to prevent slippage backward – so that the newly invented artifact or practice may preserve its new and improved form at least somewhat faithfully until a further modification or improvement comes along. Perhaps surprisingly, for many animal species it is not the creative component, but rather the stabilizing ratchet component, that is the difficult feat. Many non-human primate individuals regularly produce intelligent behavioral innovations and novelties, but then their groupmates do not engage in the kinds of social learning that would enable, over time, the cultural ratchet to do its work (Kummer and Goodall, 1985).

4. Chimpanzees learn socially not by imitation but by emulation and ritualization

Although environmental shaping is likely a part of the explanations for group differences of behavior for all species, experimental studies have demonstrated that more than this is going on in chimpanzee culture. Tomasello (1996) reviewed all of the experimental evidence on chimpanzee imitative learning of tool use and concluded that chimpanzees are very good at learning about the dynamic affordances of objects that they discover through watching others manipulate them, but they are not skillful at learning from others a new behavioral strategy per se (but see Whiten and Cunstance, 1996). For example, if a mother rolls over a log and eats the insects underneath, her child will very likely follow suit. This is simply because the child learned from the mother's act that there are insects under the log – a fact she did not know and very likely would not have discovered on her own. But she did not learn how to roll over a log or to eat insects; these are things she already knew how to do or could learn how to do on her own. (Thus, the youngster would have learned the same thing if the wind, rather than its mother, had caused the log to roll over and expose the ants). This is what has been called emulation learning because it is learning that focuses on the environmental events involved – the changes of state in the environment that the other produced – not on a conspecific's behavior or behavioral strategy (see also Nagell, Olguin and Tomasello, 1993).

Chimpanzees are thus very intelligent and creative in using tools and understanding changes in the environment brought about by the tool use of others, but they do not seem to understand the instrumental behavior of conspecifics in the same way as do humans. For humans the goal or intention of the demonstrator is a central part of what they perceive, and indeed the goal is understood as something separate from the various behavioral means that may be used to accomplish the goal. Observers' ability to separate goal and means serves to highlight for them the demonstrator's method or strategy of tool use as an independent entity – the behavior she is using in an attempt to accomplish the goal, given the possibility of other means of accomplishing it. In the absence of this ability to understand goal and behavioral means as separable in the actions of others, chimpanzee observers focus on the changes of state (including changes of spatial position) of the objects involved during the demonstration, with the motions of the demonstrator being, in effect, just other motions. The intentional states of the demonstrator, and thus her behavioral methods as distinct behavioral entities, are simply not a part of their experience.

In terms of communicative behavior, as alluded to above, virtually no chimpanzee vocalizations are learned, and the gestures that are learned are not learned by imitation but rather by a process of ritualization in which individuals mutually shape one another's behavior over repeated social interactions (Tomasello, 1996). For example, it is likely that the 'arm-raise' gesture to initiate play originates as follows:

(1) an initiating chimpanzee youngster begins rough-and-tumble play with another by play hitting;

(2) after repeated instances of this the recipient begins to anticipate the impending hit on the basis of the first part of the sequence (the raising of the arm) and so begins the rough-and-tumble play at that early point in the sequence;

(3) the initiator notices the recipient's anticipation and connects the raising of its own arm with the beginning of the play; and

(4) on some future occasion the initiator comes to use its 'arm-raise' in order to elicit play – often in a stylized manner with no attempt to actually hit, waiting for a response from the recipient.

The 'arm-raise', which was originally a preparation for instrumental action, has become an intentional communicative signal used to elicit play from others. Evidence for ritualization as the major, if not exclusive, process of chimpanzee gesture learning was presented by Tomasello, Call, Warner, Carpenter, Frost, and Nagell (1997) who found that: (i) some individuals used gestures that no other group member used (thus precluding imitation as a means of acquisition), (ii) individual variability in types of gestures used by individuals of the same group was very high, whereas many gestures were shared between isolated groups; and (iii) in an experiment, no chimpanzees in a captive group copied a novel gesture used (frequently) by one of their groupmates.

Non-human primate communication and tool use thus provide us with two very different sources of evidence about social learning. In the case of gestural signals, it is very likely that chimpanzees acquire their communicative gestures through a process of ontogenetic ritualization. In the case of tool use, it is very likely that they acquire the tool use skills they are exposed to by a process of emulation learning. Both ontogenetic ritualization and emulation learning require skills of cognition and social learning, each in its own way, but neither requires skills of imitative learning in which the learner comprehends both the demonstrator's goal and the strategy she is using to pursue that goal – and then in some way aligns this goal and strategy with her own. Indeed, emulation learning and ontogenetic ritualization are precisely the kinds of social learning one would expect of organisms that are very intelligent and quick to learn, but that do not understand others as intentional agents with whom they can align themselves.

It may be objected that there are a number of convincing observations of chimpanzee imitative learning in the literature, and indeed there are a few. However, basically all of the clear cases in the exhaustive review of Whiten and Ham (1992) concern chimpanzees that have had extensive amounts of human contact. In many cases this has taken the form of intentional instruction involving human encouragement of behavior and attention, and even direct reinforcement for imitation for many months; for example, Hayes and Hayes (1952) provided their chimpanzee Vicki with 7 months of systematic training, and Whiten and Custance

(1996) provided their two chimpanzees with 4 months of systematic training. This raises the possibility that imitative learning skills may be influenced, or even enabled, by certain kinds of social interaction during early ontogeny, a point suggested especially strongly by the study of Tomasello, Savage-Rumbaugh, and Kruger (1993) in which it was found that the imitative learning abilities of so-called enculturated chimpanzees (raised like human children) were much better than those of mother-reared captive chimpanzees, who imitated almost never. This fact may have some implications for the co-evolution of skills of teaching and imitative learning.

5. Human children integrate an understanding of intentions into social learning

In contrast – and as a foundation for the ratchet effect – human children reproduce rather faithfully the intentional actions of others, which of course requires an understanding of their intentions. For example, Meltzoff (1988) had 14-month-old children observe an adult bend at the waist and touch his head to a panel, thus turning on a light. Most infants then performed more or less this same behavior – even though it was an unusual and awkward behavior and even though it would have been easier and more natural for them simply to push the panel with their hand. One interpretation of this behavior is that infants understood: (a) that the adult had the goal of illuminating the light; (b) that he chose one means for doing so, from among other possible means; and (c) that if they had the same goal they could choose the same means. Imitative learning of this type thus relies fundamentally on infants' ability to distinguish in the actions of others the underlying goal and the different means that might be chosen to achieve it. Otherwise, the infants might have engaged in emulation learning in which they simply turn on the light with their hands (which they did not), or else they would have just mimicked the action, like a parrot, without any regard for its goal-directed nature.

Two other recent studies have tested more directly what infants understand about others' intentional actions in the context of imitative learning. In the first, Meltzoff (1995) presented 18-month-old infants with two types of demonstrations (along with some control conditions). Infants in one group saw the adult perform actions on objects, much as in previous studies. Infants in the other group, however, saw the adult try but fail to achieve the end results of the target actions; for example, the adult tried to pull two parts of an object apart but never succeeded in separating them. Infants in this group thus never saw the target actions actually performed. Meltzoff found that infants in both of these groups reproduced the target actions equally well, that is, they appeared to understand what the adult intended to do and performed that action instead of mimicking the adult's actual surface behavior. (And they were much better in both of these conditions than in the control conditions in which the adult just manipulated the objects randomly and the like.)

In the second study, Carpenter, Akhtar, and Tomasello (1998) investigated infants' imitation of accidental versus intentional actions. In this study, 14- to 18-month-old infants watched an adult perform some two-action sequences on objects that made interesting results occur. One action of the modeled sequences was marked vocally as intentional ("There!") and one action was marked vocally as accidental ("Woops!") – with order systematically manipulated across sequences. Infants were then given a chance to make the result occur themselves. Overall, infants imitated almost twice as many of the adult's intentional actions as her accidental ones regardless of the order in which they saw them, indicating that they differentiated between the two types of actions and that they were able to reproduce, again, what the adult meant to do and not only her surface behavior.

Imitative learning thus represents infants' initial entry into the cultural world around them in the sense that they can now begin to learn from adults, or, more accurately, through adults, in cognitively significant ways. This learning is not just about the affordances of objects that are revealed when others manipulate them, or just about the surface behavior in the sense of precise motor movements. Instead, from around their first birthdays, human infants begin to tune into and attempt to reproduce both the adult's intentional acts, including her acts of intentional communication.

Child Language Acquisition

6. Symbols emerge in human ontogeny from more basic skills of joint attention

Pre-linguistic human infants are able to discriminate sounds and associate particular experiences with them. But they do not comprehend and produce linguistic symbols until about their first birthdays. An interesting question is thus: Why not? One hypothesis is that they do not acquire language so early because they do not yet understand the communicative intentions of others. From about their first birthdays, however, infants begin to understand that when other persons are making funny noises at me they are trying to manipulate my attention with respect to some external entity. This understanding is one manifestation of a momentous shift in the way human infants understand other persons – which occurs at around 9 to 12 months of age, as indicated by the near simultaneous emergence of a wide array of joint attentional skills involving outside objects. These include such things as following into the gaze direction and pointing gestures of others, imitating the actions of others on objects, and manipulating the attention of others by pointing or holding up objects to 'show' them to others declaratively (Tomasello, 1995).

The first language emerges on the heels of these non-linguistic triadic behaviors (involving you, me, and it) and is highly correlated with them – in the sense that children with earlier emerging skills of non-linguistic joint attention begin to acquire linguistic skills at an earlier age as well (Carpenter, Nagell and Tomasello, 1998). Similarly, children with autism have problems with joint attention and language in a correlated fashion, that is, those who have the poorest non-linguistic joint attentional skills are those who have the poorest language skills (Sigman and Capps, 1997). When children begin to understand the actions of others as intentional in general, they also begin to understand the communicative actions of others as intentional in the sense that they are aimed at directing attention. This understanding is the sine qua non of true language acquisition conceived as learning conventional symbols for manipulating the attentional and mental states of others.

7. Children's early language is not based on abstract categories

In terms of grammar, recent research suggests that most of young children's early language is not based on any abstract linguistic categories or schemas, and so description of child language in terms of abstract grammars created for adults is fundamentally misguided (Tomasello, 2000a). For example, in a detailed diary study Tomasello (1992) found that most of his English-speaking daughter's early multi-word speech revolved around specific verbs and other predicative terms. That is to say, at any given developmental period each verb was used in its own unique set of utterance level schemas, and across developmental time each verb began to be used in new constructions on its own developmental timetable irrespective of what other verbs were doing during that same time period. There was thus no evidence that once the child mastered the use of, for example, a passive construction with one verb that she could then automatically use that same construction with other semantically appropriate verbs. Tomasello (1992) hypothesized that children's early grammars could be characterized as an inventory of 'verb-island constructions', which then defined the first syntactic categories as lexically based things such as 'hitter', 'thing hit, and 'thing hit with' (as opposed to subject/agent, object/patient, and instrument; see also Tomasello and Brooks, 1999).

Experiments using novel verbs have also found that young children's early language is almost totally concrete – with the exception that they control from early on some very local and item-based structures with abstract but highly constrained 'slots'. For example, Tomasello and Brooks (1998) exposed 2 to 3-year-old children to a novel verb used to refer to a highly transitive and novel action in which an agent was doing something to a patient. In the key condition the novel verb was used in an intransitive sentence frame such as *The sock is tamming* (to refer to a situation in which, for example, a bear was doing something that caused a sock to "tam" – similar to the verb *roll* or *spin*). Then, with novel characters performing the target action, the adult asked children the question: *What is the doggie doing?* (when the dog was causing some new character to tam). Agent questions of this type encourage a transitive reply such as *He's tamming the car* – which would be creative since the child has heard this verb only in an intransitive sentence frame. The outcome was that very few children produced a transitive utterance with the novel verb, and in another study they were quite poor at two tests of comprehension as well (Akhtar and Tomasello, 1997). In general, it is not until they are 4 to 5 years old that young children are skillful at using novel verbs creatively in novel utterances and constructions in ways that indicate the use of abstract linguistic categories and schemas (see Tomasello, 2000b, for a review).

How do children create more abstract linguistic categories and constructions? Currently the leading hypothesis is that they invoke a relational mapping across different verb island constructions (Gentner and Markman, 1997). For example, in

English the several verb island constructions that children have with the verbs *give*, *tell*, *show*, *send*, and so forth, all share a 'transfer' meaning and they all appear in a structure: NP+V+NP+NP (the ditransitive construction). The specific hypothesis is thus that children make constructional analogies based on similarities of both form and function: two utterances or constructions are analogous if a "good" structure mapping is found both on the level of linguistic form and on the level of communicative function. Precisely how this might be done is not known at this time, but there are some proposals that a key element in the process might be some kind of "critical mass" of exemplars, to give children sufficient raw material from which to construct their abstractions (Marchman and Bates, 1994).

Overall, then, we may say that young children begin by imitatively learning specific pieces of language in order to express their communicative intentions, for example, in one-word utterances (holophrases) and other fixed expressions. As they attempt to comprehend and reproduce the utterances produced by mature speakers – along with the internal constituents of those utterances – they come to discern certain patterns of language use, and these patterns lead them to construct a number of different kinds of (at first very local and then more general) linguistic categories and schemas.

Conclusion

Following Parisi and Cangelosi (this volume), I believe that a comprehensive account of language origins and evolution requires attention to three distinct time frames. First, in human evolution what I believe we are looking for is the emergence of the human ability to use symbols (Deacon, 1997). However, my own view is that if we are looking for the major cognitive adaptation underlying language, we must look much more broadly at the human ability to understand others as intentional agents like the self – which underlies not just language but a whole suite of cultural skills (Tomasello, 1999).

Second, of crucial importance in determining the shape of modern languages is grammaticalization and other processes of historical language change. Although I have not dealt with them here (I am not an expert on these matters), these processes have both psychological and sociological dimensions – involving everything from pragmatic inferences to analogical extensions to the spread of novelties due to social status – and they provide the major alternative account (i.e., alternative to Chomskyan nativism) for why modern languages look the way they do (Croft, 2000). A number of modern linguists (e.g., Heine, 1997; Bybee *et al.*, 1994) have argued and presented evidence that the linguistic constructions of individual languages have been built up over generations by groups of people communicating with one another within the general constraints of human cognition, communication, and vocal-auditory processing.

Third, in ontogeny human children hear only concrete utterances but they end up with some abstract linguistic categories and constructions. It is conceivable that they do this with biological machinery specifically dedicated to language and its acquisition, but a variety of lines of recent research are beginning to establish the viability of an alternative view. In this alternative view (e.g., Slobin, 1997; Bates,

1999; Tomasello, 2000b), young children themselves create linguistic abstractions by using general cognitive and social-cognitive skills to find patterns in the way adults are using language when communicating with them. These pattern-finding skills emerge early in human ontogeny and can be applied to a variety of different types of experiences (Saffran *et al.*, 1996; 1999).

Computer simulations of linguistic processes – evolutionary, historical, and ontogenetic – represent an exciting new approach to the question of language origins and evolution. But it is my belief that they can only make real progress with at least a little bit of back-propagation from facts about the way these processes operate in the real world. Ideally, new insights gained from these informed simulations then will lead to novel empirical investigations, which will then lead to better simulations, and so on in one more instance of an important cultural ratchet.

References

Akhtar N, Tomasello M (1997) Young children's productivity with word order and verb morphology. *Developmental Psychology*, 33: 952-965

Bates E (1999) On the nature and nurture of language. In: Levi-Montalcini R (ed) *Frontiers of biology*. Giovanni Treccani, Rome

Boesch C (1993) Towards a new image of culture in wild chimpanzees? *Behavioral and Brain Sciences*, 16: 514-515

Bybee J, Perkins R, Pagliuca W (1994) *The evolution of grammar*. University of Chicago Press, Chicago

Carpenter M, Akhtar N, Tomasello, M. (1998) Sixteen-month-old infants differentially imitate intentional and accidental actions. *Infant Behavior and Development*, 21: 315-30

Carpenter M, Nagell K, Tomasello M (1998) Social cognition, joint attention, and communicative competence from 9 to 15 months of age. *Monographs of the Society for Research in Child Development*, 63: 1-143

Cheney DL, Seyfarth RM (1990) *How monkeys see the world*. University of Chicago Press

Cheney DL, Wrangham RW (1987) Predation. In: Smuts BB, Cheney DL, Seyfarth RM, Wrangham RW, Struhsaker TT (eds) *Primate societies*. The University of Chicago Press, Chicago, pp 227-239

Croft W (2000) *Explaining language change: An evolutionary approach*. Longmans, London

Deacon T (1997) *The symbolic species*. Norton, New York

Galef B. (1992) The question of animal culture. *Human Nature*, 3: 157-178

Gentner D, Markman A (1997) Structure mapping in analogy and similarity. *American Psychologist*, 52: 45-56

Hayes K, Hayes C (1952) Imitation in a home-raised chimpanzee. *Journal of Comparative and Physiological Psychology*, 45: 450-59

Heine B (1997) *Cognitive foundations of grammar*. Oxford University Press

Marler P, Evans C, Hauser M (1992) Animal signals: Motivational, referential, or both? In: Papousek H (ed) *Nonverbal vocal communication: Comparative and developmental approaches*. Cambridge University Press, pp 66-86.

Kummer H, Goodall J (1985) Conditions of innovative behaviour in primates. *Phil. Trans. Royal Soc. Lon, B*, 308: 203-214

Marchman V, Bates E (1994) Continuity in lexical and morphological development: A test of the critical mass hypothesis. *Journal of Child Language*, 21: 339-366

McGrew WC (1992) *Chimpanzee material culture*. Cambridge University Press, Cambridge

Meltzoff A (1988) Infant imitation after a one week delay: Long term memory for novel acts and multiple stimuli. *Developmental Psychology*, 24: 470-76

Meltzoff A (1995) Understanding the intentions of others: Re-enactment of intended acts by 18-month-old children. *Developmental Psychology*, 31: 838-50

Nagell K, Olguin K, Tomasello M (1993) Processes of social learning in the tool use of chimpanzees (Pan troglodytes) and human children (Homo sapiens). *Journal of Comparative Psychology*, 107: 174-186

Owings D, Morton E (1998) *Animal vocal communication*. Cambridge Univeristy Press

Owren M, Rendell D (2001) Sound on the rebound. *Evolutionary Anthropology*, 10: 58-71

Saffran JR, Aslin RN, Newport EL (1996) Statistical learning by 8-month-old infants. *Science*, 274: 1926-1928

Saffran JR, Johnson EK, Aslin RN, Newport EL (1999) Statistical learning of tone sequences by human infants and adults. *Cognition*, 70: 27-52

Sigman M, Capps L (1997) *Children with autism: A developmental perspective*. Harvard University Press, Cambridge MA

Slobin D (1997) On the origin of grammaticalizable notions – beyond the individual mind. In: Slobin D (ed) *The crosslinguistic study of language acquisition (Volume 5)*. Erlbaum, Hillsdale NJ

Tinbergen N (1951) *The study of instinct*. Oxford University Press, New York

Tomasello M (1992) *First verbs: A case study in early grammatical development*. Cambridge University Press

Tomasello M (1995) Joint attention as social cognition. In: Moore C, Dunham P (eds) *Joint attention: Its origins and role in development*, Erlbaum, Hillsdale NJ, pp 103-130

Tomasello M (1996) Do apes ape? In: Galef BG Jr, Heyes CM (eds), *Social learning in animals: The roots of culture*, Academic Press, New York, pp 319-346

Tomasello M (1999) The cultural origins of human cognition. Harvard University Press, Cambridge MA

Tomasello M (2000a) First steps in a usage based theory of language acquisition. *Cognitive Linguistics*, 11: 61-82

Tomasello M (2000b) Do young children have adult syntactic competence? *Cognition*, 74: 209-253

Tomasello M, Brooks P (1998) Young children's earliest transitive and intransitive constructions. *Cognitive Linguistics*, 9: 379-395

Tomasello M, Brooks P (1999) Early syntactic development. In: Barrett M (ed), *The development of language*. UCL Press

Tomasello M, Call J (1997) *Primate cognition*. Oxford University Press

Tomasello M, Call J, Warren J, Frost GT, Carpenter M, Nagell K (1997) The ontogeny of chimpanzee gestural signals: A comparison across groups and generations. *Evolution of Communication*, 1: 223-253

Tomasello M, Call, J Nagell K, Olguin K, Carpenter M. (1994) The learning and use of gestural signals by young chimpanzees: A trans-generational study. *Primates*, 35: 137-154

Tomasello M, George B, Kruger AC, Farrar J, Evans E. (1985) The development of gestural communication in young chimpanzees. *Journal of Human Evolution*, 14: 175-86

Tomasello M, Gust D, Frost GT (1989) The development of gestural communication in young chimpanzees: A follow up. *Primates*, 30: 35-50

Tomasello M, Kruger AC, Ratner HH (1993) Cultural learning. *Behavioral and Brain Sciences*, 16: 495-552

Tomasello M, Savage-Rumbaugh ES, Kruger AC (1993) Imitative learning of actions on objects by children, chimpanzees, and enculturated chimpanzees. *Child Development*, 64: 1688-1705

Tomasello M, Zuberbüler K (in press) Primate vocal and gestural communication. In: Bekoff M, Allen C, Burghardt G (eds), *The cognitive animal: Empirical and theoretical perspecitives on animal cognition*. MIT Press

Visalberghi E, Fragaszy DM (1990) Food-washing behaviour in tufted capuchin monkeys, Cebus apella, and crabeating macaques, Macaca fascicularis. *Animal Behaviour*, 40: 829-836

Whiten A, Custance D (1996) Studies of imitation in chimpanzees and children. In: Heyes C, Galef B (eds) *Social learning in animals: The roots of culture*. Academic Press, San Diego, pp 291-318

Whiten A, Ham R (1992) On the nature and evolution of imitation in the animal kingdom: Reappraisal of a century of research. In: Slater PJB, Rosenblatt JS, Beer C, Milinsky M (eds), *Advances in the study of behavior*. Academic Press, New York, pp 239-283

Author Index[1]

---

[1] Pages in italic in the Author Index refer to authors' own chapters, and in the Subject Index to the most important discussion(s) of a topic.

Subject Index